INTRODUCTORY
ELECTRIC CIRCUIT ANALYSIS

— chapter 15, 16 — phasors and imaginary numbers

power factor 19 page 477

INTRODUCTORY
ELECTRIC CIRCUIT ANALYSIS

David E. Johnson

and

Johnny R. Johnson

Department of Electrical Engineering
Louisiana State University

PRENTICE-HALL, INC. *Englewood Cliffs, New Jersey 07632*

Library of Congress Cataloging in Publication Data

Johnson, David E
  Introductory electric circuit analysis.

  Includes index.
    1. Electric circuits. 2. Electric circuit analysis.
I. Johnson, Johnny Ray, joint author. II. Title.
TK454.J57   621.319′2   80–16459
ISBN 0–13–500835–2

Editorial production supervision
and interior design by: JAMES M. CHEGE

Cover design by: Wanda Lubelska

Manufacturing buyer: JOYCE LEVATINO

Printed in the United States of America

10   9   8   7   6   5   4   3   2   1

PRENTICE-HALL INTERNATIONAL, INC., *London*
PRENTICE-HALL OF AUSTRALIA PTY. LIMITED, *Sydney*
PRENTICE-HALL OF CANADA, LTD., *Toronto*
PRENTICE-HALL OF INDIA PRIVATE LIMITED, *New Delhi*
PRENTICE-HALL OF JAPAN, INC., *Tokyo*
PRENTICE-HALL OF SOUTHEAST ASIA PTE. LTD., *Singapore*
WHITEHALL BOOKS LIMITED, *Wellington New Zealand*

CONTENTS

PREFACE   *xiii*

1. INTRODUCTION   *1*

    1.1 History of Electricity   *2*
    1.2 Nature of Electricity   *6*
    1.3 Definitions   *10*
    1.4 System of Units   *11*
    1.5 Scientific Notation   *13*
    1.6 Summary   *16*
         Problems   *17*

2. ELECTRIC CIRCUIT ELEMENTS   *19*

    2.1 Charge and Current   *19*
    2.2 Voltage   *25*
    2.3 Power   *27*

2.4   Energy   29

2.5   Voltage Sources—Batteries   31

2.6   Other Sources of EMF   36

2.7   Summary   39

      Problems   40

3.   RESISTANCE   42

3.1   Ohm's Law   43

3.2   Conductance   45

3.3   Power Absorbed by a Resistor   47

3.4   Physical Resistors   50

3.5   Variable Resistors   54

3.6   Resistor Color Coding   58

3.7   Summary   62

      Problems   62

4.   SIMPLE RESISTIVE CIRCUITS   64

4.1   Kirchhoff's Current Law   65

4.2   Kirchhoff's Voltage Law   68

4.3   Series Circuits   70

4.4   Parallel Circuits   76

4.5   Power in Simple Circuits   85

4.6   Summary   88

      Problems   89

5.   SERIES–PARALLEL CIRCUITS   92

5.1   Equivalent Resistance   92

5.2   Strings and Banks of Resistances   97

5.3   Series–Parallel Circuit Examples   100

5.4   Open and Short Circuits   105

5.5   Summary   108

      Problems   109

6. VOLTAGE AND CURRENT DIVISION **112**

6.1 Voltage Division *112*

6.2 Current Division *117*

6.3 Examples Using Voltage and Current Division *122*

6.4 Wheatstone Bridge *126*

6.5 Summary *129*

Problems *129*

7. GENERAL RESISTIVE CIRCUITS **132**

7.1 Loop Analysis Using Element Currents *133*

7.2 Determinants *137*

7.3 Mesh Currents *140*

7.4 Node-Voltage Analysis *146*

7.5 Other Nodal Analysis Examples *150*

7.6 Summary *154*

Problems *154*

8. NETWORK THEOREMS **157**

8.1 Superposition *158*

8.2 Thévenin's Theorem *162*

8.3 Norton's Theorem *166*

8.4 Voltage and Current Source Conversions *169*

8.5 Millman's Theorem *173*

8.6 Y and Δ Networks *176*

8.7 Summary *181*

Problems *182*

9. DIRECT-CURRENT METERS **185**

9.1 D'Arsonval Movement *186*

9.2 Ammeters *192*

9.3 Voltmeters *196*

9.4 Ohmmeters *201*

9.5 Other Meters *205*

9.6 Summary *209*

Problems *209*

10. CONDUCTORS AND INSULATORS   212

　　10.1   Resistance   *213*

　　10.2   Wire Conductors   *216*

　　10.3   Effect of Temperature   *220*

　　10.4   Insulators   *223*

　　10.5   Switches and Fuses   *224*

　　10.6   Summary   *231*

　　　　 Problems   *232*

11. CAPACITORS   233

　　11.1   Definitions   *234*

　　11.2   Circuit Relationships   *239*

　　11.3   Series and Parallel Capacitors   *244*

　　11.4   Types of Capacitors   *248*

　　11.5   Properties of Capacitors   *257*

　　11.6   Summary   *260*

　　　　 Problems   *260*

12. *RC* CIRCUITS   262

　　12.1   Charging and Discharging a Capacitor   *263*

　　12.2   Source-free *RC* Circuits   *266*

　　12.3   Driven *RC* Circuits   *274*

　　12.4   More General Circuits   *278*

　　12.5   Shortcut Procedure   *281*

　　12.6   Summary   *286*

　　　　 Problems   *287*

13. MAGNETISM   290

　　13.1   The Magnetic Field   *291*

　　13.2   Magnetic Flux   *294*

　　13.3   Ohm's Law for Magnetic Circuits   *297*

　　13.4   Magnetic Field Intensity   *301*

　　13.5   Simple Magnetic Circuits   *305*

　　13.6   Summary   *311*

　　　　 Problems   *312*

14. INDUCTORS   *315*

14.1   Definitions   *316*

14.2   Inductance and Circuit Relationships   *319*

14.3   Series and Parallel Inductors   *324*

14.4   Source-free *RL* Circuits   *326*

14.5   Driven *RL* Circuits   *331*

14.6   Summary   *335*

     Problems   *336*

15. ALTERNATING CURRENT   *339*

15.1   AC Generator Principle   *340*

15.2   The Sine Wave   *345*

15.3   Frequency   *349*

15.4   Phase   *351*

15.5   Average Values   *354*

15.6   RMS Values   *360*

15.7   AC Circuits   *363*

15.8   Summary   *366*

     Problems   *366*

16. PHASORS   *368*

16.1   Imaginary Numbers   *369*

16.2   Complex Numbers   *373*

16.3   Operations with Complex Numbers   *381*

16.4   Phasor Representations   *388*

16.5   Impedance and Admittance   *391*

16.6   Kirchhoff's Laws and Phasor Circuits   *398*

16.7   Summary   *401*

     Problems   *401*

17. AC STEADY–STATE ANALYSIS   405

17.1   Impedance Relationships   *406*
17.2   Phase Relationships   *412*
17.3   Voltage and Current Division   *417*
17.4   Nodal Analysis   *425*
17.5   Loop Analysis   *430*
17.6   Phasor Diagrams   *433*
17.7   Summary   *439*
        Problems   *440*

18. AC NETWORK THEOREMS   445

18.1   Superposition   *445*
18.2   Thévenin's and Norton's Theorems   *451*
18.3   Voltage and Current Phasor Source Conversions   *455*
18.4   Y and Δ Phasor Networks   *458*
18.5   Summary   *465*
        Problems   *466*

19. AC STEADY–STATE POWER   470

19.1   Average Power   *470*
19.2   Power Factor   *477*
19.3   Power Triangle   *482*
19.4   Power-Factor Correction   *488*
19.5   Maximum Power Transfer   *494*
19.6   Power Measurement   *496*
19.7   Summary   *498*
        Problems   *499*

20. THREE-PHASE CIRCUITS   503

20.1   Three-Phase Generator   *504*
20.2   Y-Connected Generator   *507*
20.3   Y–Y Systems   *512*
20.4   Delta-Connected Load   *517*

20.5 Power Measurement   *525*

20.6 Unbalanced Loads   *530*

20.7 Summary   *532*

Problems   *533*

21. TRANSFORMERS   **535**

21.1 Mutual Inductance   *536*

21.2 Transformer Properties   *543*

21.3 Ideal Transformers   *550*

21.4 Equivalent Circuits   *556*

21.5 Types of Transformers   *562*

21.6 Summary   *567*

Problems   *568*

22. FILTERS   **571**

22.1 Amplitude and Phase Responses   *572*

22.2 Resonance   *576*

22.3 Bandpass Filters   *582*

22.4 Low-Pass Filters   *589*

22.5 Other Types of Filters   *592*

22.6 Summary   *601*

Problems   *601*

ANSWERS TO ODD-NUMBERED PROBLEMS   **605**

INDEX   **617**

PREFACE

This book is a basic text for beginning students in electric circuits. No previous experience in electricity is needed in order to read the book, and the highest level of mathematics that the reader needs is an introductory course in algebra. Some knowledge of complex numbers and trigonometry will make it easier to understand the book, but these topics are not required background because they are presented as they are needed.

The order of topics follows a typical one–year course in electric circuits. The book is divided into two parts, with the first half devoted to dc circuits and the second half to ac circuits. The first ten chapters cover an introduction to electric circuits and their elements, to resistance, and to circuits containing resistors and sources. Ohm's law and Kirchhoff's laws are given and applied to series circuits, parallel circuits, series–parallel circuits, and more general circuits. Also voltage and current division and network theorems are presented and used in solving circuits. The first part is concluded with a discussion of dc meters and the properties of conductors and insulators.

Chapters 11 through 22 primarily deal with ac circuits. Capacitors and inductors, and their related properties of electric and magnetic fields, are considered in an introductory manner. Alternating currents are defined and ac circuits are solved using phasors and impedances. The book concludes with chapters on ac steady–state power, three–phase circuits, transformers, and electric filters. The chapters on three–phase circuits and filters are written so that they may be omitted in a shorter course.

The standard International System (SI) of units is used wherever practical throughout the book, and emphasis is placed on the use of the electronic hand calculator in solving the problems. There are many worked–out examples in each chapter, and practice exercises with answers are given at the end of almost every section. Problems, some more difficult and some less difficult than the exercises, are given at the end of every chapter. Answers to the odd–numbered problems are provided at the end of the book.

There are many people who have provided invaluable assistance and advice concerning this book. In particular, we are especially grateful to the many manufacturers who have provided us with photographs of their products, as noted in each legend. Finally, a special note of thanks is due Mrs. Marie Jines for her rapid and expert typing of the manuscript.

Louisiana State University   DAVID E. JOHNSON
Baton Rouge, Louisiana   JOHNNY R. JOHNSON

INTRODUCTORY
ELECTRIC CIRCUIT ANALYSIS

1

INTRODUCTION

Electricity is a form of energy that can be used to produce light, heat, motion, and many other commonly observed effects. It lights, heats, and cools our homes, cooks our food, and brings us radio, motion pictures, television, and the telephone. Modern society would be very different without electricity. Our homes would be without electric irons, stoves, washing machines, toasters, grills, refrigerators, freezers, dishwashers, and all kinds of power tools and equipment. Modern industry could not exist without electricity, communication would be by word of mouth or by written messages delivered by hand, and transportation would be by horse and buggy.

Equipment that uses electricity, or which is used in its generation and distribution, ranges in size from such devices as hand calculators that can be carried in a pocket to motors and generators weighing many tons. As examples, the giant Westinghouse motor-generator set of Fig. 1.1 weighs 51 tons and the Westinghouse extra-high-voltage transformer shown in Fig. 1.2 is 31 feet high, 31 feet long, 25 feet wide, and weighs over 347 tons.

At the other extreme is Hewlett-Packard's HP-67 programmable hand calculator shown in Fig. 1.3 with a number of magnetic cards used for entering the program. Another example is that of Fig. 1.4, which is the HP-01 "wrist instrument" with an insert showing the *integrated circuitry* that made it possible. The HP-01 is much more than a digital watch. It also gives the day of the week, the day of the year, makes calculations, functions as a stopwatch, and performs many other tasks.

In this book, we are interested primarily in solving, or *analyzing,* electric circuits.

1

FIGURE 1.1   *Giant motor-generator set (Courtesy, Westinghouse Electric Corporation).*

We shall be concerned with what an electric circuit is, what we mean by its solution, or analysis, and how quantities associated with electric circuits are measured and in what units. These topics, the nature of electricity itself, and a short discussion of its history are the subjects of this chapter.

1.1   *HISTORY OF ELECTRICITY*

Everything in nature is basically electrical because all matter consists of atoms, which contain electrons. Electrons, in turn, are the basic ingredients of electricity, as we shall see. However, people did not know about electricity until about 600 B.C., when the ancient Greeks discovered that amber, when rubbed with a cloth, exhibited the basic electrical effect of attracting other light objects, such as bits of feathers or the stems of plants. In fact, our word for electricity comes from *elektron,* the Greek word for amber.

Not much more was learned about electricity from the time of the Greeks to around the eighteenth and nineteenth centuries, when great advances were made.

FIGURE 1.2 Extra-high-voltage transformer (Courtesy, Westinghouse Electric Corporation).

During this time it was discovered that in addition to the "static" electricity produced in the amber, there was also "flowing" or "current" electricity in which electrons could be made to move in a wire or, as we would say, in an *electric circuit*. The American statesman Benjamin Franklin performed his famous kite experiment in 1752 and demonstrated that lightning is electricity. He also coined the terms *positive*

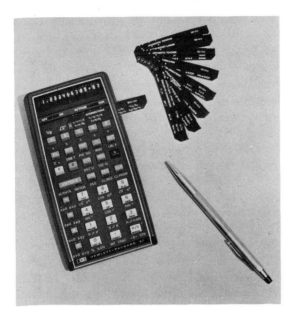

FIGURE 1.3 *HP-67 programmable hand calculator (Courtesy, Hewlett-Packard).*

FIGURE 1.4 *HP-01 wrist instrument (Courtesy, Hewlett-Packard).*

and *negative,* describing the two types of *electrically charged* bodies. Charles Augustin de Coulomb, a French physicist, in 1785 worked out the laws of force between charged bodies, and in 1800 an Italian physics professor, Alessandro Volta, built the first battery.

In 1819 a Danish scientist, Hans Christian Oersted, showed that an electric *current* (electrons moving in a wire) caused a nearby compass needle to move and concluded that electric current has a magnetic effect. In the same year, the French physicist André Marie Ampère showed that two wires carrying current attract or repel each other just as magnets do, and he was able to measure the magnetic effect of an electric current. In 1831 the English physicist Michael Faraday, and Joseph Henry, an American, showed in separate experiments that a moving magnet would produce an electric current in a coil of wire. Thus electricity can produce magnetism and magnetism can produce electricity. All electric generators and transformers work by means of the principles of Faraday and Henry.

One of the last great advances of the nineteenth century came in the late 1880s when Heinrich Rudolph Hertz, a German physicist, produced electromagnetic waves, or radio waves, that move at the speed of light. This achievement of Hertz paved the way for our modern communication system of telephones, telegraphs, radios, television, and earth satellites.

The advances of the eighteenth and nineteenth centuries continued in the twentieth century at an accelerated pace. By 1907 vacuum tubes were invented that detected and amplified radio waves, or *signals.* These tubes made the radio possible. The development of other tubes led to the invention of television in the 1920s, of radar in the 1930s, and of electronic computers in the 1940s. The transistor, which was invented in 1948; integrated circuits, developed in the 1960s; and the microprocessors of the 1970s have brought about revolutionary changes in the way we use electricity, and the end is nowhere in sight.

The transition from vacuum tubes to integrated circuits was accompanied by a fantastic shrinkage in the size of the electric elements, as may be seen in Fig. 1.5. Various examples of transistors are shown in Fig. 1.6, and Fig. 1.7 illustrates two *operational amplifiers,* which are integrated circuits containing many transistors.

Another example of an integrated circuit is the "standard calculator on a chip" of Texas Instruments, shown in Fig. 1.8. It incorporates all of the logic and memory circuits necessary to perform complete eight-digit calculator functions. This chip contains the equivalent of 6000 transistors and occupies only a small fraction of the total space needed in the calculator. The small size of the chip may be seen also from Fig. 1.9, where it is being placed for soldering onto the metal structure.

One revolutionary development, based on the existence of integrated circuits, that is having an immediate effect on students reading circuits books such as this one is that of the hand calculator. With this marvelous instrument, there is no longer any need to use a slide rule for multiplications and divisions or an adding machine for additions and subtractions. Trigonometric tables, logarithmic tables, exponential tables, and so on, are all unnecessary. The tedium of arithmetic has almost been removed, with a tremendous improvement in accuracy.

FIGURE 1.5 *Vacuum tube and integrated circuits (Courtesy, RCA Solid State Division).*

FIGURE 1.6 *Various power transistor packages (Courtesy, RCA Solid State Division).*

FIGURE 1.7 *Two operational amplifiers (Courtesy, RCA Solid State Division).*

1.2 NATURE OF ELECTRICITY

Electricity produces heat, light, motion, and so on, in such diverse systems as vacuum cleaners, television sets, motors, and lightning. All these systems, therefore, have something in common that is electrical in nature. What does lightning have in common with a television set? The answer is that both their distinguishing characteristics are produced by forces due to the movement of electric *charges,* or *particles,* which are associated with the atom, the basic element in all matter.

Atomic Structure: Atoms are so tiny that they can be seen only with powerful electron microscopes. For example, a single drop of water contains more than 100 billion billion atoms. Even so, atoms are made up of even smaller particles. The *nucleus,* or center core, of the atom consists of several different kinds of particles,

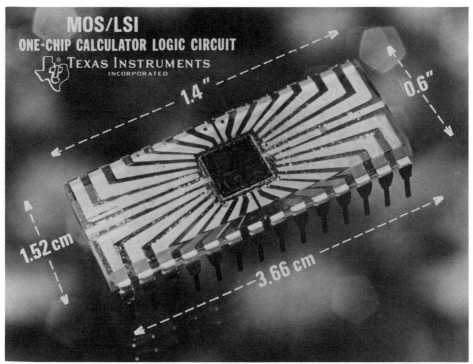

FIGURE 1.8   *One-chip calculator circuit (Courtesy, Texas Instruments, Incorporated).*

packed closely together. The basic particles of the nucleus are *protons* and *neutrons,* the protons being positively charged and the neutrons having no charge. The rest of the atom consists of *electrons* whirling around the nucleus in *orbits,* or *shells,* at fantastic speeds. Electrons have a negative charge equal and opposite to the positive charge of a proton, but they have almost no mass. To illustrate, a proton has the same mass as a neutron and has about 1836 times the mass of an electron.

The simplest atom is hydrogen, with a single proton constituting its nucleus and one electron spinning about the nucleus. The hydrogen atom is represented by Fig. 1.10(a). A more complicated atom is carbon, or carbon 12, represented in Fig. 1.10(b). Its nucleus contains 6 protons and 6 neutrons and there are 6 electrons in two orbits, 2 in the shell closer to the nucleus and 4 in the outer shell. The most complex atom is uranium 238, with a nucleus of 92 protons and 146 neutrons, and 92 electrons arranged in seven shells.

In general, the first shell, which is nearest the nucleus, can contain no more than 2 electrons; the second shell can contain a maximum of 8; the third 8 or 18; the fourth 8, 18, or 32; and so on, depending on the particular atom. However, if there are at least two shells, the maximum number of electrons in the outermost shell is always 8.

The number of electrons in the outermost shell determines, to a great extent, the atom's electrical stability. The electrons in the outermost shell, being farther

FIGURE 1.9 *Integrated circuit being soldered on a metal structure (Courtesy, Texas Instruments, Incorporated).*

away, are not as tightly bound to the nucleus as are the electrons in the inner shells. Thus if the outermost shell is not filled (and thus not stable), there is a greater tendency for its electrons to be pulled loose by an external force or by the force of a neighboring atom. For example, copper has 29 electrons in its shells: 2 in the first, 8 in the second, 18 in the third, and only 1 in the outermost shell. This single outermost electron can move easily from one atom to another in a piece of copper and for this reason is called a *free* electron. This movement of free electrons provides the electrical current, which is the principal characteristic of electricity.

Conductors, Insulators, and Semiconductors: A material with many free electrons, which can move easily from atom to atom, is called a *conductor*. Most metals are generally conductors, with silver the best and copper second. Both silver and gold atoms, like copper, have one free electron in the outermost shell, but because copper is much less expensive it is most often used as an electrical conductor. Because

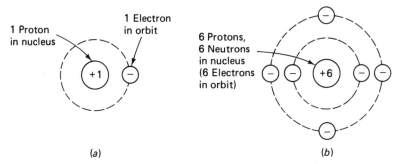

FIGURE 1.10 *Representation of (a) hydrogen and (b) carbon atoms.*

of the cost factor, only aluminum has been a competitor with copper for commercial use.

A material with very stable atoms (the electrons tend to stay in their orbits) is a very poor conductor of electricity. Such a material is called an *insulator* or a *dielectric*. Conductors are needed when current is to flow easily with minimum opposition and insulators are required when it is necessary to prevent current flow. Some common insulating materials are air, rubber, paper, glass, and mica.

Materials with conducting abilities in between conductors and insulators are called *semiconductors*. They are both very poor conductors and very poor insulators, but they are extremely important materials in modern *solid-state* electronic devices. Examples of some such semiconductor devices are diodes, transistors, integrated circuits, and so on. Typical semiconductor atoms are silicon and germanium, each with four electrons in the outermost shell.

Positive and Negative Charges: In its stable state an atom is electrically neutral, which means that the number of positively charged protons in its nucleus is equal to the number of negatively charged electrons in its orbits. However, if an electron, such as the free electron of a copper atom, is pulled loose from the outer orbit, the atom is a *positively charged* particle because it has more positive charges in the nucleus than negative charges in orbit. The free electron is, of course, an example of a *negatively charged* particle. The opposite situation may also occur. That is, an atom may acquire additional electrons in its outer orbit (free electrons from other atoms, for example), and thus become a negatively charged particle. Atoms that have acquired or lost electrons are called *ions* and are good examples of positive or negative *charges*. The usual example of a negative charge is, of course, the electron itself.

Forces Due to Charges: The essence of electricity and the property that makes it work for us is the *force* exhibited between two charged bodies. Two particles that have opposite charges, or opposite *polarities*, attract each other, as illustrated in Fig. 1.11(a), and two particles with like charges, or the same polarities, repel each other, as in the case of the two positive charges of Fig. 1.11(b) and the two negative charges of Fig. 1.11(c). For example, we may produce a negative charge on a rubber balloon

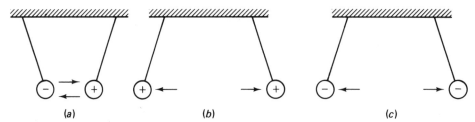

FIGURE 1.11 *Representation of force between charges which have (a) opposite, (b) both positive, and (c) both negative polarities.*

by rubbing it on our hair. (The balloon acquires its negative polarity from the electrons rubbed off the hair in the process.) The balloon will then stick to the wall because of the force of attraction exerted on the neutral wall by the negatively charged balloon.

1.3 DEFINITIONS

Before we start on any course of study we should define the terms we are going to use, the quantities we are going to measure, the units of measurement we are going to use, and so forth. This is especially true of circuit analysis, and we shall begin in this section by giving certain basic definitions and in the following section we shall describe the system of units that we shall use throughout the book.

Electric Circuit and Elements: An electric *circuit,* or electric *network,* is simply a collection of electrical *elements* connected in some specified way. There are many common examples of electrical elements, which we will define formally in later chapters. Some of these are resistors, capacitors, inductors, batteries, and generators, which may already be familiar to you. These are all elements with two *terminals,* represented in the general case by the rectangular shape with terminals *a* and *b* in Fig. 1.12. There are other, more complicated, elements with three or more terminals, such as transistors, vacuum tubes, operational amplifiers, and so on, which are studied in a later course in electronics.

Electric circuits are formed by connecting the terminals of electrical elements to those of other electrical elements. An example is the electric circuit of Fig. 1.13. The wires used to connect terminals *a, b, c,* and *d* of the electrical elements 1, 2, 3, 4, and 5 are conductors such as copper. Thus if at least one of the elements is a source of energy, such as a battery, free electrons, or current, can readily flow around *closed paths,* such as path *abca,* containing elements 1, 2, and 3.

Circuit Analysis: An electrical element is often defined, as we shall see, by considering quantities associated with it, such as current, voltage, and power. The *analysis* of the circuit, which is our main concern in this book, is the determination of one or more of these quantities for certain elements, or perhaps for every element, in the circuit. This can be done only if we have a standard system of units so that

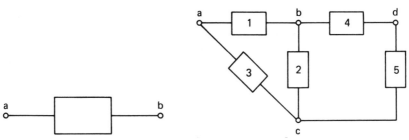

FIGURE 1.12 General two-terminal electrical element.

FIGURE 1.13 Example of an electric circuit.

when a quantity is described by measuring it, everyone will agree on what the measurement means. Such a system of units is described in Section 1.4.

1.4 SYSTEM OF UNITS

In the past, one system of units that was used was the English system of pounds, inches, feet, seconds, and so on. However, this system is extremely cumbersome to use in any scientific application, or indeed in any application where conversions must be made from one unit to another. For example, there are 12 inches (in.) in a foot, 3 feet (ft) in a yard (yd), 5280 feet in a mile (mi), and so on, so that the number of yards in 25 miles is much more difficult to find than, say, the number of cents in 25 dollars in the U.S. monetary system. The reason for this, of course, is that the monetary system is based on *powers* of 10: 10 cents in a dime, 10 dimes in a dollar, and so on.

International System of Units: In studying electric circuits we need a standard system of units that is easy to use, such as the U.S. monetary system. Fortunately, there is such a system that is used today by almost all technicians and engineers. This system, which we shall use, is the metric system, called the *International System of Units,* abbreviated SI, which was adopted in 1960 by the General Conference on Weights and Measures.

The basic unit of length in the SI is the *meter,* abbreviated m, which is related to the English system by the fact that 1 in. is 0.0254 m. Those of us who are more familiar with the English system may get a feeling for a meter by noting that it is almost equivalent to a yard (1 yd = 0.914 m). The basic SI unit of mass is the *kilogram* (kg). A kilogram is 1000 *grams* (g) and is slightly more than 2 pounds of mass in the English system (1 kg = 2.205 lb). In the case of time, the SI and the English system both use the *second* (s) as the basic unit.

Prior to the adoption of the SI units there were two versions of the metric system that were widely used, and in some cases are still being used, in scientific work. These are the MKS system, in which the basic length, mass, and time units are meters, kilograms, and seconds, and CGS system, which uses centimeters (cm), grams, and seconds. Since 1 m = 100 cm, the CGS system is closely related to the

MKS system, which in turn is almost identical to the SI. All three are, of course, variations of the metric system.

The unit of force in the SI is the *newton* (N), named for the great English scientist, astronomer, and mathematician Sir Isaac Newton (1642–1727). One newton is the force required to accelerate a 1-kg mass by 1 meter per second per second (1 m/s²). One pound of force in the English system is equivalent to 4.45 N.

Work or energy is force times distance and the fundamental SI unit is the *joule* (J), named for the British physicist James P. Joule (1818–1889). A joule is the work done by a 1-N force applied through a 1-m distance. One foot-pound in the English system equals 1.356 joules.

The rate at which work is being done, or energy is being expended, is *power*. Its basic SI unit is the *watt* (W), named for James Watt (1736–1819), the Scottish engineer whose engine design first made steam power practicable. One watt is defined to be 1 joule per second. We may get a feeling for the size of a watt by noting that a person performing vigorous manual labor averages 25 to 50 W output.

Temperature in the English system is measured in degrees Fahrenheit (°F). In the MKS and CGS systems, Celsius (°C) is used, and in the SI, Kelvin (K) is the standard. We may convert °F to °C by the formula

$$C = \frac{5}{9}(F - 32) \tag{1.1}$$

and °C to °F by

$$F = \frac{9}{5}C + 32 \tag{1.2}$$

Kelvin and Celsius are related by

$$K = 273.15 + C \tag{1.3}$$

water freezes at 32°F, which is equivalent to 0°C or 273.15 K. It boils at 212°F = 100°C = 373.15 K. Absolute zero, the temperature at which all motion ceases, including that of the electrons orbiting the nucleus, is −459.7°F = −273.15°C = 0 K.

Example 1.1: Convert 300 km to yards.

Solution: Since there are 0.914 m/yd and 1000 m/km, the number of yards N is

$$N = \frac{(300 \text{ km})(1000 \text{ m/km})}{0.914 \text{ m/yd}}$$

$$= \frac{300 \times 1000}{0.914} \text{ yd}$$

$$= 328{,}227.6 \text{ yd}$$

Note that in the second step all the units cancel except the one we want.

Example 1.2: Convert 50°F to Celsius and Kelvin.

Solution: By (1.1) we have the Celsius equivalent, given by

$$C = \frac{5}{9}(50 - 32)$$

$$= \frac{5}{9}(18)$$

$$= 10°C$$

By (1.3), with $C = 10$, we have the Kelvin equivalent, given by

$$K = 273.15 + 10$$

$$= 283.15 \text{ K}$$

PRACTICE EXERCISES

1-4.1   Find the number of miles in 20 km.   *Ans.* 12.4274 mi

1-4.2   Find the work in foot-pounds done by a force of 2 lb for a distance of 4 yd.
            Ans. 24 ft-lb

1-4.3   Change the answer in Exercise 1-4.2 to joules. (*Suggestion:* Recall that 1 ft-lb = 1.356 J.)
            Ans. 32.544 J

1-4.4   Convert 68°F to (a) Celsius and (b) Kelvin.   *Ans.* (a) 20°C, (b) 293.15 K

1.5 SCIENTIFIC NOTATION

One of the greatest advantages that the SI has over the English system is its use of powers of 10 to indicate larger and smaller amounts of the basic unit. The SI is thus perfectly suited for the *scientific notation* of representing numbers of extreme size, either large or small, by means of powers of 10. For example, suppose that we have under consideration a distance of 2,136,000.0 m. In the scientific notation this number is written 2.136×10^6 m, which is a number with one digit to the left of the decimal point multiplied by an appropriate power of 10. In the case of positive powers, the power of 10 is the number of places the decimal point must be moved to the left in the number to arrive at a single place to the left of the decimal point.

This follows from the definition of powers of 10, such as

$$1 = 10^0$$
$$10 = 10^1$$
$$100 = 10^2$$
$$1000 = 10^3$$

and so on.

In the case of negative powers of 10, we have

$$1/10 = \quad 0.1 = 10^{-1}$$
$$1/100 = \quad 0.01 = 10^{-2}$$
$$1/1000 = 0.001 = 10^{-3}$$

and so on. Thus the scientific notation for 0.002136 is 2.136×10^{-3}. The negative exponent is the number of places we must move the decimal point to the right to arrive at a single digit to the left of the decimal point.

Example 1.3: Write in scientific notation the numbers (a) 2.3 million and (b) 0.00243.

Solution: For (a) we may write 2.3 million as 2,300,000., which has the scientific notation

$$2,300,000. = 2.3 \times 10^6$$

Note that the decimal point is moved six places to the left. For (b) the scientific notation for 0.00243 is

$$0.00243 = 2.43 \times 10^{-3}$$

because the decimal point is moved three places to the right.

Multiplication and division are also easily performed in the scientific notation. For example, to multiply two numbers, such as 10^a and 10^b, we simply add the exponents a and b. That is,

$$10^a \times 10^b = 10^{a+b}$$

In performing division we subtract the exponents, as in

$$\frac{10^a}{10^b} = 10^{a-b}$$

Example 1.4: Perform the operations indicated in the number N given by

$$N = \frac{(2.3 \times 10^4)(12 \times 10^{-14})}{2 \times 10^{-7}}$$

and express the result in scientific notation.

Solution:

$$N = \frac{2.3 \times 12 \times 10^{4-14+7}}{2} = 13.8 \times 10^{-3}$$

$$= 1.38 \times 10^{-2}$$

Prefixes in the SI: The use of prefixes in the International System eliminates the need for powers of 10 in representing numbers. Some standard prefixes and their abbreviations are shown in Table 1.1. For example, 2725 m = 2.725×10^3 m = 2.725 km (kilometers). As another example, it was once thought that a second was a short time, and that fractions such as 0.1 or 0.01 s were unimaginably short. Nowadays, in applications such as digital computers, the second is an impractically large unit, and times such as 1 microsecond (1 μs or 10^{-6} s) or 1 nanosecond (1 ns or 10^{-9} s) are in common use. (Note: The symbol μ is the lowercase Greek letter *mu*.)

TABLE 1.1 *Prefixes in the SI.*

Power of 10	Prefix	Abbreviation
10^{12}	Tera	T
10^9	Giga	G
10^6	Mega	M
10^3	Kilo	k
10^{-3}	Milli	m
10^{-6}	Micro	μ
10^{-9}	Nano	n
10^{-12}	Pico	p

Example 1.5: Convert (a) 2 MW (megawatts) to watts and (b) 26 μs to s.

Solution: In case (a), 2 MW = 2×10^6 W = 2,000,000 W. In case (b), 26 μs = 26×10^{-6} s = 0.000026 s.

Example 1.6: Find the work done in millijoules (mJ) by a force of 25 micronewtons (μN) applied to a mass for a distance of 100 m.

Solution: Since 25 μN $= 25 \times 10^{-6}$ N and work is force times distance, we have the work in joules given by

$$(25 \times 10^{-6} \text{ N})(100 \text{ m}) = 25 \times 10^{-6} \times 10^2 \text{ Nm}$$

$$= 2.5 \times 10^{-3} \text{ J}$$

Thus, since 10^{-3} J $= 1$ mJ, the work is 2.5 mJ.

The solution may be carried out in much more compact form by noting that

$$\text{Work} = (25 \ \mu\text{N})(100 \text{ m}) = 2500\mu\text{J} = 2.5 \text{ mJ}$$

where we have used the fact that 1 μNm $= 1$ μJ and 1000 μJ $= 1$ mJ. The prefixes μ and m are used in place of the powers of 10 they represent.

Example 1.7: Find the work done by a force of 25 mN applied to a mass for a distance of 2 km.

Solution: Work $= (25 \text{ mN})(2 \text{ km}) = 50$ Nm $= 50$ J. Note that in the product the prefixes m (milli) and k (kilo) cancel each other, since they represent the multiples 10^{-3} and 10^3.

PRACTICE EXERCISES

1-5.1 Write in scientific notation the numbers (a) 25,320,000 and (b) 0.0000101.

Ans. (a) 2.532×10^7, (b) 1.01×10^{-5}

1-5.2 Make the following conversions:

(a) 1250 g to kg
(b) 0.0136 kg to g
(c) 0.00517 s to ms
(d) 0.00517 s to μs   *Ans.* (a) 1.25, (b) 13.6, (c) 5.17, (d) 5170

1-5.3 Find the work in nJ done by a force of 200 μN applied to a mass for a distance of 50 mm.   *Ans.* 10,000 nJ

1.6 SUMMARY

Electricity was discovered in 600 B.C. and thus has had a long history. However, during most of its history very few advancements were made until the eighteenth, nineteenth, and twentieth centuries, during which time almost everything we know about electricity was discovered.

The essence of electricity is the existence of charged bodies which exert forces on one another and which may move, constituting electric current. A typical negative charge is the electron of an atom, and typical positive charges are atoms that have lost electrons.

Materials such as copper, silver, aluminum, and so on, made up of atoms having free electrons, are good conductors of electricity. Materials with few free electrons are insulators and materials that are neither good conductors nor good insulators are called semiconductors.

An electric circuit is a connection of two or more electric elements, such as resistors, capacitors, inductors, batteries, and so on. The analysis of a circuit is the determination of certain quantities associated with the elements, such as voltage, current, and power. This task is made easier by the existence of a standard universally accepted system of units known as the International System of Units, or SI. All the quantities associated with electric circuits are measured in standard units, and the use of prefixes in the SI makes it easy to use the scientific notation.

PROBLEMS

1.1 Find all the closed paths of the electric circuit of Fig. 1.13.

1.2 Add an element between terminals *a* and *d* in Fig. 1.13 and find all the closed paths of the resulting circuit.

1.3 Make the following conversions:

　　(*a*) 1 km to mi
　　(*b*) 1 mi to km
　　(*c*) 50 mJ to ft-lb

1.4 Make the following conversions:

　　(*a*) 10 kN to lb
　　(*b*) 10 lb to mN
　　(*c*) 0.002 ft-lb to μJ

1.5 Make the following conversions:

　　(*a*) 70°F to Celsius
　　(*b*) 50°C to Fahrenheit
　　(*c*) 60°F to Kelvin

1.6 Find the work done in μJ by a force of 100 μN applied to a mass for a distance of (a) 20 mm and (b) 100 in.

1.7 Solve Problem 1.6 if the force is 50 nN.

1.8 Make the following conversions:

　　(*a*) 0.02 s to μs
　　(*b*) 0.02 s to ms

(c) 50 mm to km

(d) 50 mm to μm

1.9 Perform the following operations and express the result in the scientific notation:

(a) $\dfrac{(10^{-6})(10{,}000)(500)}{0.002}$

(b) $\dfrac{(100)^3(20)}{10^{12}}$

(c) $\sqrt{10{,}000} = 10{,}000^{1/2}$

(d) $\dfrac{\sqrt{1000}}{0.001}$

1.10 Repeat Problem 1.9 for the following operations:

(a) $\dfrac{(200)^2(100)}{10^8}$

(b) $\dfrac{(2000)^3(10^{-4})}{2 \times 10^2}$

(c) $(0.01)^2(4000)(0.2)^3$

(d) $\dfrac{(50)(0.002)(10^{-3})}{(0.0002)^3}$

2

ELECTRIC CIRCUIT ELEMENTS

The two-terminal elements of an electric circuit are typically resistors, capacitors, inductors, and voltage sources, and they come in a variety of sizes and shapes. *Discrete* components (as opposed to integrated-circuit components) may range in size from the tiny resistors and capacitors of Figs. 2.1 and 2.2, respectively, to the large high-voltage power capacitor of Fig. 2.3. In the latter case the container housing the capacitor is over 1 foot high. Other examples of rather large circuit elements are the motor-generator set and the transformer shown in Figs. 1.1 and 1.2, respectively.

We shall devote individual chapters to a discussion of resistors, capacitors, and inductors, and in this chapter we will consider voltage sources, particularly batteries, in some detail in Section 2.5. However, the main purpose of the chapter is to consider two-terminal circuit elements in general and give a brief discussion of their associated quantities, such as current, voltage, and power. These quantities and their interrelationships are of prime importance in the study of circuits.

2.1 CHARGE AND CURRENT

As we have stated in Chapter 1, electric current is the movement of charges. In this section we shall formally define these quantities and their SI units, and be more specific about current in electric circuits.

19

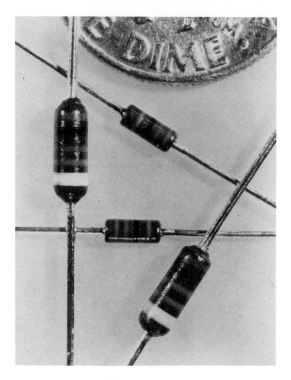

FIGURE 2.1 *Tin oxide film resistors (Courtesy, Corning Glass Works).*

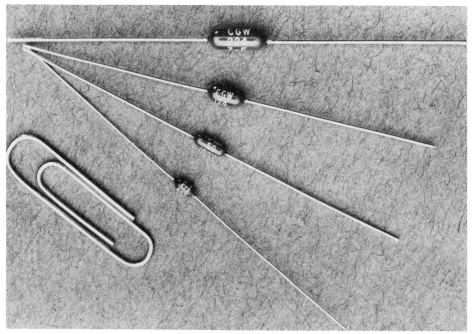

FIGURE 2.2 *Ceramic capacitors (Courtesy, Corning Glass Works).*

FIGURE 2.3 High-voltage capacitor (Courtesy, Westinghouse Electric Corporation).

Coulomb: The unit for charge is the *coulomb* (C), named for Charles Augustin de Coulomb (1736–1806), the French scientist, inventor, and army engineer, whom we mentioned earlier. We shall use Q or q as the symbol for charge, so that a statement such as

$$Q = 2 \text{ C}$$

means that we are discussing a positive charge Q with a magnitude of 2 coulombs. In most cases, capital letters such as Q are reserved for *steady*, or *constant*, quantities and lowercase letters such as q denote *instantaneous* values of quantities that are changing with time.

An electron has a negative charge of approximately 1.6×10^{-19} coulomb, which means that a charge of 1 C is that of $1 \div (1.6 \times 10^{-19}) = 6.25 \times 10^{18}$ electrons. This was first shown by the American physicist Robert A. Millikan (1868–1953) in his famous oil-drop experiment. The charge of an electron and the number of electrons in a coulomb are unimaginable figures to any ordinary human being, but their sizes enable us to use more manageable numbers, such as 2 C, in circuit theory. In almost any practical application of electricity the charge of many billions of electrons are required, and it is certainly easier to say 4 mC, for example, than its equivalent of 25 million billion electrons.

Example 2.1: A charge Q is positive with a magnitude of 25×10^{18} electrons. Find its charge in coulombs.

Solution: $Q = \dfrac{25 \times 10^{18} \text{ electrons}}{6.25 \times 10^{18} \text{ electrons/C}}$

$\qquad\quad = \dfrac{25}{6.25} \text{ C}$

$\qquad\quad = 4 \text{ C}$

Electric Current: The primary purpose of an electric circuit is to move or transfer charges along specified paths. In a conductor, such as a piece of copper wire, in the absence of an external force, the free electrons are moving at random from one atom to another. However, when an external force, such as from a battery, is applied, the free electrons are caused to move in a definite direction, like water flowing through a pipe. This motion of charges is the *current,* which we denote by the letters I or i, taken from the French word "intensité."

The formal definition of current is the *rate of flow* of charge. Thus if Q is the amount of charge passing some arbitrary point in the conductor in a length of time t, the current is given by

$$I = \frac{Q}{t} \tag{2.1}$$

(We are considering the case where Q is constant. If Q varies with time, we must use a small value of t to get an accurate measurement of current.)

If Q is measured in coulombs and t in seconds, the unit of current is coulombs/second (C/s). This unit is called the *ampere* (A), in honor of the French mathematician and physicist André Marie Ampère (1775–1836), mentioned earlier. Thus $1 \text{ A} = 1 \text{ C/s}$.

Example 2.2: A charge of 10 C moves past a given point in a conductor every second. Find the current I in the conductor.

Solution: The current I may be found from (2.1), since the charge is $Q = 10$ C in the time $t = 1$ s. Therefore, we have

$$I = \frac{Q}{t} = \frac{10 \text{ C}}{1 \text{ s}} = 10 \text{ A}$$

Example 2.3: Suppose that the charge in Example 2.2 moves past the point in 0.1 s instead of 1 s. Find *I*.

Solution: Again by (2.1), we have

$$I = \frac{Q}{t} = \frac{10 \text{ C}}{0.1 \text{ s}} = 100 \text{ A}$$

Electron and Conventional Current: As we have described the current in a conductor, it is a flow of free electrons. Thus we are considering *negative* charges as moving in the circuit to constitute the current. This is actually what happens in a wire, and such a current is sometimes called *electron* current. However, in circuit analysis, current is generally thought of as the movement of positive charges. This convention stems from Benjamin Franklin (1706–1790), who with his kite experiment thought that electricity traveled from positive to negative. This current of positive charges is called *conventional* current, to distinguish it from the electron current. Throughout the book we will use conventional current, but it should be noted that conventional current in one direction is the same as electron current in the other direction.

There are, of course, cases where current is actually the movement of positive charges. For example, positive ions (atoms that have lost electrons) may be induced to flow in certain media, such as liquids or gases. Also, certain semiconductor materials with a deficiency of electrons have what are called *holes* or *hole* charges, which are places where electrons are missing. These holes have the opposite polarity of electrons and thus are positive. As electrons fill holes, leaving other holes, the effect is a motion of holes, or positive charges.

To see why electric current in a conductor is similar to water flowing in a pipe, let us consider the effect of the external force. When one end of the conductor is made positive with respect to the other end, which is negative, the positive charges near the negative end are attracted toward it. This leaves the position they vacated with a deficiency of positive charges, so that it is negatively charged. This attracts positive charges farther down the conductor and their movement attracts positive charges still farther down. Since all these movements are incredibly fast, the effect is a virtually simultaneous movement of charges all along the conductor, like the flow of water from a higher pressure to a lower pressure in a pipe.

In summary, electron current is the movement, as in a conductor, of electrons, whereas conventional current is the movement of positive charges. These currents are illustrated in Fig. 2.4, where I_e represents the movement of electrons, symbolized as shown, across imaginary planes from negative to positive. The conventional current I_c is of course in the opposite direction, from positive to negative.

In any case, whether the current is electron or conventional, the current in one direction in a conductor is the negative of that in the other direction. Suppose, for example, that we have 2 A in the direction shown in Fig. 2.5(a). This situation is identical to that of Fig. 2.5(b), where the current is −2 A in the opposite direction.

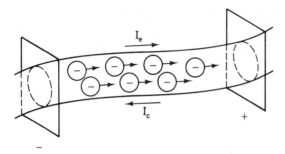

FIGURE 2.4 *Electron and conventional current in a conductor.*

(a)   (b)

FIGURE 2.5 *Two representations of the same current.*

Direct Current (dc): If in Fig. 2.4 the polarity of the external source is always the same, then the current is always in the same direction. It may be steady, or constant, if the external source is steady, as is the case with a battery, or it may vary in magnitude with the source. Constant current is called *direct current* or dc. An example of a direct current is shown in Fig. 2.6(a).

Alternating Current (ac): Current that changes direction periodically is called *alternating current,* or ac. The polarity continually reverses, or alternates, and the magnitude of the current builds up through zero to a peak value in one direction and then through zero to a peak value in the other direction. A graph of a typical alternating current is shown in Fig. 2.6(b). The pattern from $t = 0$ to $t = T$, which is continually repeated, is called a *cycle,* and the number of cycles executed in a second is the *frequency.* The unit of frequency (cycles/s) is the *hertz* (Hz), named for the German physicist Heinrich Rudolph Hertz (1857–1894), referred to earlier. The 60-hertz current used in most homes in the United States is a common ac example.

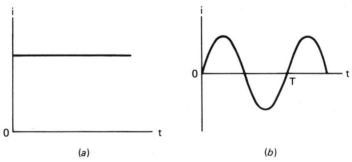

(a)   (b)

FIGURE 2.6 *(a) dc and (b) ac current.*

PRACTICE EXERCISES

2-1.1 If 18.75×10^{18} electrons pass an arbitrary point in a conductor every half-second, find the current in amperes.   *Ans.* 6 A

2-1.2 Find the charge in picocoulombs represented by 100,000 electrons.   *Ans.* 0.016 pC

2-1.3 If the current in a conductor is 4 A, find the charge in coulombs that passes an arbitrary point in (a) 10 s and (b) 0.1 s. [*Suggestion:* Note from (2.1) that $Q = I \times t$.]
Ans. (a) 40, (b) 0.4 C

2.2 VOLTAGE

The external force, considered in the previous section, which causes current through an electrical element is an *electromotive force* (emf) that provides a *voltage,* or *potential difference,* "across" the element. When two charges that want to stay together (that is, unlike charges) are physically separated, force must be exerted and work is done. As long as the charges are held apart they have the *potential* to do work. That is, if they are released, they will attract each other and try to go back together. This phenomenon is somewhat like that of moving a rock uphill. Work is done in the process and *potential energy* (energy with the potential to do work once the rock is released) is stored by moving the rock. There is, of course, a *difference* of potential in two rocks if one is farther uphill than the other.

The Volt: We define the voltage across an element as the work done in moving a coulomb of charge through the element from one terminal to the other. The unit of voltage is the *volt,* abbreviated V, which is named for the Italian physicist Alessandro Guiseppe Antonio Anastasio Volta (1745–1827), who, as we have said, invented the first battery. The convention we will use for labeling the voltage on an element is shown in Fig. 2.7. The voltage across the element is V volts and the $+$, $-$ polarity indicates that terminal a is at a higher potential than terminal b. That is, terminal a is at a potential of V volts higher than terminal b.

Since the voltage is the number of joules of work performed on 1 coulomb of charge, we may say that 1 volt is 1 joule per coulomb. That is,

$$1 \text{ V} = 1 \text{ J/C} \tag{2.2}$$

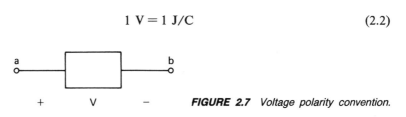

FIGURE 2.7 Voltage polarity convention.

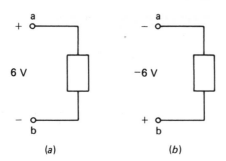

FIGURE 2.8 Two equivalent voltage representations.

(a)   (b)

For example, if it takes 12 J of work to transfer 2 C of charge through an element, the voltage across the element is

$$V = \frac{12 \text{ J}}{2 \text{ C}} = 6 \text{ J/C} = 6 \text{ V}$$

Equivalent Voltage Representations: As examples illustrating the voltage polarity convention, consider Fig. 2.8, which shows two versions of exactly the same voltage. In Fig. 2.8(a), terminal *a* is at a potential +6 V higher than that of terminal *b*. In Fig. 2.8(b), the potential of *b* is −6 V above *a* (or +6 V below *a*). We will use the double-subscript notation V_{ab} for the potential of point *a* with respect to point *b*. That is, in Fig. 2.8(a) we have

$$V_{ab} = 6 \text{ V}$$

Using this notation, we have $V_{ba} = -V_{ab}$, which in this case is

$$V_{ba} = -6 \text{ V}$$

This is more clearly seen in Fig. 2.8(b).

In analyzing a circuit, as we will see later, we may not know beforehand which terminal is at a higher potential. Thus we may simply label the voltage as *V* and determine the true situation depending on whether *V* turns out to be positive or negative.

PRACTICE EXERCISES

2-2.1 If it takes 12 J of work to transfer a charge of 4 C through an element, find the voltage across the element.   *Ans.* 3 V

2-2.2 The voltage across an element is 12 V. Work is done in moving a charge *Q* through the element. If the work is 48 J, find *Q*.   *Ans.* 4 C

2.3 POWER

From the units displayed in (2.2) we may discover another important quantity associated with a circuit element. For example, the element shown in Fig. 2.9 has a voltage V across it and a current I through it, with the polarities as shown. Let us consider what we have, as far as the units are concerned, if we form the product VI. Since the units of V, by (2.2), are J/C and those of I, as we know from the definition of current, are C/s, the units of VI are (J/C)(C/s). Canceling C, we have units of J/s. Thus VI is the rate at which energy (joules) is being expended. Since this is by definition power, which we denote by P or p, we have in general

$$P = VI \qquad\qquad (2.3)$$

That is, the power associated with an element is simply the product of the voltage and current of the element. If V is in volts and I is in amperes, P is in watts (W).

If V and I are not constants, (2.3) should be written

$$p = vi$$

where v and i are varying values of voltage and current and p is the *instantaneous* power (the power delivered at a particular instant of time).

Example 2.4: Suppose that a 12-V source of voltage (such as a battery) produces 2 A in a circuit. Find the power generated by the source.

Solution: By (2.3) the power is

$$P = VI = 12 \times 2 = 24 \text{ W}$$

Example 2.5: A light bulb connected to a 120-V power line draws a current of 0.5 A. How much power is used?

Solution: Again by (2.3), we have

$$P = 120 \times 0.5 = 60 \text{ W}$$

Absorbing and Delivering Power: The last two examples illustrate the difference between an element's providing power or using power. In Example 2.4 the source provides power, or *delivers* power, to the external circuit, or *load*. In Example 2.5, however, the light bulb is *absorbing* power, which is delivered by some external source of emf. In the general case of Fig. 2.9, the element is absorbing power, given by VI. The current is *entering* the positive terminal of the element. If either the polarity

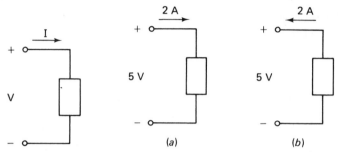

FIGURE 2.9 *Element with indicated voltage and current polarities.*

FIGURE 2.10 *Element (a) absorbing and (b) delivering power.*

of V or I (but not both) is reversed, the current enters the negative terminal, or leaves the positive terminal, in which case the element is delivering power, given by VI, to the external circuit. Thus the element in this case is behaving like a source of emf.

As an example, in Fig. 2.10(a) the element is absorbing power of $P = 5 \times 2 = 10$ W because the current is entering its positive terminal. In Fig. 2.10(b) the element is delivering power of 10 W to the external circuit since its current enters the negative terminal (and thus leaves the positive terminal). Equivalently, we may say in Fig. 2.10(b) that the current entering the positive terminal is $I = -2$ A. Thus the power *absorbed* by the element is

$$P = (5)(-2) = -10 \text{ W}$$

Since this number is negative, the element is actually delivering $+10$ W to the external circuit.

Example 2.6: An element connected to a 12-V line is absorbing 30 W of power. Find the amount and direction of the current drawn by the element.

Solution: From (2.3) we may write

$$I = \frac{P}{V}$$

and therefore the amount of the current is

$$I = \frac{30}{12} = 2.5 \text{ A}$$

Since the element is absorbing power, the direction of the current is *into* the positive voltage terminal.

Horsepower: In the English system the unit for power is *horsepower* (hp), which is equal to 550 ft-lb/s. It may help to visualize a watt to know that

$$1 \text{ hp} = 746 \text{ W} \qquad (2.4)$$

Power used in applications may range from a few picowatts in applications such as satellite communications to millions of watts in supplying the needs of a city. Even in an ordinary household the kilowatt (1000 watts) is a more practical amount than the watt. Since 1 W = 0.001 kW, we have, from (2.4),

$$1 \text{ hp} = 0.746 \text{ kW}$$

Thus we see that a horsepower is roughly $\frac{3}{4}$ kilowatt.

PRACTICE EXERCISES

2-3.1 In Fig. 2.9, $V = 12$ V and $I = 4$ A. Find the power associated with the element. Is it delivering or absorbing power?   *Ans.* absorbing 48 W

2-3.2 Repeat Exercise 2.3.1 if $I = -4$ A.   *Ans.* delivering 48 W

2-3.3 An element drawing a current of 3 mA is absorbing power of 0.012 W. Find the voltage across the element.   *Ans.* 4 V

2-3.4 Convert the answer to Exercise 2-3.1 to horsepower.   *Ans.* 0.0643 hp

2.4 *ENERGY*

Since power is the *rate* of doing work or expending energy, it must be used over a period of time if energy is to be lost or gained. Of course, the longer power is used, the more energy is expended.

If a force, such as that supplied by an electric motor, is providing power of P watts for a time t seconds, the work done or energy used, say W, is given in joules by

$$W = Pt \qquad (2.5)$$

The power, of course, is

$$P = \frac{W}{t} \qquad (2.6)$$

with units of J/s. (Again we are considering small values of t if P is not constant.)

FIGURE 2.11 Kilowatthour meter (Courtesy, Westinghouse Electric Corporation).

From (2.5) we see that a joule, the SI unit for energy, is a wattsecond (Ws). The wattsecond, however, is much too small for most practical purposes, such as the energy consumed in an ordinary home. A more practical amount is the *kilowatthour* (kWh), which is used by the power company in preparing our electric bills. The instrument used for measuring kWh is the kilowatthour meter, an example of which is shown in Fig. 2.11.

Example 2.7: A light bulb uses 500 W for 3 h. Find the energy expended in kWh.

Solution: Since 500 W = 0.5 kW, we have

$$W = Pt = (0.5 \text{ kW})(3 \text{ h}) = 1.5 \text{ kWh}$$

Example 2.8: An element takes 10 A from a 120-V line. Find (a) the power in kW and (b) the number of kWh if the power is absorbed for 45 minutes.

Solution: For case (a), the power is given by

$$P = VI = 120 \times 10 = 1200 \text{ W}$$
$$= 1.2 \text{ kW}$$

For case (b), since 45 minutes is $\frac{3}{4}$ hour, the number of kWh is

$$(1.2 \text{ kW})(\tfrac{3}{4} \text{ h}) = 0.9 \text{ kWh}$$

PRACTICE EXERCISES

2-4.1 Find the number of kWh used by a 1200-W toaster which is on for 20 min.

Ans. 0.4 kWh

2-4.2 Find the number of kWh used by a 5100-W electric clothes dryer for 20 min.

Ans. 1.7 kWh

2-4.3 How long must a 200-W television set be on to use 4 kWh of energy?   *Ans.* 20 h

2.5 VOLTAGE SOURCES—BATTERIES

A *voltage source* is a two-terminal element that maintains a specific voltage between its terminals. The voltage may be a steady, or constant value, such as that supplied by a battery, or it may vary with time, as is the case with an ac generator. A standard symbol for a voltage source is that shown in Fig. 2.12, where the polarity convention indicates that terminal *a* is *v* volts above terminal *b*. That is, if *v* is positive ($v > 0$), terminal *a* is at a higher potential than terminal *b*. The opposite is true if *v* is negative ($v < 0$).

Batteries: The most common dc voltage source is the battery, with a terminal voltage produced by chemical action. That is, the work of separating the positive and negative charges to provide the potential difference is done by chemical forces. This may be illustrated by considering the *voltaic cell* of Fig. 2.13, which consists of two different conducting materials, or *electrodes*, immersed in another material, called an *electrolyte*. The electrolyte is a chemical compound such as an acid, base, or salt, that decomposes into positive and negative ions when it is placed in a solution such as water. The chemical action of forming a new solution causes the separation of charges, resulting in positive charges on one electrode, called the *anode*, and negative

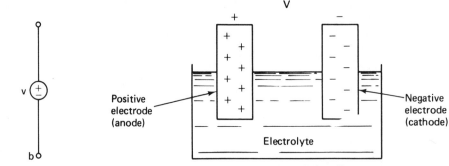

FIGURE 2.12 Volt-
age source.

FIGURE 2.13   Voltaic cell.

charges on the other electrode, called the *cathode*. As current is drawn from one
electrode, or terminal, to the other through an external load, the chemical action
continuously separates charges to maintain the terminal voltage V.

A battery (or battery of cells) is a combination of cells connected in such a
manner that the individual cell voltages add to yield the total voltage of the battery.
A typical circuit symbol for a battery with a terminal voltage of V volts is shown
in Fig. 2.14(a), and is interpreted as having more than one cell, each one like that
of Fig. 2.14(b). The polarity marks may be left off if we understand the longer and
shorter lines in the symbol to represent positive and negative polarities.

To meet the design requirements in a wide variety of uses, such as calculators,
cameras, radios, watches, toys and games, flashlights, smoke alarms, and hearing
aids, batteries are manufactured in numerous shapes, sizes, and voltages. Some examples are the alkaline, mercury, and silver–oxide batteries shown in Figs. 2.15 and
2.16.

Primary and Secondary Cells: Batteries or cells may be classified generally as
of *primary* or *secondary* types. The secondary type is rechargeable, whereas the primary
type is not. In the rechargeable type the current may be reversed to charge the
battery by making the solution build up the electrodes to more or less the original
potential difference. A good example of a primary cell is the ordinary carbon–zinc

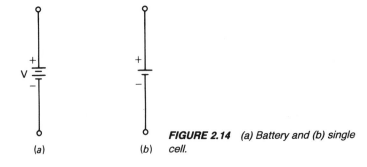

(a)

(b)

FIGURE 2.14   (a) Battery and (b) single
cell.

FIGURE 2.15 *Various types of batteries (Courtesy, P. R. Mallory & Co. Inc.).*

flashlight battery in which the cathode is carbon (mixed with manganese dioxide), the anode is zinc, and the electrolyte is a solution of ammonium chloride. Common secondary batteries are those in automobiles and electronic hand calculators. A cutaway view of a typical lead–acid 6-V truck battery is shown in Fig. 2.17. Its electrodes are lead and lead peroxide and the electrolyte is sulfuric acid. The car battery is a "wet"-cell battery, as opposed to the "dry"-cell flashlight battery. Another example of a rechargeable battery is the lantern battery of Fig. 2.18. The most common rechargeable dry battery is the nickel–cadmium battery used in hand calculators as well as in many other devices.

Specific Gravity: The condition of a lead–acid battery is generally checked by measuring the *specific gravity* of the electrolyte. This is the ratio of the weight of the electrolyte to the weight of an equal volume of water. Concentrated sulfuric acid is 1.835 times as heavy as water and thus its specific gravity is 1.835. In a fully charged cell the specific gravity of the electrolyte at room temperature is approximately 1.280, and in a completely discharged cell it is down to about 1.150.

The battery is checked with a battery hydrometer, which has a calibrated float that rests at a level proportional to the specific gravity. The decimal point is usually omitted, so that a hydrometer reading of 1250 means 1.250 and indicates a cell that is about half-charged.

FIGURE 2.16 *Other types of batteries (Courtesy, P. R. Mallory & Co. Inc.).*

FIGURE 2.17 *Six-volt truck battery (Courtesy, Globe Battery Division of Globe Union, Inc.).*

FIGURE *2.18 Rechargeable lantern battery (Courtesy, Globe Battery Division of Globe Union, Inc.).*

Life of a Battery: Batteries have a capacity rating measured in ampere-hours (Ah). For example, a typical 12-V automobile battery may be rated at 70 Ah at 3.5 A, which means that if the current were 3.5 A, the life of the battery would be $70/3.5 = 20$ h. In general, the life (in hours) is given by

$$\text{life (h)} = \frac{\text{ampere-hour rating (Ah)}}{\text{amperes drawn (A)}} \qquad (2.7)$$

Example 2.9: Find the life of a 70-Ah battery if the current is a steady value of 2 A.

Solution: By (2.7), we have

$$\text{life} = \frac{70 \text{ Ah}}{2 \text{ A}} = 35 \text{ h}$$

PRACTICE EXERCISES

2-5.1 A battery has an ampere-hour rating of 100 Ah. What current will it theoretically provide for 20 h?   *Ans.* 5 A

2-5.2 Ampere-hour ratings are affected by the rate of discharge, or current drawn from a battery. The higher the discharge rate, the lower the Ah rating. Suppose that a battery is rated at 70 Ah at 3.5 A and at 60 Ah at 15 A. Find its life if the discharge rate is (a) 3.5 A and (b) 15 A.   *Ans.* (a) 20 h, (b) 4 h

2-5.3 Temperature also affects the Ah rating of a battery. Suppose that the 70-Ah rating at 3.5 A of the battery of Exercise 2-5.2 is for a temperature of 80°F. If the Ah rating drops by 20% if the temperature is 30°F, find its life at 30°F with a discharge rate of 3.5 A.   *Ans.* 16 h

2-5.4 Find (a) the power supplied by the battery of Exercise 2-5.1 and (b) the energy it supplies during its lifetime of 20 h. Assume that its terminal voltage is 12 V.
                    Ans. (a) 60 W, (b) 1.2 kWh

2.6 OTHER SOURCES OF EMF

There are many sources of emf other than batteries. All of them work on a common principle, that of converting other forms of energy, such as mechanical, light, or heat energy, into electrical energy. In the case of the battery, the conversion, of course, is from chemical to electrical energy.

Generators: As we observed in Chapter 1, the experiments of Faraday and Henry in the nineteenth century demonstrated that current and voltage may be produced by the motion of a coil of wire in the vicinity of a magnet. The movement of the wire through the magnetic field produced by the magnet induces a voltage across the ends of the wire. Therefore, if wires are systematically wound on a rotating cylinder, or *rotor,* and made to revolve around a shaft in a magnetic field, a voltage will be produced across the ends of the wires. If the wires are connected so that their voltages add in a prescribed way, the result is a *generator.* An example is, of course, the generator of Fig. 1.1.

The voltage generated is an ac voltage because it continuously changes with the position of the rotor. However, dc voltages may also be generated by using a device called a *commutator* to reverse the polarity of every other half-cycle of the output voltage. In this case the rotating cylinder is called an *armature,* and the process of changing ac to dc is called *rectification.*

As an illustration, the small laboratory-type generator of Fig. 2.19 may be used by hand to generate an ac voltage. A battery (not shown) is used with a coil of wire to produce the magnetic field.

We shall discuss generators in more detail when we consider ac voltages in general. For the time being we shall simply note that they are represented by the same voltage source symbol as that given in Fig. 2.12. To emphasize the alternating nature of an ac generator, an ac cycle is sometimes shown in the symbol, as illustrated in Fig. 2.20.

FIGURE 2.19 *Laboratory-type hand operated generator (Courtesy, Sargent-Welch Scientific Co.).*

Other Voltage Sources: Batteries and generators are by far the most common voltage sources, but there are many others as well. One example is the *photovoltaic* cell, or *solar* cell, which uses light to produce a potential difference. Another example uses *thermal emission,* which is the release of electrons from certain materials by the application of heat. Finally, the most common dc source used in the laboratory is a *power supply,* which provides a dc voltage from an ac voltage by rectification. Examples of dc laboratory power supplies are shown in Fig. 2.21.

FIGURE 2.20   *Ac generator symbol.*

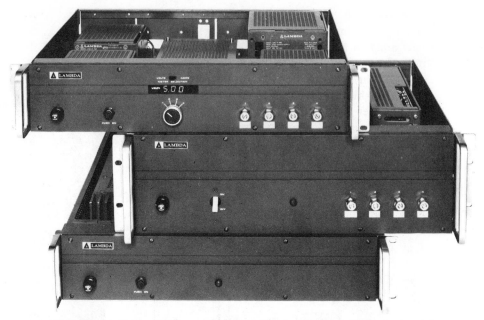

FIGURE 2.21 *Rack adapter mounted power supply assemblies (Courtesy, Lambda Electronics, Division of Veeco Instruments, Inc.).*

Current Sources: A *current source* is a two-terminal element that maintains a specific current through its terminals. Like the voltage of a voltage source, the current may be constant or it may vary with time. In either case we shall use as a representation of a current source the standard symbol of Fig. 2.22(a). The polarity convention shown indicates a current of i amperes in the direction of the arrow regardless of any external load connected to the terminals. Occasionally, in the case of an ac source, we may use the symbol shown in Fig. 2.22(b).

Current sources are not constructed as primary elements in the way that voltage sources are formed. That is, we normally require a voltage source for use in constructing a current source. For example, a battery or generator connected with a resistor

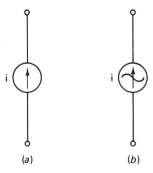

(a)   (b)

FIGURE 2.22 *(a) Standard current source and (b) ac current source symbols.*

may furnish an approximately fixed current over a wide range of operating conditions. In another manner, a transistor properly connected with three resistors and a battery may function as a constant-current source with a terminal current of the order of 2 mA and a terminal voltage varying from −6 to 30 V.

We may also have laboratory power supplies which are current sources. These may resemble the devices of Fig. 2.21. They will provide a remarkably constant current over a wide range of terminal voltages. Again, the ultimate source of power is the voltage source of the ac power line.

PRACTICE EXERCISES

2-6.1 A current source will supply a nearly constant current of 2 mA as long as its terminal voltage is between 2 and 20 V. Find the power it delivers to a load if the load voltage is 15 V. *Ans.* 30 mW

2-6.2 Find the maximum and minimum values of power the source of Exercise 2-6.1 can supply with a nearly constant current. *Ans.* 40 mW, 4 mW

2.7 SUMMARY

Circuit elements may be defined in terms of their associated quantities, such as current and voltage. Current is the rate of flow of charge and is measured in amperes, an ampere being a coulomb of charge per second. Electron current is the flow of electrons and conventional current, with which we shall concern ourselves, is the flow of positive charges. Voltage, measured in volts, is the work done in moving a coulomb of charge through the element. Thus a volt is a joule per coulomb.

Power is the rate of doing work or expending energy. Its unit is the watt, which is a joule per second. In the case of an electric element with voltage v and current i, the power associated with it is the product vi. The element may be absorbing power if i enters the positive terminal of v, or delivering power otherwise.

Energy is expended when power is supplied over a period of time. The unit of energy, the joule, is thus a wattsecond. A more practical unit is the kilowatthour, which is the unit used in monitoring the electricity used by our houses.

Sources of emf may be of the voltage-source or current-source variety. A voltage source supplies a specified voltage across its terminals and a current source maintains a specified current through its terminals. In the ideal case, sources are not influenced by externally connected loads. Typical voltage sources are batteries, generators, solar cells, and so on, and typical current sources are constructed with voltage sources together with other elements.

PROBLEMS

2.1 Find the current in amperes if 1800 C of charge passes through a wire in (a) 90 s and (b) 3 min.

2.2 Find the current in amperes if a charge equivalent to 25×10^{18} electrons passes through a wire every 2 s.

2.3 If a current in a wire is 6 A, find the charge in coulombs that passes an arbitrary point in the wire in (a) 25 s and (b) 0.01 s.

2.4 If 100,000 electrons pass a point in a wire every 2 s, find the current in microamperes.

2.5 How many coulombs of charge pass through an element in 5 min if the current is 20 mA?

2.6 If the voltage across an element is 20 V, how much work is required to move 4 C of charge through the element?

2.7 If 50 J of work are required to move 4 C of charge through an element, find the voltage across the element.

2.8 The terminal voltage of a battery is 1.5 V. If it supplies 15 J in moving a charge Q, find the value of Q in coulombs.

2.9 A 12-V battery produces 3 A in a circuit. Find the power delivered by the battery.

2.10 If 60 mW is supplied by a 1.5-V battery, find the current it provides.

2.11 If 48 J of work is done in 20 s in supplying energy to an element carrying a current of 30 mA, find the terminal voltage of the element.

2.12 If 60 J of work is done in 40 s in supplying energy to an element whose terminal voltage is 30 V, find the current through the element.

2.13 Charge is flowing through an element at the rate of 100 C/min and the element is absorbing energy at the rate of 40 J/min. Find the voltage across the element.

2.14 An element drawing a current of 3.73 A has a terminal voltage of 60 V. Find the power it absorbs in horsepower.

2.15 If the element of Problem 2.14 absorbs power for 100 s, how much energy in joules is delivered to it?

2.16 Change the answer to Problem 2.15 to kWh.

2.17 An element draws 5 A from a source of 12 V. Find (a) the power it absorbs and (b) the number of kWh if the power is absorbed for 20 h.

2.18 How long must a 1200-W toaster be on to use 6 kWh of energy?

2.19 For how many hours will a battery with a rating of 100 Ah theoretically provide a current of 4 A?

2.20 A battery is rated at 70 Ah at 3.5 A and at 15 A it has a life of 3 h. Find (a) its life at a discharge rate of 3.5 A and (b) its ampere-hour rating at 15 A.

2.21 A battery has an ampere-hour rating of 100 Ah and a life of 40 h at rated current. Find (a) the power supplied by the battery at rated current and 12 V terminal voltage and (b) the energy it supplies during its lifetime under these conditions.

2.22 A current source of 6 A supplies power to an element for 20 min at a constant voltage V. If 0.1 kWh is delivered, find V.

3

RESISTANCE

The simplest and most common electric circuit element is the *resistor,* which we will consider in this chapter. A good example is a wire conductor in which a current can be produced by applying a voltage across its terminals. The moving electrons, which constitute the current, collide with other electrons and atoms in the conductor, causing heat and a *resistance* to the current. Thus resistance is a measure of the resistor's opposition to current and may be thought of as the electrical equivalent of friction. For a fixed terminal voltage, the higher the resistance, the lower will be the current, and vice versa. A copper conductor has relatively low resistance because copper contains many free electrons which are easily made to move by an emf. On the other hand, a material such as carbon, with a lower number of free electrons that are more tightly bound to their nuclei, has a very high resistance.

In this chapter we shall describe resistors and consider their properties, such as resistance and its unit of measurement, their voltage–current relationship (known as Ohm's law), their power and energy relationships, and certain practical features such as their physical makeup and color codes. In Chapter 4 we will apply our knowledge of resistors to analyze *resistive* circuits, which are electric circuits containing only resistors and sources.

3.1 OHM's LAW

We define a *resistor* as a two-terminal element whose voltage is directly proportional to its current. The standard circuit symbol for a resistor is that shown in Fig. 3.1, with an associated voltage V and current I as shown. The statement that V is proportional to I may be represented by

$$V = RI \qquad (3.1)$$

where R, the constant of proportionality, is the *resistance* of the resistor.

The voltage–current relationship (3.1) is known as *Ohm's law,* in honor of the German physicist Georg Simon Ohm (1787–1854), who formulated it in 1826 and published his results the following year. The resistance R is given by

$$R = \frac{V}{I} \qquad (3.2)$$

and thus has the standard unit of volt/ampere, which is defined as the *ohm*. That is, 1 ohm is 1 volt/ampere.

Resistance Symbol: The symbol used to represent the ohm is the capital Greek letter *omega* (Ω). Thus we have

$$1\ \Omega = 1\ \text{V/A} \qquad (3.3)$$

An example is the resistor of Fig. 3.2, whose resistance is $R = 5\ \Omega$.

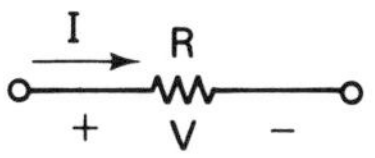

FIGURE 3.1 *Circuit symbol for a resistor with associated voltage and current and resistance R.*

FIGURE 3.2 *A 5-ohm resistor.*

Example 3.1: A toaster is essentially a resistor that becomes hot when it carries a current. If a toaster has a current of 5 A with a voltage of 120 V, find its resistance.

Solution: By (3.2) the resistance is

$$R = \frac{V}{I}$$

$$= \frac{120}{5}$$

$$= 24\ \Omega$$

Current in a Resistor: Ohm's law (3.1) may be solved for the current, resulting in

$$I = \frac{V}{R} \tag{3.4}$$

Thus an ampere is a volt/ohm (1 A = 1 V/Ω). Also we see that for a fixed voltage, the higher the resistance, the lower the current, and vice versa.

> *Example 3.2:* The voltage across a resistor is 12 V. Find the current if the resistance is (a) 3 Ω and (b) 1 kΩ (1000 Ω, or 1 *kilo-ohm,* usually shortened to 1 *kilohm*).

Solution: In case (a) we have, by (3.4),

$$I = \frac{V}{R} = \frac{12 \text{ V}}{3 \text{ } \Omega} = 4 \text{ A}$$

In case (b),

$$I = \frac{12 \text{ V}}{1 \text{ k}\Omega}$$

$$= \frac{12 \text{ V}}{10^3 \text{ } \Omega}$$

$$= 12 \times 10^{-3} \text{ A}$$

$$= 12 \text{ mA}$$

Other Units: In Example 3.2 we see that resistance, like any other electrical quantity, may come in large sizes, which are easier to handle by means of the prefixes of the SI. Thus a volt/kilohm is a milliampere, and so forth. In many applications, volts and amperes are practical units, and thus ohms are practical units. This is true, for example, with power circuits like those in our home. In other cases, such as in solid-state electronic devices, milliamperes are more practical current units, and thus kΩ and MΩ (*mega-ohms,* or simply *megohms*) are common units of resistance.

Polarities Associated with Ohm's Law: Equation (3.1) applies to Fig. 3.1, where the current enters the positive voltage terminal. If either the current or the voltage polarities (but not both) are reversed, the current enters the negative voltage terminal, as shown in Fig. 3.3. Since this is equivalent to changing the sign of either I or V in (3.1), Ohm's law for this case is

$$V = -RI \tag{3.5}$$

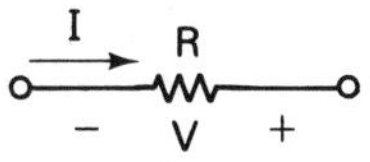

FIGURE 3.3 *Resistor with a reversed voltage polarity.*

Equations (3.1) and (3.5), together with Figs. 3.1 and 3.3, illustrate that in the case of resistor current the charges flow from high to low potential, in a manner like that of a rock which falls from a high to a low place on a hill. The resistance R is a positive number, so that in Fig. 3.1 V is positive when I is positive and thus I is from high to low potential. On the other hand, in Fig. 3.3, if I is positive, then by (3.5) V is negative and the current is still from high to low potential.

Linear Resistors: The resistors that we are considering are called *linear* resistors because the voltage–current relationship (3.1) is the equation of a straight line. That is, the variables V and I appear to the first degree, and the ratio of V to I is the constant value R. Resistors for which the ratio V/I is not constant are *nonlinear* (not linear). A common example of a nonlinear resistor is an incandescent lamp.

In reality, all resistors are nonlinear because the electrical characteristics of all conductors are affected by environmental factors such as temperature. Many materials, however, closely approximate an ideal linear resistor over a desired operating region. We shall consider only these types of elements, or linear resistors, and refer to them simply as resistors.

PRACTICE EXERCISES

3-1.1 The terminal voltage of a 20-Ω resistor is 100 V. Find the current carried by the resistor.   *Ans.* 5 A

3-1.2 Find the voltage that must be applied across a 2-kΩ resistor in order for the resistor current to be 6 mA.   *Ans.* 12 V

3-1.3 For the resistor of Fig. 3.3 the resistance is $R = 10\ \Omega$ and the current is $I = 4$ A. Find the voltage V.   *Ans.* −40 V

3.2 CONDUCTANCE

Another important quantity that is associated with a resistor is its *conductance,* denoted by G and given by

$$G = \frac{1}{R} \qquad (3.6)$$

That is, the conductance is the reciprocal of the resistance. Thus for a given resistor voltage, the higher the conductance, the lower the resistance and, consequently, the higher the current.

The unit of conductance is the *mho*, which is ohm spelled backward. The symbol for the mho is an inverted omega ($\mho$). As an example, a 10-Ω resistor has a conductance of 0.1 $\mho$ ($1/R = 1/10$). (Another term that is sometimes used for the conductance unit is *siemens.*)

Duals: In terms of conductance, Ohm's law,

$$V = RI$$

may be written, by (3.6), as

$$I = GV \qquad (3.7)$$

Thus we see that

$$G = \frac{I}{V}$$

and therefore that $1 \; \mho = 1 \; \text{A/V}$.

We may note from the two versions of Ohm's law that one may be obtained from the other by replacing V by I, I by V, and R by G, or G by R. For this reason we say that one version is the *dual* of the other, and that V and I are duals and R and G are duals. This method of obtaining duals of statements is often extremely useful in circuit theory.

Example 3.3: A resistor with $G = 2 \; \text{m}\mho$ has a current of 6 mA. Find the voltage across the resistor.

Solution: By (3.7) we have

$$V = \frac{I}{G} = \frac{6 \times 10^{-3} \; \text{A}}{2 \times 10^{-3} \; \mho} = \frac{6 \; \text{mA}}{2 \; \text{m}\mho} = 3 \; \text{V}$$

Example 3.4: Find the resistance of the resistor of Example 3.3.

Solution: From (3.6) we have

$$R = \frac{1}{G} = \frac{1}{2 \times 10^{-3} \; \mho} = 0.5 \times 10^3 \; \Omega = 0.5 \; \text{k}\Omega$$

or more compactly,

$$R = \frac{1}{G} = \frac{1}{2 \; \text{m}\mho} = 0.5 \; \text{k}\Omega$$

PRACTICE EXERCISES

3-2.1 Find the conductance of a resistor for which (a) $R = 5\ \Omega$ and (b) $R = 2\ k\Omega$.

Ans. (a) 0.2 ℧, (b) 0.5 m℧

3-2.2 Find the current in a resistor that has a terminal voltage of 10 V and a conductance of (a) 2 ℧ and (b) 2 m℧. *Ans.* (a) 20 A, (b) 20 mA

3.3 POWER ABSORBED BY A RESISTOR

As we have noted, current in a resistor produces heat because of the collisions of the moving electrons with other electrons and with the nuclei of the atoms in the resistor material. Thus resistor current converts electrical energy into heat and therefore power is being absorbed by, or *dissipated* in, the resistor. Dissipation is a good description of this process because the joules of heat that are produced are lost to the surrounding atmosphere and cannot be returned to the circuit as electrical energy. The heat that is lost quite often serves a useful purpose, however, since it may be used to provide light from a bulb, warmth from a heater or an iron, or to open a fuse by melting its metal link when the current becomes excessive. Other times, of course, the heat produced is undesirable. Desirable or not, however, it is always present.

Relationship for Power: We know from Chapter 2 that the power delivered to any electric element with voltage V and current I is given by

$$P = VI \tag{3.8}$$

In the case of power dissipated in a resistor, it is convenient to express this result in terms of the resistance R. This is readily done by replacing V by RI from Ohm's law, resulting in

$$P = I^2 R \tag{3.9}$$

For another form we may replace I by V/R, which yields

$$P = \frac{V^2}{R} \tag{3.10}$$

In each case, if V is in volts, I in amperes, and R in ohms, the power is in watts.

Example 3.5: A 50-Ω resistor carries a current of 4 A. Find the power dissipated by the three methods of (3.8) to (3.10).

Solution: By Ohm's law the voltage is

$$V = RI = (50)(4) = 200 \text{ V}$$

By (3.8) we have the power given by

$$P = VI = (200)(4) = 800 \text{ W}$$

by (3.9) we have

$$P = I^2 R = (4)^2(50) = 800 \text{ W}$$

and by (3.10) the power is

$$P = \frac{V^2}{R} = \frac{(200)^2}{50} = 800 \text{ W}$$

Power Rating: Electrical elements that utilize heat produced by a resistor, such as light bulbs, toasters, and irons, are usually rated in terms of the power they dissipate. The power rating is at the normal operating voltage. For example, a light bulb to be operated at a voltage of 120 V may be rated 300 W. In this case the resistance R of the bulb is such that at 120 V the current is

$$I = \frac{P}{V} = \frac{300}{120} = 2.5 \text{ A}$$

Therefore, the resistance is

$$R = \frac{P}{I^2} = \frac{300}{(2.5)^2} = 48 \ \Omega \tag{3.11}$$

In (3.11) we have R given in terms of P and I. It may also be found in terms of P and V by means of (3.10). The result is

$$R = \frac{V^2}{P} \tag{3.12}$$

In the same manner we may obtain expressions for I and V from (3.9) and (3.10), given by

$$I = \sqrt{\frac{P}{R}} \tag{3.13}$$

and

$$V = \sqrt{RP} \tag{3.14}$$

Energy: As we saw in Chapter 2, if an electric element is absorbing a constant power of P watts for a time of t seconds, the total energy in joules used by the element is

$$W = Pt \tag{3.15}$$

In the case of a resistor with resistance R and current I, the energy, which is converted to heat, is

$$W = I^2Rt \tag{3.16}$$

Equivalently, if $V = RI$ is the voltage across the resistor, (3.15) may also be written

$$W = \frac{V^2t}{R} \tag{3.17}$$

Example 3.6: A toaster with a resistance of 24 Ω operates at a voltage of 120 V. If the toaster is on for 20 s, find the energy it uses.

Solution: By (3.17) we have

$$W = \frac{V^2t}{R} = \frac{(120)^2(20)}{24} \text{ J} = 12 \text{ kJ}$$

PRACTICE EXERCISES

3-3.1 A source of 100 V produces 4 A in a resistor. Find (a) the power dissipated and (b) the resistance.   *Ans.* (a) 400 W, (b) 25 Ω

3-3.2 A toaster is rated at 600 W at a voltage of 120 V. Find (a) the current it normally draws and (b) its resistance.   *Ans.* (a) 5A, (b) 24 Ω

3-3.3 A 20-Ω resistor dissipates 180 W of power. Find the current it carries.   *Ans.* 3 A

3-3.4 Electric shock is a painful contraction of the muscles caused by current passing through the body, and a current as low as 20 mA could be fatal. If the body resistance is typically 25 kΩ, find the voltage that it must be subjected to for a body current of 20 mA.   *Ans.* 500 V

3-3.5 A factor as important in electric shock as high voltage is the amount of power the source of the voltage can supply. In the example of Exercise 3-3.4, find the power provided by the source.   *Ans.* 10 W

3.4 PHYSICAL RESISTORS

Physically, resistors are made from a variety of materials and are available in many sizes, values, and shapes, as may be seen in Figs. 3.4 to 3.6. Their resistance may range in value from a fraction of an ohm to many megohms and be capable of dissipating power from a fraction of a watt to several hundred watts.

Resistor Characteristics: The two main characteristics of a resistor are its resistance, of course, and its *wattage rating,* or *power rating.* The nominal resistance value is identified on the resistor by means of a number or a code, and its actual resistance may vary from this nominal value by no more than some specified amount, known as its *tolerance.* For example, a resistor may have a nominal value of 1000 Ω with a tolerance of ±5%. In this case its actual value may deviate from the nominal

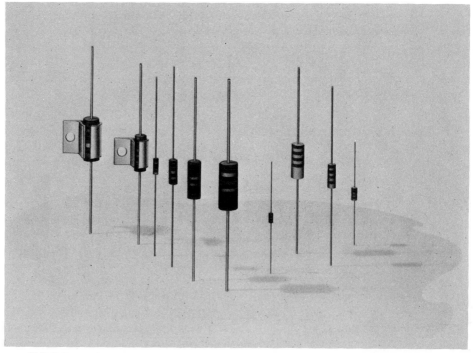

FIGURE 3.4 *Carbon-composition resistors of various sizes (Courtesy, Allen-Bradley Co.).*

FIGURE 3.5　*Thin resistors in a stack of three (Courtesy, Ohmite Manufacturing Co.).*

FIGURE 3.6　*High-current, low-resistance resistor (Courtesy, Ohmite Manufacturing Co.).*

value by as much as 5% of 1000, or

$$(0.05)(1000) = 50 \ \Omega$$

Thus the actual resistance may lie between $1000 - 50 = 950 \ \Omega$ and $1000 + 50 = 1050 \ \Omega$.

The wattage rating is the maximum wattage the resistor can dissipate without damage to it. For example, if a 100-Ω resistor has a wattage rating of $\frac{1}{4}$ W, the maximum current I that it could safely carry may be found from

$$100I^2 = 0.25$$

which yields $I = 0.05$ A $= 50$ mA.

Carbon-Composition Resistors:　One of the two most common types of resistors used in electric circuits is the *carbon-composition* resistor. As shown in Fig. 3.7, it is made of hot-pressed carbon granules mixed with an insulating material in the right proportion to yield the desired resistance. The resistance material is enclosed in a plastic case with wire leads connected to form the two terminals. As examples, the resistors of Fig. 3.4 are carbon-composition resistors.

Carbon-Film Resistors:　The other most common resistor type is the *carbon-film* type, which consists of carbon powder deposited on an insulating material and packaged like the carbon-composition resistor. These two types of carbon resistors are the least expensive of all the resistor types, which accounts for their popularity. However, they have the disadvantage of a relatively high variation of resistance with temperature and thus are not as desirable in many applications as the more expensive resistors. Values of carbon resistors range from 2.7 Ω to 22 MΩ, with wattages from $\frac{1}{8}$ to 2 W.

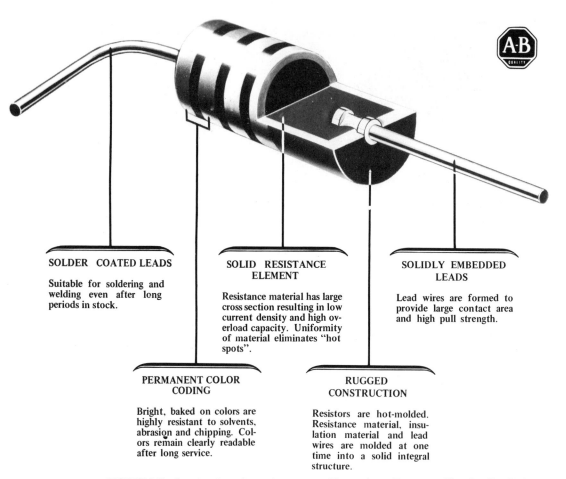

SOLDER COATED LEADS

Suitable for soldering and welding even after long periods in stock.

SOLID RESISTANCE ELEMENT

Resistance material has large cross section resulting in low current density and high overload capacity. Uniformity of material eliminates "hot spots".

SOLIDLY EMBEDDED LEADS

Lead wires are formed to provide large contact area and high pull strength.

PERMANENT COLOR CODING

Bright, baked on colors are highly resistant to solvents, abrasion and chipping. Colors remain clearly readable after long service.

RUGGED CONSTRUCTION

Resistors are hot-molded. Resistance material, insulation material and lead wires are molded at one time into a solid integral structure.

FIGURE 3.7   Construction of a carbon-composition resistor (Courtesy, Allen-Bradley Co.).

Wire-wound Resistors: In applications where high performance is desired or where temperature is an important factor, resistors should be used that are of higher quality than the carbon resistors. One such resistor is the *wire-wound* type, an example of which is shown in Fig. 3.8. This type consists of a metallic wire, usually a nickel–cadmium alloy, wound on a ceramic core. Low-temperature coefficient wire permits the fabrication of very precise resistors with accuracy of the order of $\pm 1\%$ to $\pm 0.001\%$. Wire-wound resistors have wattage ratings from 5 W to several hundred watts and range in resistance from a fraction of an ohm to several thousand ohms.

Metal-Film Resistors: Another valuable and useful resistor type is the *metal-film* resistor, which is made of a thin metal layer on an insulating material. Accuracy and stability for these resistors approaches that of wire-wound types, and high resistance values are much easier to attain.

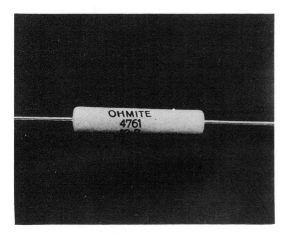

FIGURE 3.8 Wire-wound resistor (Courtesy, Ohmite Manufacturing Co.).

Integrated-Circuit Resistors: The resistors we have considered thus far are discrete resistors, as opposed to resistors in *integrated-circuit* form. An integrated circuit, consisting of a single chip of semiconductor material, may have a large number of resistors fabricated in it. A chip about $\frac{1}{8}$ in. square could contain hundreds of resistors. Three examples of integrated circuits containing a number of resistors are shown in Fig. 3.9.

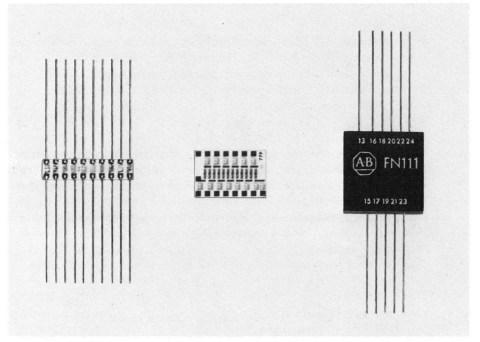

FIGURE 3.9 Integrated-circuit resistor networks (Courtesy, Allen-Bradley Co.).

PRACTICE EXERCISES

3-4.1 A resistor with a nominal value of 820 Ω has a $\pm 10\%$ tolerance. Find the range in which its actual resistance lies. *Ans.* 738 to 902 Ω

3-4.2 A 100-Ω carbon resistor is to carry a current of 0.1 A. Find the wattage rating that should be used if the safety factor is 2. (That is, the wattage rating used is to be twice the calculated value.) *Ans.* 2 W

3-4.3 A 100-Ω resistor has a wattage rating of 1 W. (a) Find the voltage that can safely be applied to the resistor. (b) Find the maximum voltage that the resistor is selected for if the safety factor involved in calculating the wattage rating is 2. *Ans.* (a) 10 V, (b) 7.07 V

3.5 VARIABLE RESISTORS

The resistors discussed in Section 3.4 were all *fixed* resistors. That is, their resistances were fixed when they were made and could not be changed by the user. *Variable* resistors are those whose resistance can be adjusted by turning a knob, moving a slider, or by applying a screwdriver to yield a value from 0 to some specified R. The symbol for a two-terminal variable resistor, or *rheostat,* is shown in Fig. 3.10.

Potentiometer: A *potentiometer,* or *pot,* is a variable resistor with three terminals, or *lugs,* symbolized by Fig. 3.11. The center lug, or center tap, c is movable on the continuous resistor *a-b,* providing a resistance kR between points a and c. The parameter k varies from 0 to 1 and represents the fraction R_{ac} of the total available resistance R that is seen between terminals *a-c*. That is, when the center tap is at a, we have $k = 0$ and $R_{ac} = 0$, and when the center tap is at b, we have $k = 1$ and $R_{ac} = R$. In between, of course, we have $R_{ac} = kR$.

A potentiometer may function as a rheostat by simply leaving terminal b open (that is, disconnected), as shown in Fig. 3.12. In this case we have a two-terminal variable resistor with terminals a and c whose resistance varies from 0 to R as the center tap is moved from a to b.

The name "potentiometer" derives from the fact that the moving center tap c

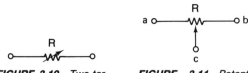

FIGURE 3.10 *Two-terminal variable resistor, or rheostat.*

FIGURE 3.11 *Potentiometer symbol.*

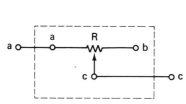

FIGURE 3.12 *Potentiometer functioning as a rheostat.*

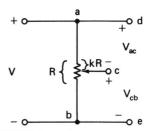

FIGURE 3.13 *Potentiometer used to control output voltages.*

is often used to control by its position the potential differences V_{ac} and V_{cb} between terminals a and c and between terminals c and b, respectively. This is shown in Fig. 3.13, where V is an applied, or *input*, voltage across terminals a and b. As we shall see later when we consider *voltage division*, if there is no current in the *output* leads d, c, and e, then

$$V_{ac} = kV \tag{3.18}$$

and

$$V_{cb} = (1 - k)V \tag{3.19}$$

Example 3.7: Determine k for the potentiometer of Fig. 3.13 if the output voltage V_{ac} is to be twice the output voltage V_{cb}. Find the ratio V_{ac}/V in this case.

Solution: From (3.18) and (3.19) we have

$$\frac{V_{ac}}{V_{cb}} = \frac{kV}{(1 - k)V} = \frac{k}{1 - k}$$

Since $V_{ac} = 2V_{cb}$, this becomes

$$2 = \frac{k}{1 - k}$$

from which $k = \frac{2}{3}$. By (3.18) we have

$$\frac{V_{ac}}{V} = k = \frac{2}{3}$$

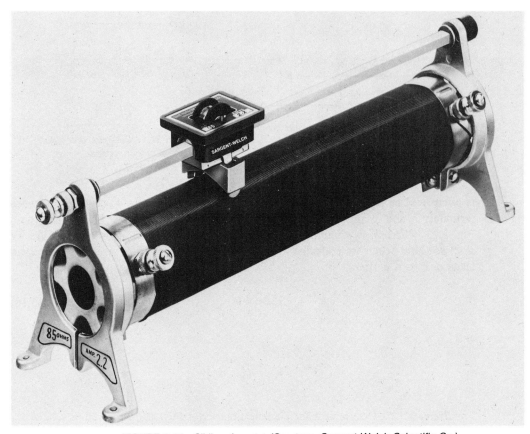

FIGURE 3.14   *Sliding rheostat (Courtesy, Sargent-Welch Scientific Co.).*

An example of a rheostat, whose resistance is determined by the position of a slider on a coil of wire, is shown in Fig. 3.14, and five types of potentiometers are illustrated in Fig. 3.15.

Decade Resistance Box: Another type of variable resistor is a *decade resistance box,* examples of which are shown in Figs. 3.16 and 3.17. The term *decade* stems from the fact that by turning dials resistance values may be obtained which are sums of multiples of powers of 10, like decimal numbers. In this sense the box of Fig. 3.17 is not truly a decade box but a resistance *substitution* box, as noted. Its available resistance values are standard values, and thus it is very useful in applications where resistors on the shelf are to be substituted for the resistance box, or vice versa.

In each of the boxes of Figs. 3.16 and 3.17 there are two dials, one for a low resistance range and one for a high resistance range. In Fig. 3.16(a), for example, we may obtain a resistance of 25 Ω by setting the left dial (the units dial) at 5 and

FIGURE 3.15 *Five types of potentiometers (Courtesy, Centralab Electronics Division, Globe-Union, Inc.).*

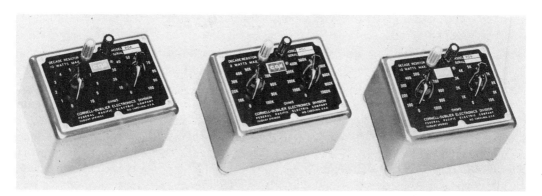

FIGURE 3.16 *Decade resistance boxes with resistance ranges of (a) 1 to 110Ω, (b) 10 kΩ to 1.1 MΩ, and (c) 1 to 11 kΩ (Courtesy, Cornell-Dubilier Electronics/ Subsidiary of Federal Pacific Electric Company).*

FIGURE 3.17 Resistance substitution box (Courtesy, Heath Company).

the right dial (the tens dial) at 2. In Fig. 3.17 we may obtain resistance values with either the low-range or the high-range dial by setting the switch to LO or HI.

PRACTICE EXERCISES

3-5.1 Determine k for the potentiometer of Fig. 3.13 if V_{ac} is to be (a) 3 times V_{cb} and (b) $\frac{1}{2}$ of V_{cb}.   *Ans.* (a) $\frac{3}{4}$, (b) $\frac{1}{3}$

3-5.2 Find k for Fig. 3.13 if V_{cb} is to be $\frac{1}{2}$ the total available voltage V. [*Suggestion:* See (3.19).]   *Ans.* 0.5

3.6 RESISTOR COLOR CODING

Because of their small size it is inconvenient to label the carbon resistors with their numerical values of resistance. For this reason a standard *color code* is used instead of numbers to represent their resistance in ohms.

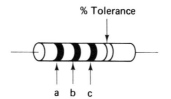

FIGURE 3.18 *Carbon resistor.*

Color Bands: A typical carbon resistor is shown in Fig. 3.18, with four bands labeled *a, b, c,* and % tolerance. These are *color bands,* which are painted on one end of the resistor body to indicate the nominal value of the resistance. Bands *a, b,* and *c* give the nominal resistance and the % tolerance, or *tolerance,* band gives the percent that the resistance may deviate, either up or down, from its nominal value.

TABLE 3.1 *Color code for carbon resistors*

	Bands a, b, and c		
Color	*Value*	*Color*	*Value*
Silver[a]	−2	Yellow	4
Gold[a]	−1	Green	5
Black	0	Blue	6
Brown	1	Violet	7
Red	2	Gray	8
Orange	3	White	9
	% tolerance band		
Gold	±5%		
Silver	±10%		

[a] These colors apply to band *c* only.

The bands are colored in accordance with Table 3.1 to form the color code for the resistance. The first band, *a,* always nearest one end of the resistor, indicates the first digit in the numerical value of the resistance. The second band, *b,* indicates the second digit, and the third band, *c,* indicates the power of 10 that multiplies the two-digit number represented by bands *a* and *b* to give the resistance. For example, if band *a* is red (representing 2 by Table 3.1), band *b* is violet (representing 7), and band *c* is orange (representing 3, or the multiple 10^3), the nominal resistance is 27×10^3 Ω, or 27 kΩ.

Resistance Formula: If we let *a, b,* and *c* denote the values assigned to their bands, we may write the formula for the nominal resistance as

$$R = (10a + b) \times 10^c \qquad (3.20)$$

To illustrate the use of the formula, in the example just cited we have $a = 2$, $b = 7$, and $c = 3$, so that by (3.20) the nominal resistance is

$$R = [10(2) + 7] \times 10^3$$
$$= 27 \times 10^3 \ \Omega \qquad\qquad (3.21)$$
$$= 27 \ k\Omega$$

Tolerance: As we have noted, the actual resistance may deviate from the nominal value, and its greatest allowed deviation is prescribed by the tolerance band. As indicated in Table 3.1, the color of the tolerance band may be either gold (for $\pm 5\%$ tolerance) or silver (for $\pm 10\%$ tolerance). For example, if the resistor described in (3.21) has a silver tolerance band, its resistance may vary as much as 10% of 27 $k\Omega$, or 2.7 $k\Omega$, up or down. The actual resistance, therefore, should be between $27 - 2.7 = 24.3$ and $27 + 2.7 = 29.7 \ k\Omega$.

Resistances under 10 Ohms: By Table 3.1 we may note that band c, but not bands a or b, may be silver or gold. This is to accommodate the negative powers $10^{-2} = 0.01$ (for the silver band) and $10^{-1} = 0.1$ (for the gold band), needed for resistances under 10 Ω. For example, suppose that the a, b, and c bands are green, blue, and gold, respectively. Then we have $a = 5$, $b = 6$, and $c = -1$, and the resistance is

$$R = 56 \times 10^{-1} = 5.6 \ \Omega$$

Other Resistor Markings: Wire-wound and metal-film resistors are usually large enough to have their resistance value and tolerance numbers printed on the resistor body. For example, the resistance value of 10 Ω appears below the manufacturer's code number on the resistor of Fig. 3.8. In some cases, small wire-wound resistors are color-coded like the carbon resistors. In these cases the first band is double the width of the others to distinguish it from the carbon resistors.

Standard Resistance Values: Resistors that are mass-produced come in standard values of resistance. These standard values, however, are selected so that they, along with their allowable deviations, completely cover the spectrum of values from a fraction of an ohm to many megohms. Table 3.2 shows the standard nominal values from 0.1 Ω to 22 MΩ.

The resistors of Table 3.2 are all available with 5% tolerances. Resistors with 10% tolerance are available only in values of 10, 12, 15, 18, 22, 27, 33, 39, 47, 56, 68, and 82 multiplied by powers of 10. For example, a 10% resistor of 6800 Ω is standard, since $6800 = 68 \times 10^2$. A resistor of 16 Ω is available with a 5% tolerance but not with a 10% tolerance.

TABLE 3.2. *Available standard values of resistors.*

Ohms (Ω)					Kilohms (kΩ)		Megohms (MΩ)	
0.10	1.0	10	100	1000	10	100	1.0	10.0
0.11	1.1	11	110	1100	11	110	1.1	11.0
0.12	1.2	12	120	1200	12	120	1.2	12.0
0.13	1.3	13	130	1300	13	130	1.3	13.0
0.15	1.5	15	150	1500	15	150	1.5	15.0
0.16	1.6	16	160	1600	16	160	1.6	16.0
0.18	1.8	18	180	1800	18	180	1.8	18.0
0.20	2.0	20	200	2000	20	200	2.0	20.0
0.22	2.2	22	220	2200	22	220	2.2	22.0
0.24	2.4	24	240	2400	24	240	2.4	
0.27	2.7	27	270	2700	27	270	2.7	
0.30	3.0	30	300	3000	30	300	3.0	
0.33	3.3	33	330	3300	33	330	3.3	
0.36	3.6	36	360	3600	36	360	3.6	
0.39	3.9	39	390	3900	39	390	3.9	
0.43	4.3	43	430	4300	43	430	4.3	
0.47	4.7	47	470	4700	47	470	4.7	
0.51	5.1	51	510	5100	51	510	5.1	
0.56	5.6	56	560	5600	56	560	5.6	
0.62	6.2	62	620	6200	62	620	6.2	
0.68	6.8	68	680	6800	68	680	6.8	
0.75	7.5	75	750	7500	75	750	7.5	
0.82	8.2	82	820	8200	82	820	8.2	
0.91	9.1	91	910	9100	91	910	9.1	

PRACTICE EXERCISES

3-6.1 Find the nominal resistance and the range in which the actual resistance lies of a carbon resistor with color bands *a, b, c,* and % tolerance which are, respectively,

    (a) Yellow, orange, red, gold
    (b) Brown, green, black, silver
    (c) Gray, red, gold, silver

                                     Ans. (a)   4300 Ω, 4085 to 4515 Ω
                                            (b)   15 Ω, 13.5 to 16.5 Ω
                                            (c)   8.2 Ω, 7.38 to 9.02 Ω

3-6.2 Find the color code (bands *a, b, c,* and % tolerance in that order) for a 5% resistance of (a) 0.43 Ω, (b) 5100 Ω, (c) 24 kΩ, and (d) 10 MΩ.

                                     Ans. (a)   Yellow, orange, silver, gold
                                          (b)   Green, brown, red, gold
                                          (c)   Red, yellow, orange, gold
                                          (d)   Brown, black, blue, gold

3.7 SUMMARY

Resistors are the most common circuit elements and are characterized by their resistance R, measured in ohms, and their voltage–current relation, known as Ohm's law and given by

$$V = RI$$

If I is in amperes and V is in volts, then R is in ohms. Other practical units are kilohms when I is in milliamperes and megohms when I is in microamperes. In both cases V is in volts.

The conductance G is defined as the reciprocal of resistance, and its standard units are mhos. In terms of conductance, Ohm's law may be written

$$I = GV$$

The power P dissipated in a resistor is given by

$$P = I^2 R$$

or by

$$P = \frac{V^2}{R}$$

A current-carrying resistor dissipates power by converting electrical energy into heat. The amount of power a resistor can safely dissipate without damage to itself is its wattage rating.

Resistors are manufactured in a variety of standard resistance sizes, and actual resistance values may deviate from the stated value by a percentage known as tolerance. The least expensive resistors, with the higher tolerances, are carbon-composition and carbon-film resistors. Higher-quality resistors are of the wire-wound or metal-film type. Resistors with variable resistances are also available. These include rheostats, decade boxes, and potentiometers.

Carbon resistance values and tolerances are identified on the resistors by means of a color code. This consists of four colored bands located at one end of the resistor, with each color representing a numerical value.

PROBLEMS

3.1 Find the current carried by a resistor having a terminal voltage of 10 V if the resistance is (a) 20 Ω and (b) 20 kΩ.

3.2 Find the voltage that must be applied across a resistor to produce a current of 2 mA if the resistance is (a) 4 kΩ, (b) 600 Ω, and (c) 1 MΩ.

3.3 The terminal current of a 20-Ω resistor is 6 A. Find (a) the conductance, (b) the terminal voltage, and (c) the power dissipated.

3.4 If the current in a resistor is 5 mA, find the resistance and the power absorbed for the cases where the terminal voltage is (a) 1 V and (b) 100 V.

3.5 A 2-kΩ resistor is connected to a battery with a resulting current of 10 mA. Find the current if the battery is connected across a 400-Ω resistor.

3.6 A 100-V source is connected across a 2-kΩ resistor. Find (a) the current and (b) the current that would result if the resistance were doubled.

3.7 Find the power dissipated in the resistor in Problem 3.6 in both cases.

3.8 A 10-Ω resistor carries a current of 2 A. Find (a) the power dissipated and (b) the power dissipated if the current is doubled.

3.9 Find the current in a 2-kΩ resistor if the power dissipated is (a) 0.8 W and (b) 320 W.

3.10 Find the current and the power absorbed in a resistor if the voltage is 12 V and the conductance is (a) 6 $\mho$ and (b) 4 m$\mho$.

3.11 How long must a constant current of 5 A flow in a 10-Ω resistor to dissipate 1 kJ?

3.12 A 12-Ω resistor dissipates 192 W of power. Find the current it carries.

3.13 What are the resistance and current rating of a 120-V, 150-W bulb?

3.14 If a 3000-W, 240-V clothes dryer is considered as a resistor, find its resistance and rated current.

3.15 A $\frac{1}{4}$-W resistor has a resistance of 10 kΩ. Find the current it can safely carry.

3.16 Find the wattage rating of a 100-Ω resistor that can safely carry a maximum current of 100 mA.

3.17 Find V_{ac} for the potentiometer of Fig. 3.13 if

$$\frac{V_{ac}}{V_{cb}} = 4$$

and $V = 50$ V.

3.18 Find k for Fig. 3.13 if V_{cb} is 0.4 times the total available voltage V.

3.19 A carbon resistor has color-code bands a, b, c, and % tolerance as follows:

   (a) Orange, black, orange, silver
   (b) Red, yellow, silver, gold
   (c) Blue, gray, black, silver

Find in each case the nominal resistance and range in which the actual resistance lies.

3.20 Find the color code (bands a, b, c, and % tolerance in that order) for a 10% resistance of (a) 82 Ω, (b) 4700 Ω, and (c) 18 MΩ.

4

SIMPLE RESISTIVE CIRCUITS

At this point we have defined a circuit as a connection of electrical elements with voltages and currents, and we have discussed in some detail current and voltage sources and resistors. In particular, we have considered Ohm's law and how it may be used to find the current, voltage, or power associated with a resistor. However, we are still not able to analyze even the simplest electric circuit because, in addition to Ohm's law, we must have laws that take into account the way in which the elements are connected.

There are two such laws that we will consider in this chapter. They are known as *Kirchhoff's current law* and *Kirchhoff's voltage law,* and were first given in 1847 by the German physicist Gustav Robert Kirchhoff (1824–1887). With Kirchhoff's two laws and Ohm's law it is theoretically possible to analyze any *resistive* circuit, which we define as a circuit whose elements are resistors and sources. However, we shall restrict ourselves for the present to relatively *simple* circuits, which may be described by a single expression of one of Kirchhoff's laws. Such simple circuits are classified as *series* circuits, which consist of elements connected terminal to terminal around a single loop, and *parallel* circuits, which consist of elements all connected between a common pair of terminals. As we shall see, it is always possible, in the case of series and parallel circuits, to combine a number of resistors into a single *equivalent* resistor and a number of sources into a single equivalent source. Using these equivalences we may often analyze the circuit by a simple application of Ohm's law.

4.1 KIRCHHOFF'S CURRENT LAW

Before discussing Kirchhoff's laws we should first consider the concepts of *nodes* and *loops*. A node is simply a point of connection of two or more circuit elements. Examples of nodes are terminals *a* and *b* of the 2-Ω resistor of Fig. 4.1(a). Point *a* is also a terminal of the 6-V source and point *b* is a common terminal of the 4-Ω resistor and the 3-A source. The circuit of Fig. 4.1(a) has only three nodes, since points *c* and *d* are not separate nodes. This may be seen in Fig. 4.1(b), where the nodes are shown enclosed by dashed lines. The three nodes are more clearly shown in the redrawn equivalent circuit of Fig. 4.1(c).

A loop is a closed path of elements such as path *abcda* in Fig. 4.1(a), which consists of the 2-Ω resistor, the 4-Ω resistor, and the 6-V source. Another loop in Fig. 4.1(a) is path *bcb,* consisting of the 3-A source and the 4-Ω resistor. Still another loop is the outside loop of the 2-Ω resistor, the 3-A source, and the 6-V source.

Kirchhoff's current law (KCL) states that for any circuit,

The algebraic sum of the currents entering any node is zero.

To illustrate this law let us consider the node shown in Fig. 4.2, having four elements connected to it with currents i_1, i_2, i_3, and i_4 as indicated. By KCL we have

$$i_1 + (-i_2) + i_3 + i_4 = 0 \tag{4.1}$$

We note that in the figure i_2 is leaving the node, which is equivalent to $-i_2$ entering the node. This accounts for the term $-i_2$ in (4.1).

Kirchhoff's current law may seem plausible by recalling that current is the rate of movement of charge. Therefore, since a point or node has no volume, there is no place for charges to accumulate in the node and thus the net rate, or total current, must add to zero.

If we multiply (4.1) through by -1, we have the equally correct statement

$$-i_1 + i_2 - i_3 - i_4 = 0$$

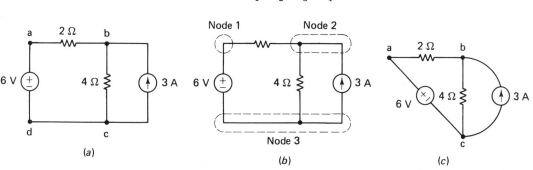

(a)

(b)

(c)

FIGURE 4.1 *Three versions of a circuit with three nodes.*

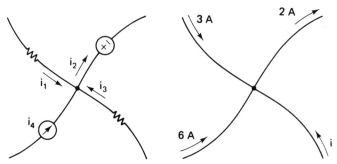

FIGURE 4.2 *Currents entering a node.*   **FIGURE 4.3** *Example of KCL.*

Since $-i_1$, i_2, $-i_3$, and $-i_4$ are the currents leaving the node, it is clear that their sum is also zero. This is an example of an equivalent form of Kirchhoff's current law, which states that

The algebraic sum of the currents leaving any node is zero.

Finally, let us rearrange (4.1) in the form

$$i_1 + i_3 + i_4 = i_2 \tag{4.2}$$

which is obtained by transposing the term $-i_2$ to the right side of the equation. Referring back to Fig. 4.2, we see that (4.2) is an example of still another correct form of Kirchhoff's current law, which states that

The sum of the currents entering any node equals the sum of the currents leaving the node.

Example 4.1: To illustrate KCL with a numerical example, let us find the current i in Fig. 4.3.

Solution: Summing the currents entering the node, we have

$$3 + (-2) + i + 6 = 0$$

which yields

$$i = -7 \text{ A}$$

We note that -7 A entering the node is equivalent to $+7$ A leaving the node. Therefore, the unknown current, which was guessed to be entering, is actually 7 A leaving the node. Thus it is not necessary to guess the correct current direction prior to solving the problem.

(In a complicated problem there may be no way to do so, anyhow.) We still arrive at the correct answer in the end.

Example 4.2: Solve Example 4.1 using the third form of KCL.

Solution: Equating the currents entering to the currents leaving, we have

$$3 + i + 6 = 2$$

or, as before,

$$i = -7 \text{ A}$$

PRACTICE EXERCISES

4-1.1 Identify the nodes in the figure.

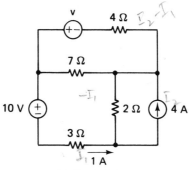

EXERCISE 4.1.1

4-1.2 Five elements are connected to a node. Three element currents that leave are 2 A, 5 A, and 7 A. The other two element currents enter and are 4 A and I. Find I.

Ans. 10 A

4-1.3 Find i_1 and i_2.   *Ans.* −4 A, 11A

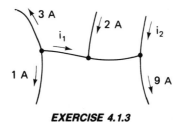

EXERCISE 4.1.3

4.2 KIRCHHOFF'S VOLTAGE LAW

The other law of Kirchhoff's in which we are interested is *Kirchhoff's voltage law* (KVL), which states that

The algebraic sum of the voltages around any loop is zero.

As a justification for this law we might note that if we start at any point at a given potential and follow, or traverse, the loop back to the same point and potential, the difference of potential, or the net voltage around the loop, must be zero.

Procedure: In determining the algebraic signs to be used on the voltages around the loop, we may select a point in the loop and traverse the loop in a given direction, assigning the polarity of the terminal reached first in traversing an element. For example, let us consider the loop *abcda* in Fig. 4.4, which has other elements not shown that are connected to it at points *a*, *b*, and *d*. Starting at point *a* and going clockwise, we encounter v_1, v_2, v_3, and $-v_4$ as we traverse the loop. The signs used are the first polarity marks encountered with each element. Thus KVL is

$$v_1 + v_2 + v_3 - v_4 = 0 \qquad (4.3)$$

The application of Kirchhoff's voltage law is independent of the direction we take around the loop. For example, if the loop of Fig. 4.4 is traversed in a counterclockwise direction, we have, by KVL starting at point *a*,

$$v_4 - v_3 - v_2 - v_1 = 0$$

This result is equivalent to (4.3). In fact, it is (4.3) multiplied through by -1.

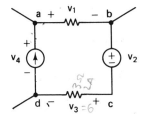

FIGURE 4.4 Voltages around a loop.

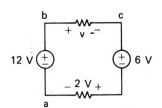

FIGURE 4.5 Example illustrating KVL.

Example 4.3: Let us illustrate the application of Kirchhoff's voltage law by finding the voltage *v* in the circuit of Fig. 4.5.

Solution: Starting at point *a* and traversing the circuit in a clockwise direction, we have

$$-12 + v + 6 + 2 = 0 \qquad (4.4)$$

from which $v = 4$ V.

68

Voltage Rises and Drops: Some authors prefer to discuss Kirchhoff's voltage law in terms of voltage *rises* and voltage *drops* around a loop. Referring to Fig. 4.5, a voltage rise of *v* volts occurs in moving from point *c* to point *b*, whereas a voltage drop of *v* volts occurs in moving from *b* to *c*. If we transpose all the terms with negative signs to the other side in a KVL equation, we have the equivalent statement of Kirchhoff's voltage law:

Around any loop the sum of the voltage rises equals the sum of the voltage drops.

For example, we may write (4.4) in the form

$$v + 6 + 2 = 12 \tag{4.5}$$

by transposing the negative terms to the right side of the equation. This last expression is evidently a statement that the sum of the voltage rises 12 (only one in this case) equals the sum of the voltage drops $v + 6 + 2$.

The second version of Kirchhoff's law is especially useful in a circuit containing a voltage source and a number of resistors. In such a case we may mark the resistor polarities so that they represent drops, in which case the applied voltage rise of the source equals the sum of the resistor voltage drops. The voltage drops in this case are sometimes called *RI* drops, or *IR* drops, because they are resistor drops for which $V = RI$. For example, there is an *IR* drop across the 3-Ω resistor in Practice Exercise 4-1.1 of $3 \times 1 = 3$ V.

PRACTICE EXERCISES

4-2.1 Find v_1 in Fig. 4.4 if $v_2 = 12$ V, $v_4 = 24$ V, and v_3 is across a 3-Ω resistor carrying a current of 2 A directed to the left. *Ans.* 6 V

4-2.2 Find *v*. *Ans.* 2 V

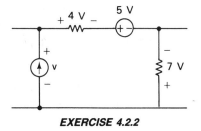

EXERCISE 4.2.2

4-2.3 Find i_1, i_2, v_1, and v_2.   *Ans.* 4 A, 2 A, 12 V, 22 V

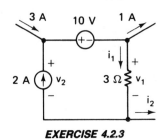

EXERCISE 4.2.3

4-2.4 To illustrate the truth of the statement earlier in this chapter that Kirchhoff's laws are required in the analysis of even the simplest circuit, find I in the given simple circuit. Note that I follows from Ohm's law if we know V_1, the voltage across the resistor, and $V_1 = V = 6$ V only as a consequence of KVL around the loop.

Ans. 2 A

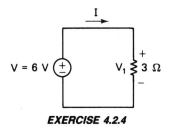

EXERCISE 4.2.4

4.3 SERIES CIRCUITS

One of the simplest types of electric circuits is one in which all the elements are connected successively around a single loop with a terminal of one element connected to a terminal of the next element, and so forth. By Kirchhoff's current law, there is only one current in the circuit and it flows through every element. That is, the current leaving the terminal of any element must enter the terminal of the next element in the loop. Elements connected in this manner (with a common current) are called *series* elements or are said to be *connected* in *series,* and a circuit consisting of a single loop of series-connected elements is called a *series* circuit. An example is that of Fig. 4.6(a), with a battery and two resistors connected in series. A schematic diagram of the circuit is shown in Fig. 4.6(b). We shall, of course, refer to either figure as a series circuit, but in most cases we shall use the schematic diagram.

The series circuit of Fig. 4.6 consists of two resistors in series with a battery. A similar example is the circuit previously considered in Fig. 4.5, which is a slightly more complicated example in that it is a series connection of *two* voltage sources and two resistors.

Analysis of Series Circuits: To analyze a series resistive circuit we need only find the current that is common to every element. With this knowledge of the current we may then find any other characteristic of the circuit. For example, the source

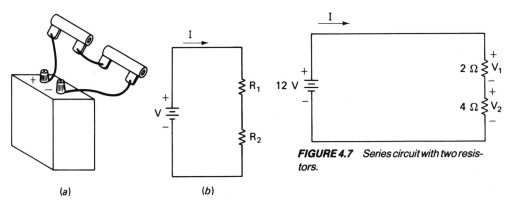

FIGURE 4.7   *Series circuit with two resistors.*

FIGURE 4.6   *(a) A series circuit and (b) its schematic diagram.*

voltages will be known, since they are presumed to be given, and the resistor voltages may be found from the current and resistances by Ohm's law.

To illustrate the analysis, let us find the current I and the resistor voltages V_1 and V_2 in the circuit of Fig. 4.7. By KVL around the circuit we have

$$V_1 + V_2 = 12 \qquad\qquad (4.6)$$

and by Ohm's law we have the IR drops

$$V_1 = 2I$$
$$V_2 = 4I$$

Substituting these values into (4.6), we have

$$2I + 4I = 12$$

or

$$6I = 12$$

Thus we have the current given by

$$I = 2 \text{ A}$$

General Case of a Two-Resistor Series Circuit: In the general case of Fig. 4.8(a), with a voltage source and two resistors, we may use Kirchhoff's voltage law and Ohm's law to equate the source voltage to the sum of the IR drops. The result is

$$V = V_1 + V_2 = R_1 I + R_2 I$$

or

$$V = (R_1 + R_2)I$$

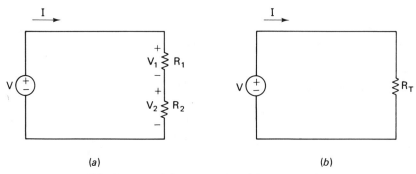

FIGURE 4.8   (a) Series circuit and (b) its equivalent.

The current is therefore

$$I = \frac{V}{R_1 + R_2} \tag{4.7}$$

We define the circuit of Fig. 4.8(b) to be the *equivalent* circuit, as far as V and I are concerned, of the circuit of Fig. 4.8(a) if the resistance R_T is such that for the same source V we have the same current I for both circuits. Since the current I in Fig. 4.8(b) is given by

$$I = \frac{V}{R_T} \tag{4.8}$$

we see by comparing (4.7) and (4.8) that the two circuits are equivalent (I is the same if V is the same) if we have

$$R_T = R_1 + R_2 \tag{4.9}$$

The resistance R_T is the *equivalent resistance,* or the *total* resistance in this case, of the two series resistances R_1 and R_2, and is obtained by simply adding the series resistances.

> *Example 4.4:* Find the equivalent resistance of the series circuit of Fig. 4.7 and use the result to find the current I.

> *Solution:* By (4.9) the equivalent resistance is

$$R_T = 2 + 4 = 6 \ \Omega$$

> and by (4.8) the current is

$$I = \frac{V}{R_T} = \frac{12}{6} = 2 \text{ A}$$

The fact that the equivalent resistance R_T of two series resistances R_1 and R_2 is their sum is represented in Fig. 4.9. The arrow indicates that R_T is the resistance seen at the terminals.

Equivalent Resistance in the General Case: The result given in (4.9) for the two-resistor series circuit may be extended to any number of resistors in series. For example, let us consider the series circuit of Fig. 4.10, which contains N resistors. [For $N = 2$ this case reduces to Fig. 4.8(a), of course.] By KVL we have

$$V = V_1 + V_2 + \cdots + V_N$$

which by Ohm's law may be written

$$V = R_1 I + R_2 I + \cdots + R_N I$$
$$= (R_1 + R_2 + \cdots + R_N)I$$

Therefore, the current is

$$I = \frac{V}{R_1 + R_2 + \cdots + R_N} \tag{4.10}$$

Comparing this result with (4.8), we see that Fig. 4.8(b) is an equivalent circuit of Fig. 4.10 provided that

$$R_T = R_1 + R_2 + \cdots + R_N \tag{4.11}$$

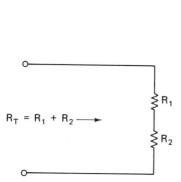

FIGURE 4.9 Representation of equivalent resistance.

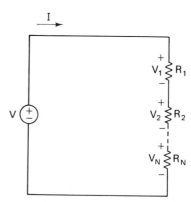

FIGURE 4.10 Series circuit with N resistors.

That is, the equivalent resistance of a series connection of any number of resistances is simply their sum.

Example 4.5: A series circuit consists of a 36-V source and resistances $R_1 = 2 \; \Omega$, $R_2 = 3 \; \Omega$, and $R_3 = 7 \; \Omega$. Find the equivalent resistance R_T and the current I.

Solution: By (4.11) we have

$$R_T = R_1 + R_2 + R_3$$
$$= 2 + 3 + 7$$
$$= 12 \; \Omega$$

The current is given by (4.8) to be

$$I = \frac{V}{R_T} = \frac{36}{12} = 3 \; \text{A}$$

General Analysis Method: As a final example in this section of a series circuit, let us consider the circuit of Fig. 4.11(a). Applying KVL around the loop, we have

$$-12 + V_1 + 6 + V_2 + V_3 - 14 + V_4 = 0$$

or

$$V_1 + V_2 + V_3 + V_4 = 12 - 6 + 14 \tag{4.12}$$

We note that the left member of this equation is the sum of the IR drops across the four resistors, given by

$$V_1 = 2I$$
$$V_2 = 3I$$
$$V_3 = 1I$$
$$V_4 = 4I$$

$$\tag{4.13}$$

The right member of (4.12) is the algebraic sum of the source voltages around the loop with a positive sign on those sources *aiding* the current (that is, those whose polarities are such that the current leaves the positive terminal), and with a negative sign on those sources *opposing* the current (those with currents entering the positive terminal). If we denote this algebraic sum of source voltages by V_T, then V_T is the

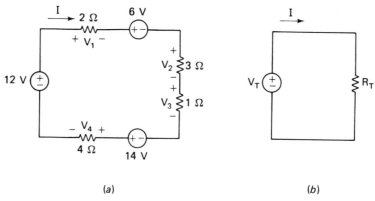

(a)   (b)

FIGURE 4.11 (a) Series circuit and (b) general equivalent circuit.

net or, in this case, the total source voltage around the loop and is given in Fig. 4.11(a) by

$$V_T = 12 - 6 + 14 = 20 \text{ V} \qquad (4.14)$$

By substituting (4.13) into (4.12), we may obtain

$$(2 + 3 + 1 + 4)I = 12 - 6 + 14 \qquad (4.15)$$

from which we see that the coefficient of I is the equivalent series resistance,

$$R_T = 2 + 3 + 1 + 4 = 10 \text{ } \Omega$$

Using this result and (4.14), we may write (4.15) in the form

$$R_T I = V_T \qquad (4.16)$$

This is a general series circuit result, which may be illustrated by the equivalent circuit of Fig. 4.11(b).

In the case of Fig. 4.11(a), (4.16) is given by

$$10I = 20$$

so that the current is $I = 2$ A.

This example illustrates a general procedure for analyzing series circuits. We simply find R_T, the equivalent resistance, as the sum of all the resistances in the loop, and find V_T, the net voltage aiding the current around the loop, and use (4.16) to find the current I. The general situation, as far as I is concerned, is, of course, that of Fig. 4.11(b).

PRACTICE EXERCISES

4-3.1 Find the equivalent resistance R_T, the current I, and the IR drops. Note that the sum of the IR drops equals the applied source voltage.

Ans. 12 Ω; 1.5 A; 4.5, 7.5, 6 V

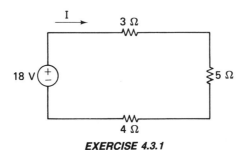

EXERCISE 4.3.1

4-3.2 A series circuit consists of resistances of 2 Ω, 3 Ω, 5 Ω, and 6 Ω, and a voltage source V. If the current is to be 2 A, what must V be?   *Ans.* 32 V

4-3.3 Find V_T, R_T, and I.   *Ans.* 6 V, 12 kΩ, 0.5 mA

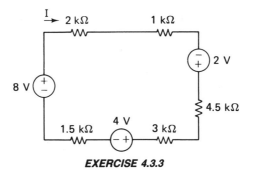

EXERCISE 4.3.3

4.4 PARALLEL CIRCUITS

The other simple type of circuit, besides the series circuit, is the *parallel* circuit, in which all the elements are connected between the same two nodes. In other words, every element in the circuit shares its two terminals with every other element. By Kirchhoff's voltage law all the elements in the circuit therefore have the same voltage, and are said to be *parallel* elements, or elements *connected* in *parallel*.

An example of a parallel circuit with three elements—a voltage source and two resistors—is shown in Fig. 4.12(a), with a schematic diagram shown in Fig. 4.12(b). In this case the common terminals are terminals *a* and *b*. Another example is that of two electrical elements or appliances, such as a lamp and a television set, which are plugged into a 120-V wall receptacle. The 120-V power line connected to the receptacle is the source and the two elements are the connected parallel loads.

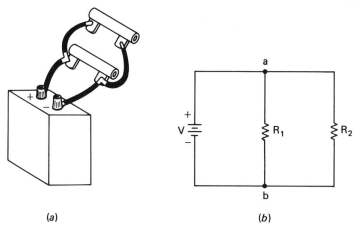

(a)　　　　　　　　　　　　　　(b)

FIGURE 4.12　(a) Parallel circuit and (b) its schematic diagram.

Analysis of Parallel Circuits:　A parallel circuit, such as that of Fig. 4.13, is analyzed when the common voltage V is determined. Knowing V, we may find all the unknown currents by Ohm's law, and then every current and voltage in the circuit is known.

For example, in Fig. 4.12 we already know V. Thus the currents in R_1 and R_2 are V/R_1 and V/R_2, respectively.

To illustrate the analysis further, let us find V in Fig. 4.13, assuming that I, R_1, and R_2 are known. By Ohm's law we have the currents I_1 and I_2 given by

$$I_1 = \frac{V}{R_1}$$

$$I_2 = \frac{V}{R_2}$$　　　　　(4.17)

Also, by Kirchhoff's current law we have

$$I = I_1 + I_2$$

which by (4.17) may be written

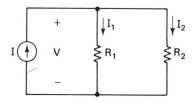

FIGURE 4.13　Parallel circuit with a current source.

$$I = \frac{V}{R_1} + \frac{V}{R_2}$$

or

$$I = \left(\frac{1}{R_1} + \frac{1}{R_2}\right) V \tag{4.18}$$

Parallel Resistances: A circuit equivalent to that of Fig. 4.13, as far as I and V are concerned, is that of Fig. 4.14(a), provided that the *equivalent* resistance R_T is such that the same current I produces the same voltage V. That is, we must have, by Ohm's law,

$$I = \frac{1}{R_T} V \tag{4.19}$$

Comparing (4.18) and 4.19), we have

$$\frac{1}{R_T} = \frac{1}{R_1} + \frac{1}{R_2} \tag{4.20}$$

or

$$R_T = \frac{1}{1/R_1 + 1/R_2}$$

Equation (4.20) may be written as

$$\frac{1}{R_T} = \frac{R_1 + R_2}{R_1 R_2}$$

or

$$R_T = \frac{R_1 R_2}{R_1 + R_2} \tag{4.21}$$

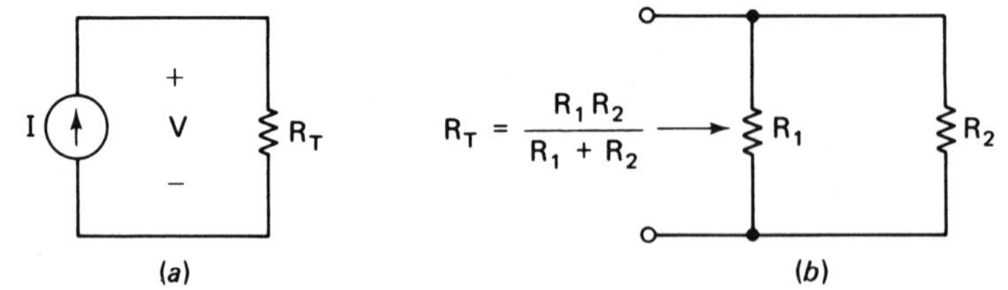

FIGURE 4.14   *(a) Equivalent circuit of Fig. 4.13 and (b) representation of equivalent resistance.*

That is, the equivalent resistance of two resistances in parallel is the product over the sum of the two resistances. This result is symbolized in Fig. 4.14(b).

Example 4.6: Find the equivalent resistance of a 3-Ω and a 6-Ω resistance in parallel.

Solution: By (4.21) we have

$$R_T = \frac{3 \times 6}{3 + 6} = \frac{18}{9} = 2 \ \Omega$$

We note in this example that the equivalent resistance 2 Ω is less than either of the two parallel resistances. This is true in general since it is easier for the current to flow through two parallel paths than through either of the paths alone. This may also be seen from (4.20), since $1/R_T$ is greater than either $1/R_1$ or $1/R_2$.

Example 4.7: Replace the two parallel resistors in Fig. 4.15(a) by their equivalent resistor and find the voltage V and the currents I_1 and I_2.

Solution: The equivalent resistance, as shown in Fig. 4.15(b), is

$$R_T = \frac{4 \times 12}{4 + 12} = \frac{48}{16} = 3 \ \Omega$$

The voltage, therefore, by Ohm's law is

$$V = (3)(5) = 15 \ V$$

Finally, from Fig. 4.15(a) and Ohm's law we have

$$I_1 = \frac{V}{4} = \frac{15}{4} = 3.75 \ A$$

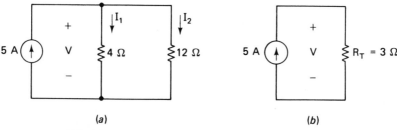

(a)   (b)

FIGURE 4.15   *(a) A parallel circuit and (b) its equivalent.*

and

$$I_2 = \frac{V}{12} = \frac{15}{12} = 1.25 \text{ A}$$

As a check, we see that KCL holds, since $I_1 + I_2 = 5$.

General Parallel Resistor Case: In the general case of N resistors connected in parallel, as shown in Fig. 4.16, we may also obtain an equivalent resistance R_T. To see this let us note that by KCL we have

$$I = I_1 + I_2 + \cdots + I_N$$

which by Ohm's law may be written

$$I = \frac{V}{R_1} + \frac{V}{R_2} + \cdots + \frac{V}{R_N}$$

or

$$I = \left(\frac{1}{R_1} + \frac{1}{R_2} + \cdots + \frac{1}{R_N} \right) V \tag{4.22}$$

If Fig. 4.14(a) is the equivalent circuit of Fig. 4.16, then by comparing (4.19) and (4.22), we have

$$\frac{1}{R_T} = \frac{1}{R_1} + \frac{1}{R_2} + \cdots + \frac{1}{R_N} \tag{4.23}$$

In other words, the reciprocal of the equivalent resistance R_T is the sum of the reciprocals of the individual parallel resistances. The case of two parallel resistances given in (4.20) is the special one where $N = 2$. It is also true in the general case, as it was in the two-resistor case, that R_T is less than any of the individual parallel resistances.

Example 4.8: Find the equivalent resistance R_T of the circuit of Fig. 4.17(a) and use the result to obtain the voltage V.

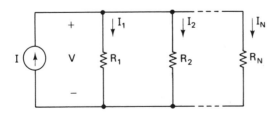

FIGURE 4.16 *Parallel circuit with N resistors.*

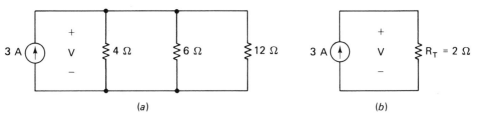

FIGURE 4.17   (a) Parallel circuit and (b) its equivalent.

Solution: By (4.23) the equivalent resistance satisfies

$$\frac{1}{R_T} = \frac{1}{4} + \frac{1}{6} + \frac{1}{12} = \frac{3+2+1}{12} = \frac{1}{2}$$

Thus we have $R_T = 2\ \Omega$, as shown in the equivalent circuit of Fig. 4.17(b). From this figure the voltage is

$$V = R_T I = (2)(3) = 6\ \text{V}$$

In terms of conductances, if $G_T = 1/R_T$, then (4.23) may be written

$$G_T = G_1 + G_2 + \cdots + G_N$$

This is the dual statement of the series case in (4.11). Thus series and parallel circuits may be considered to be duals of each other.

Special Case of Equal Resistances: If all the parallel resistances are equal, the formulas for the equivalent resistance R_T become relatively simple. For example, if in the case of two parallel resistances R_1 and R_2 we have $R_2 = R_1$, the equivalent resistance is by (4.21),

$$R_T = \frac{R_1 R_2}{R_1 + R_2}$$

$$= \frac{R_1 R_1}{2R_1}$$

or

$$R_T = \frac{R_1}{2} \tag{4.24}$$

Thus the equivalent of two parallel and equal resistances is half of one of them.

Example 4.9: Find R_T and V if $R_1 = R_2 = 12$ Ω and $I = 4$ A in Fig. 4.13.

Solution: By (4.24) we have

$$R_T = \frac{12}{2} = 6 \text{ Ω}$$

Thus using the equivalent circuit of Fig. 4.14(a), we have

$$V = R_T I = (6)(4) = 24 \text{ V}$$

In the case of N equal parallel resistances, each equal to R_1, we have by (4.23)

$$\frac{1}{R_T} = \frac{1}{R_1} + \frac{1}{R_1} + \cdots + \frac{1}{R_1}$$

with N terms in the right member. Thus we have

$$\frac{1}{R_T} = \frac{N}{R_1}$$

or

$$R_T = \frac{R_1}{N} \tag{4.25}$$

For $N = 2$ this becomes (4.24), of course.

Example 4.10: The parallel circuit of Fig. 4.16 has 10 resistors, each with a resistance of $R_1 = 200$ Ω. If $I = 3$ A, find V.

Solution: By (4.25) we have

$$R_T = \frac{200}{10} = 20 \text{ Ω}$$

so that

$$V = R_T I = (20)(3) = 60 \text{ V}$$

General Analysis Method: As a final example, let us consider the circuit of Fig. 4.18(a), where it is required to find the common voltage V across each of the five parallel elements. By KCL the sum of the currents downward from the top node is given by

$$-7 + I_1 + 2 + I_2 + I_3 = 0$$

or

$$I_1 + I_2 + I_3 = 7 - 2 \qquad (4.26)$$

The left member of this equation is the sum of the resistor currents, given by

$$I_1 = \frac{V}{6}$$

$$I_2 = \frac{V}{18} \qquad (4.27)$$

$$I_3 = \frac{V}{9}$$

Thus the left member may be written

$$I_1 + I_2 + I_3 = \left(\frac{1}{6} + \frac{1}{18} + \frac{1}{9}\right)V$$

$$\qquad (4.28)$$

$$= \frac{V}{R_T}$$

where R_T is the equivalent of the three parallel resistances, given by

$$\frac{1}{R_T} = \frac{1}{6} + \frac{1}{18} + \frac{1}{9} \qquad (4.29)$$

The right member of (4.26) is the algebraic sum of the currents into the top node through the sources. If we denote this sum by I_T, the *equivalent* of the current sources, we have

$$I_T = 7 - 2 = 5 \text{ A} \qquad (4.30)$$

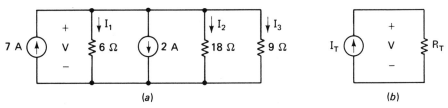

FIGURE 4.18 (a) Parallel circuit and (b) general equivalent circuit.

From (4.28) and (4.30) we see that (4.26) is equivalent to

$$\frac{V}{R_T} = I_T \qquad (4.31)$$

which is described by the circuit of Fig. 4.18(b). This circuit is, of course, an equivalent circuit, as far as V is concerned, of the original circuit of Fig. 4.18(a).

From (4.29) we may show that $R_T = 3\ \Omega$ and since $I_T = 5$ A in (4.30), we may write (4.31) as

$$\frac{V}{3} = 5$$

so that $V = 15$ V.

In summary, we may analyze general parallel circuits with current sources and resistors by finding the equivalent resistance R_T of the parallel resistances and the equivalent current source I_T as the algebraic sum of the current sources. Then the voltage V follows by Ohm's law from (4.31).

PRACTICE EXERCISES

4-4.1 Find the equivalent resistance R_T of the parallel connection of resistances of (a) 10 and 90 Ω, (b) 16 and 16 Ω, (c) 8, 12, and 24 Ω, and (d) 24, 24, 24, and 24 Ω.

Ans. (a) 9, (b) 8, (c) 4, (d) 6 Ω

4-4.2 Find *R*.   *Ans.* 800 Ω

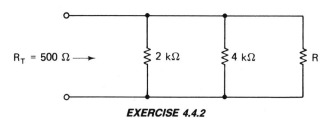

EXERCISE 4.4.2

4-4.3 Find V, I_1, and I_2.   *Ans.* 12 V, 2 A, 1 A

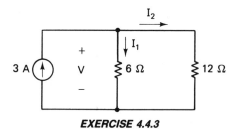

EXERCISE 4.4.3

4.5 POWER IN SIMPLE CIRCUITS

The total power dissipated in a resistive circuit must be supplied by the source, or sources. Therefore, if there is only one source, it must supply the power absorbed in all the resistors. If there are several sources, some may be absorbing power which is delivered by others, as is the case when one battery is used to charge another. However, the *net* power delivered by the sources must equal the power absorbed by the resistors. This concept is sometimes referred to as *conservation of power*.

Example 4.11: To illustrate conservation of power, let us find in Fig. 4.19 the power P_T supplied by the battery, the power P_1 delivered to the resistor R_1, and the power P_2 delivered to R_2. Finally, let us show that

$$P_T = P_1 + P_2 \tag{4.32}$$

and thus that conservation of power holds.

Solution: Since the circuit is a series circuit, the equivalent resistance seen by the source is

$$R_T = R_1 + R_2$$
$$= 2 + 4$$
$$= 6 \ \Omega$$

The current is therefore

$$I = \frac{12}{R_T} = \frac{12}{6} = 2 \ \text{A}$$

The power P_1 is

$$P_1 = R_1 I^2 = 2(2)^2 = 8 \ \text{W}$$

and the power P_2 is

$$P_2 = R_2 I^2 = 4(2)^2 = 16 \ \text{W}$$

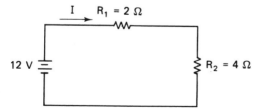

FIGURE 4.19 *Series circuit example.*

The power supplied by the battery is given by

$$P_T = 12I = 12(2) = 24 \text{ W}$$

and therefore (4.32) holds.

Example 4.12: Find the power delivered by or absorbed by each of the elements of Fig. 4.20 and show that conservation of power holds.

Solution: The circuit is a series circuit and thus the equivalent resistance is

$$R_T = 3 + 2 = 5 \text{ } \Omega$$

and the equivalent source aiding the current is

$$V_T = 20 - 5 = 15 \text{ V}$$

Therefore, the current is

$$I = \frac{V_T}{R_T} = \frac{15}{5} = 3 \text{ A}$$

The power P_1 absorbed by the 3-Ω resistor is

$$P_1 = 3I^2 = 3(3)^2 = 27 \text{ W}$$

and the power P_2 absorbed by the 2-Ω resistor is

$$P_2 = 2I^2 = 2(3)^2 = 18 \text{ W}$$

The 5-V source is also absorbing power, say P_3, because the current is entering the positive voltage terminal. This power is

$$P_3 = 5I = 5(3) = 15 \text{ W}$$

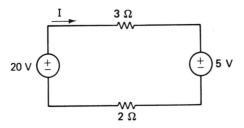

FIGURE 4.20 *Series circuit with two sources.*

The 20-V source is delivering power of

$$P_4 = 20I = 20(3) = 60 \text{ W}$$

Thus conservation of power holds, as the reader may verify by noting that

$$P_4 = P_1 + P_2 + P_3$$

Example 4.13: As a final example, let us consider the parallel circuit of Fig. 4.21, where it is required to find the power P_1 delivered to the 3-Ω resistor, the power P_2 delivered to the 6-Ω resistor, and the power P_3 delivered by the source.

Solution: The equivalent resistance is

$$R_T = \frac{(3)(6)}{3+6} = 2 \ \Omega$$

and therefore we have

$$V = R_T I = 2(6) = 12 \text{ V}$$

The powers are

$$P_1 = \frac{V^2}{3} = \frac{(12)^2}{3} = 48 \text{ W}$$

$$P_2 = \frac{V^2}{6} = \frac{(12)^2}{6} = 24 \text{ W}$$

and

$$P_3 = VI = 12(6) = 72 \text{ W}$$

Conservation of power holds since

$$P_3 = P_1 + P_2$$

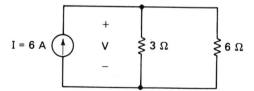

FIGURE 4.21 *Parallel circuit example.*

PRACTICE EXERCISES

4-5.1 In Fig. 4.11(a) find the power P_1 absorbed by the 2-Ω resistor, P_2 by the 3-Ω resistor, P_3 by the 1-Ω resistor, P_4 by the 4-Ω resistor, and P_5 by the 6-V source. Also find the power P_6 delivered by the 14-V source and P_7 delivered by the 12-V source. Finally, show that conservation of power holds by noting that

$$P_1 + P_2 + P_3 + P_4 + P_5 = P_6 + P_7$$

Ans. 8, 12, 4, 16, 12, 28, 24 W

4-5.2 In Fig. 4.18(a) find the powers P_1, P_2, P_3, and P_4 absorbed by the 6-Ω, 18-Ω, and 9-Ω resistors, and the 2-A source. Also find the power P_5 delivered by the 7-A source and show that

$$P_5 = P_1 + P_2 + P_3 + P_4$$

Ans. 37.5, 12.5, 25, 30, 105 W

4.6 SUMMARY

To analyze even the simplest circuit, we need, in addition to Ohm's law, two laws that take into account how the circuit elements are connected. These are Kirchhoff's current law, which says that the algebraic sum of the currents entering any node, or terminal, is zero, and Kirchhoff's voltage law, which says that the algebraic sum of the voltages around any loop is zero.

With Ohm's and Kirchhoff's laws we may analyze any resistive circuit, which is a circuit of resistors and one or more sources. In particular, simple circuits such as series and parallel circuits are especially easy to solve. A series circuit, since it consists of a single loop of elements, is analyzed when the current I, common to all the elements, is found. A parallel circuit has all its elements connected between a common pair of terminals and thus is analyzed when the voltage V across the two terminals is found.

The work is made easier by the use of equivalent resistances. The equivalent resistance of a set of series resistances is simply their sum, and the reciprocal of the equivalent resistance of a set of parallel resistances is the sum of their reciprocals. Thus in either case (series or parallel) the circuit is equivalent to a circuit with a single resistance, so that the unknown current (in the series case) or voltage (in the parallel case) may be found using Ohm's law.

Power is absorbed by some elements (such as resistances and, in some cases, certain sources) and delivered by others (sources). In every case, the power delivered is equal to that absorbed. This is the so-called conservation-of-power principle.

PROBLEMS

4.1 A 12-V battery is connected to a resistor and a current of 3 A results. Find the resistance of the resistor.

4.2 Repeat Problem 4.1 if the current is 3 mA.

4.3 Find i and v.

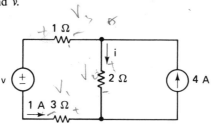

PROBLEM 4.3

4.4 Find v in Practice Exercise 4-1.1.

4.5 If the 9-A current in Practice Exercise 4-1.3 is reversed, find i_1 and i_2.

4.6 A series circuit consists of a 10-V source and a resistance of 5 Ω. Find (a) the current in the circuit and (b) the resistance that must be added in series to reduce the current by one-half.

4.7 Two resistors connected in series with a 100-V source draw 25 mA. If the voltage across one resistor is 25 V, find the value of the two resistances.

4.8 A series circuit of a 10-V source and several resistors has a current of 2 A. If a 5-Ω resistor is inserted in series in the circuit, find the new current.

4.9 Three resistors of 20, 30, and 70 Ω are connected in series. If a 12-V battery is connected across the series combination, find (a) the resistance R_T, (b) the current in the circuit, and (c) the IR drop across each resistor. Show that KVL holds by noting that the sum of the IR drops equals the battery voltage.

4.10 Three 20-Ω resistors are in series with a voltage source. If the voltage across each resistor is 60 V, find the current in the circuit and the voltage across the source.

4.11 If twenty 5-Ω resistors are in series with a 10-V source, find the current in the circuit and the voltage across each resistor.

4.12 Find I.

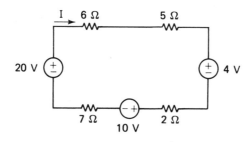

PROBLEM 4.12

4.13 Find *I* in Problem 4.12 if the 4-V source has its polarity reversed.

4.14 Find *R*.

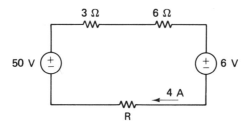

PROBLEM 4.14

4.15 If $R_1 = 3$ Ω, $R_2 = 8$ Ω, $R_3 = R_4 = 48$ Ω are connected in parallel, find R_T.

4.16 If the parallel circuit of Problem 4.15 is connected in parallel with a 12-A source, find the voltage *V* across the resistance combination and the current in each resistor.

4.17 Two 40-Ω resistors and a 60-V voltage source are all connected in parallel. Find the current in the source and the current in each resistor.

4.18 How many 100-kΩ resistors must be connected in parallel to provide an equivalent resistance of $R_T = 4$ kΩ?

4.19 A 40-Ω resistor, a 60-Ω resistor, and a current source *I* are connected in parallel. If the voltage across each element is to be 120 V, find *I*.

4.20 The conductances of four parallel resistors are 1, 2, 5, and 12 m℧. Find the equivalent resistance in ohms.

4.21 Find *V*, I_1, and I_2 by writing KCL at the top node.

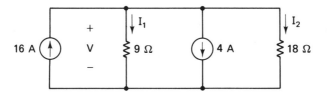

PROBLEM 4.21

4.22 Find *V* in Problem 4.21 by finding R_T and I_T.

4.23 A battery and two resistors are connected in parallel. The terminal voltage of the battery is 12 V and it delivers 3 A to the resistor combination. If one resistance is 5 Ω, find the other.

4.24 Find R_T if $R_1 = R_2 = 112$ Ω and $R_3 = R_4 = R_5 = 24$ Ω.

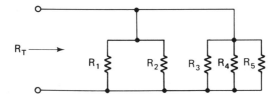

PROBLEM 4.24

4.25 Find R_T in Problem 4.24 if $R_1 = 6\ \Omega$, $R_2 = 12\ \Omega$, $R_3 = 18\ \Omega$, $R_4 = 40\ \Omega$, and $R_5 = 360\ \Omega$.

4.26 In Problem 4.7 find the power (a) delivered by the source and (b) absorbed by each of the two resistors. Show that conservation of power holds by noting that the answer in part (a) is the sum of the answers in part (b).

4.27 In Problem 4.11 find (a) the power delivered by the source and (b) the power delivered to each resistor.

4.28 In Problem 4.12 find the power absorbed by each of the 4-V source, the 10-V source, and the 6-Ω, 5-Ω, 2-Ω, and 7-Ω resistors, and delivered by the 20-V source. Show that conservation of power holds.

4.29 In Problem 4.21 find the power absorbed or delivered (and state which) by (a) the 16-A source, (b) the 4-A source, (c) the 9-Ω resistor, and (d) the 18-Ω resistor. Show that conservation of power holds.

4.30 If in Problem 4.24 the current at the outside terminals is $I = 12$ A, find the power delivered to each resistor. Check by showing that the sum of the answers is equal to $I^2 R_T$.

5

SERIES–PARALLEL CIRCUITS

In Chapter 4 we considered simple circuits of the series and parallel types and in each case we were able to obtain an equivalent circuit with a single resistor and a single source. This circuit, in turn, could be used to analyze the original circuit or to find currents and voltages that could be used in its analysis.

In this chapter we shall consider more complicated circuits in which some elements are connected terminal to terminal (that is, in series) and thus have the same current, and others are connected between common pairs of terminals (that is, in parallel) and thus have the same voltage. We shall refer to these circuits as *series–parallel* circuits and we shall see that they, too, have equivalent circuits containing a single resistor. Therefore, they may be analyzed in much the same way as series or parallel circuits.

The most general resistive circuit, of course, is not of the series, parallel, or series–parallel types. We shall give standard methods for analyzing such circuits, as well as the simpler ones of this and the previous chapter, in Chapter 7. However, these more powerful methods are often more difficult to apply and as a rule should only be used when the equivalent resistance procedures are not applicable.

5.1 *EQUIVALENT RESISTANCE*

A simple example of a series–parallel circuit is that of Fig. 5.1(a), with the schematic diagram shown in Fig. 5.1(b). Resistors R_2 and R_3 are connected in parallel (since they have the same terminals) and their parallel combination is in series with R_1.

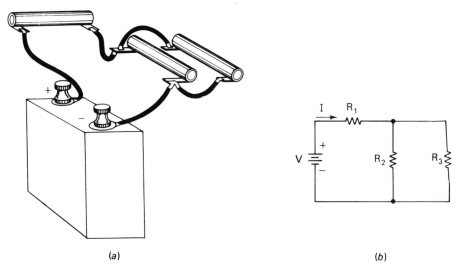

(a)   (b)

FIGURE 5.1   (a) Series-parallel circuit and (b) its schematic diagram.

Thus we may replace the parallel resistances R_2 and R_3 by their equivalent resistance, say R_4, given by

$$R_4 = \frac{R_2 R_3}{R_2 + R_3} \tag{5.1}$$

resulting in the circuit of Fig. 5.2(a). As far as V and I are concerned, this circuit is equivalent to that of Fig. 5.1.

Finally, the equivalent resistance R_T seen by the source V may be found by noting that R_1 and R_4 in Fig. 5.2(a) are in series (since they have the same current). Thus we have

$$R_T = R_1 + R_4 \tag{5.2}$$

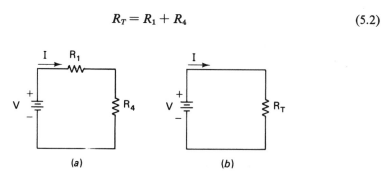

(a)   (b)

FIGURE 5.2   Steps in obtaining the equivalent resistance of Fig. 5.1.

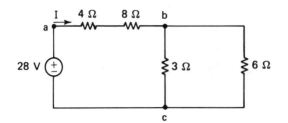

FIGURE 5.3 *Series-parallel circuit example.*

as indicated in Fig. 5.2(b). The current I is now easily found from Ohm's law,

$$I = \frac{V}{R_T} \tag{5.3}$$

We may combine (5.1) and (5.2) to obtain R_T in one step. The result is

$$R_T = R_1 + \frac{R_2 R_3}{R_2 + R_3} \tag{5.4}$$

We could have obtained this result from Fig. 5.1(b) by noting that R_2 and R_3 are in parallel, and thus equivalent to $R_2 R_3/(R_2 + R_3)$; then since the parallel combination is in series with R_1, (5.4) follows.

Example 5.1: Find the equivalent resistance R_T seen by the source in Fig. 5.3, and use the result to obtain the current I.

Solution: The 4-Ω and 8-Ω resistances are in series and may be replaced between terminals a and b by their equivalent resistance

$$R_{ab} = 4 + 8 = 12 \ \Omega$$

The 3-Ω and 6-Ω resistances between terminals b and c are in parallel and may be replaced by their equivalent resistance

$$R_{bc} = \frac{(3)(6)}{3 + 6} = 2 \ \Omega$$

These results are shown in Fig. 5.4(a), from where we see that R_{ab} and R_{bc} are in series. Therefore, their equivalent resistance is R_T, given by

$$R_T = R_{ab} + R_{bc}$$
$$= 12 + 2$$
$$= 14 \ \Omega$$

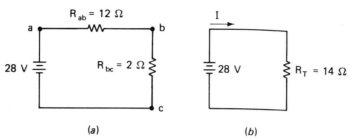

FIGURE 5.4   Steps in obtaining the equivalent circuit of Fig. 5.3.

as shown in Fig. 5.4(b). The current is therefore

$$I = \frac{28}{14} = 2 \text{ A}$$

Example 5.2: To illustrate further the procedure for finding equivalent resistances of series–parallel circuits, let us find R_T in the more complex circuit of Fig. 5.5.

Solution: To begin, we note that the 1-Ω and 5-Ω resistors are in series and may be replaced by their equivalent of $1 + 5 = 6$ Ω. Also, the 4-Ω and 12-Ω resistors are in parallel and have an equivalent of

$$\frac{(4)(12)}{4+12} = 3 \text{ Ω}$$

Making these changes we have the equivalent circuit of Fig. 5.6(a). Since in this latter circuit the 3-Ω and 6-Ω resistors are in parallel, we may replace their combination by

$$\frac{(3)(6)}{3+6} = 2 \text{ Ω}$$

resulting in the equivalent circuit of Fig. 5.6(b).

In Fig. 5.6(b) the 8-Ω and 2-Ω resistors are in series and may be replaced by one of $8 + 2 = 10$ Ω, as shown in Fig. 5.6(c). In the latter case the 15-Ω and 10-Ω resistors are in parallel with the equivalent resistance

$$\frac{(15)(10)}{25} = 6 \text{ Ω}$$

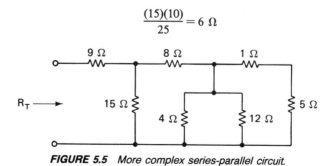

FIGURE 5.5   More complex series-parallel circuit.

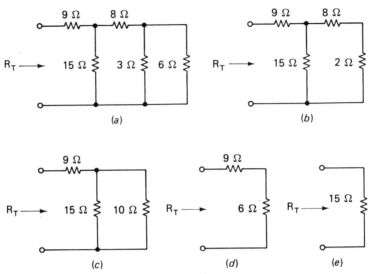

FIGURE 5.6 Steps in obtaining R_T for Fig. 5.5.

Making this replacement, we have Fig. 5.6(d) with the 9-Ω and 6-Ω resistors in series. Thus we have

$$R_T = 9 + 6 = 15 \ \Omega$$

as shown in Fig. 5.6(e).

General Case: The procedure for reducing a series–parallel combination of resistances to a single equivalent resistance R_T should be clear from the preceding example. We simply combine into one equivalent resistance any series resistances or any parallel resistances. Since these steps create more series and/or parallel resistances, we repeat the procedure until there is only the one resistance R_T left.

Not all resistive circuits are series, parallel, or series–parallel, as we noted at the beginning of the chapter. An example is the *bridge circuit* of Fig. 5.7, so called because R_6 is *bridged* across terminals *a-b*. The reader may verify by considering

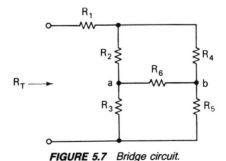

FIGURE 5.7 Bridge circuit.

each resistor one at a time that no two are in series or in parallel. We shall have a procedure for analyzing such circuits in Chapter 7.

PRACTICE EXERCISES

5-1.1   Find R_T and I in Fig. 5.1(b) if $R_1 = 2\ \Omega$, $R_2 = 9\ \Omega$, $R_3 = 72\ \Omega$, and $V = 30$ V.

Ans. 10 Ω, 3 A

5-1.2   Find R_T.   *Ans.* 12 Ω

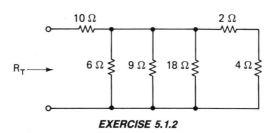

EXERCISE 5.1.2

5-1.3   Find I.   *Ans.* 4 A

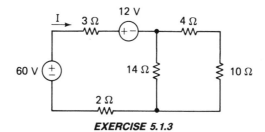

EXERCISE 5.1.3

5.2   STRINGS AND BANKS OF RESISTANCES

A common type of series–parallel circuit is one in which series resistors are connected in sets, or *strings*, of two or more, and parallel resistors occur in sets, or *banks*, of two or more. A string of three resistors is shown in Fig. 5.8(a) and consists simply of three resistors in series. A bank of four resistors, which is a parallel connection of four resistors, is shown in Fig. 5.8(b). Evidently, connections of strings and banks of resistors constitute a series–parallel circuit.

As an example, in the circuit of Fig. 5.3 the 4-Ω and 8-Ω series resistors constitute a string of two resistors and the 3-Ω and 6-Ω parallel resistors form a bank of two resistors. The circuit was analyzed by replacing the string and the bank by their equivalent resistances and finding R_T for the circuit.

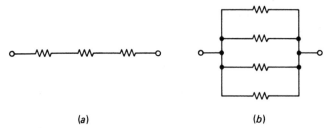

(a)   (b)

FIGURE 5.8   (a) String of three resistors and (b) bank of four resistors.

Example 5.3: Suppose that we have a number of light bulbs which are rated at 100 W at 120 V, and our voltage source is 240 V. Connecting each bulb across the source would result in too much voltage and damage to the bulb. Connecting too many bulbs in series with the source may result in too little voltage across each bulb and thus the light would be substandard. A solution is to connect the bulbs in strings so that the voltage across each bulb is the proper amount. Determine the proper connection.

Solution: Representing the bulbs as resistors, each with resistance R_1, we note that the equivalent resistance of a string of N bulbs is NR_1. Therefore, the IR drop across the string is NR_1I, if the current in the string is I. Connecting the string across the 240-V line, we must have

$$NR_1I = 240$$

But the IR drop of each bulb, given by IR_1, should be 120 V. Therefore, setting $R_1I = 120$, we have

$$120N = 240$$

or $N = 2$ bulbs per string. An example of three strings of two bulbs each is shown in Fig. 5.9. The voltage across each bulb (represented by R_1) is 120 V, as required.

Example 5.4: The circuit of Fig. 5.10 is a series–parallel circuit with two strings and two banks of resistors as shown. Find R_T.

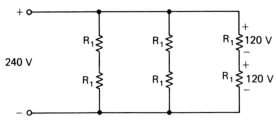

FIGURE 5.9   Three identical strings of resistors in parallel.

STRINGS AND BANKS OF RESISTANCES

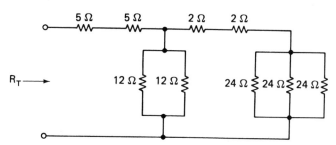

FIGURE 5.10 *Circuit of two strings and two banks of resistors.*

Solution: The string of two 5-Ω resistors has an equivalent resistance of $5 + 5 = 10$ Ω, the string of two 2-Ω resistors is equivalent to $2 + 2 = 4$ Ω, the bank of two 12-Ω resistors is equivalent to $12/2 = 6$ Ω, and the bank of three 24-Ω resistors is equivalent to $24/3 = 8$ Ω. Replacing these strings and banks by their equivalents, we have the equivalent circuit of Fig. 5.11(a). Finally, the 8-Ω and 4-Ω resistors in series are equivalent to a 12-Ω resistor, which is in parallel with the 6-Ω resistor. This combination is equivalent to $6(12)/(6 + 12) = 4$ Ω, which is then in series with the 10-Ω resistor. Thus $R_T = 10 + 4 = 14$ Ω, as shown in Fig. 5.11(b).

An example of a string of resistors is an old-fashioned string of Christmas tree lights. In this case the circuit is a series circuit of resistors (lights) connected across the 120-V wall plug. A disadvantage of this type of light string, of course, is that if one light burns out, the circuit is broken and none of the lights will burn. This is why this type of Christmas tree lights is seldom used nowadays. We shall consider this idea in more detail in Section 5.4.

Strings of Batteries: We may also "string out" a set of batteries in a series connection to provide a higher voltage than could be provided by a single battery. For example, in Fig. 5.12 we have three 1.5-V batteries in a string to provide a terminal voltage of $V = 4.5$ V. That V is the sum of the battery voltages is easily seen from KVL applied around the imaginary loop containing V and the three sources. The result is

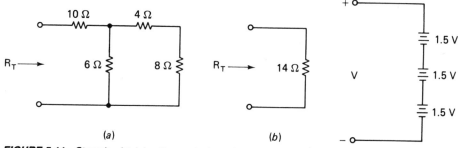

FIGURE 5.11 *Steps in obtaining the equivalent circuit of Fig. 5.10.*

FIGURE 5.12 *String of three batteries.*

$$-V + 1.5 + 1.5 + 1.5 = 0$$

from which V follows.

PRACTICE EXERCISES

5-2.1 A string of six resistors, each with a resistance R_1, is to be connected in parallel with a string of four resistors, each with a resistance of 9 Ω. Find R_1 if the equivalent of the circuit is to be $R_T = 33$ Ω.   *Ans.* 66 Ω

5-2.2 If the 100-W, 120-V bulbs (represented by R_1) in Fig. 5.9 are operating under rated conditions, find R_1 and the current drawn from the 240-V line.   *Ans.* 144 Ω, 2.5 A

5-2.3 Find R_T in Fig. 5.9 under the conditions of Exercise 5-2.2.   *Ans.* 96 Ω

5-2.4 Find I.   *Ans.* 2 A

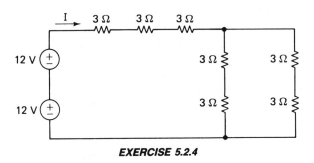

EXERCISE 5.2.4

5.3 SERIES–PARALLEL CIRCUIT EXAMPLES

To analyze a circuit completely, we must find all its currents and voltages. In a series–parallel circuit such as Fig. 5.1(b) we may find the equivalent resistance R_T seen by the source V by combining series and parallel resistors (repeating the process as necessary), and use the result to find the current I through the source. The other currents and voltages may then be found by applying Ohm's and Kirchhoff's laws to the remaining elements in a systematic way. In this section we shall illustrate the procedure by analyzing a number of series–parallel circuit examples.

Example 5.5: Find I, V_1, V_2, I_1, and I_2 in the series–parallel circuit of Fig. 5.13.

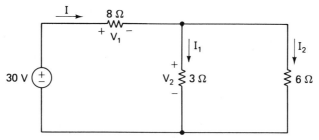

FIGURE 5.13 Series-parallel circuit example.

Solution: To find R_T we note that the 3-Ω and 6-Ω resistors are in parallel with an equivalent resistance of

$$\frac{3(6)}{3+6} = 2 \ \Omega$$

This 2-Ω resistance is in series with the 8-Ω resistance, so that we have

$$R_T = 2 + 8 = 10 \ \Omega$$

Therefore, the current I is

$$I = \frac{30}{10} = 3 \ A$$

and by Ohm's law the voltage V_1 is

$$V_1 = 8I = 8(3) = 24 \ V$$

Since by KVL, $V_1 + V_2 = 30$, we have

$$V_2 = 30 - V_1 = 30 - 24 = 6 \ V$$

Therefore, by Ohm's law,

$$I_1 = \frac{V_2}{3} = \frac{6}{3} = 2 \ A$$

We may find I_2 from Ohm's law given by

$$I_2 = \frac{V_2}{6} = \frac{6}{6} = 1 \ A$$

or from Kirchhoff's current law, $I = I_1 + I_2$, which yields

$$I_2 = I - I_1 = 3 - 2 = 1 \ A$$

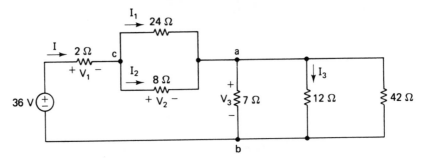

FIGURE 5.14 *Another series-parallel circuit example.*

Example 5.6: Find I_1, I_2, and I_3 in Fig. 5.14.

Solution: The equivalent resistance R_{ab} of the parallel 7-Ω, 12-Ω, and 42-Ω resistances satisfies

$$\frac{1}{R_{ab}} = \frac{1}{7} + \frac{1}{12} + \frac{1}{42}$$

$$= \frac{12 + 7 + 2}{84} = \frac{1}{4}$$

Therefore, $R_{ab} = 4\ \Omega$. The parallel 24-Ω and 8-Ω resistances are equivalent to

$$R_{ca} = \frac{(24)(8)}{32} = 6\ \Omega$$

Therefore, the resistance between terminals a and b is equivalent to $R_{ab} = 4\ \Omega$ and that between terminals c and a is equivalent to $R_{ca} = 6\ \Omega$. This is shown in Fig. 5.15, where it is seen that the resistance R_T seen by the source is

$$R_T = 2 + 6 + 4 = 12\ \Omega$$

Also, in Fig. 5.15 we have identified I, V_1, V_2 (across c-a), and V_3 (across a-b) in the original circuit.

From Fig. 5.15 we have

$$I = \frac{36}{R_T} = \frac{36}{12} = 3\ \text{A}$$

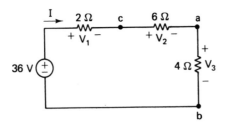

FIGURE 5.15 *Equivalent circuit of Fig. 5.14.*

and therefore

$$V_1 = 2I = 6 \text{ V}$$

$$V_2 = 6I = 18 \text{ V}$$

and

$$V_3 = 4I = 12 \text{ V}$$

Finally from Fig. 5.14 we have

$$I_1 = \frac{V_2}{24} = \frac{18}{24} = 0.75 \text{ A}$$

$$I_2 = \frac{V_2}{8} = \frac{18}{8} = 2.25 \text{ A}$$

and

$$I_3 = \frac{V_3}{12} = \frac{12}{12} = 1 \text{ A}$$

Ladder Network: One of the most important of the series–parallel circuits is the *ladder* network, an example of which is shown in Fig. 5.16. The network gets its name, of course, from its resemblance to a ladder.

To analyze the ladder network, we may find R_T seen by the source, and use Ohm's and Kirchhoff's laws to calculate successively I, V_1, V_2, I_1, I_2, and so on. To illustrate the procedure, let us analyze Fig. 5.16.

The series 4-Ω and 8-Ω resistances are equivalent to 12 Ω, which is in parallel with the 6-Ω resistance. Their equivalent, $6(12)/(6 + 12) = 4$ Ω, is in series with the 12-Ω resistor. Thus the equivalent resistance seen by I_2 is $4 + 12 = 16$ Ω. This 16-Ω resistance is in parallel with the 16-Ω resistance of the circuit, so that their equivalent is $16/2 = 8$ Ω, which is in series with the 2-Ω resistance. Therefore, we have

$$R_T = 2 + 8 = 10 \ \Omega$$

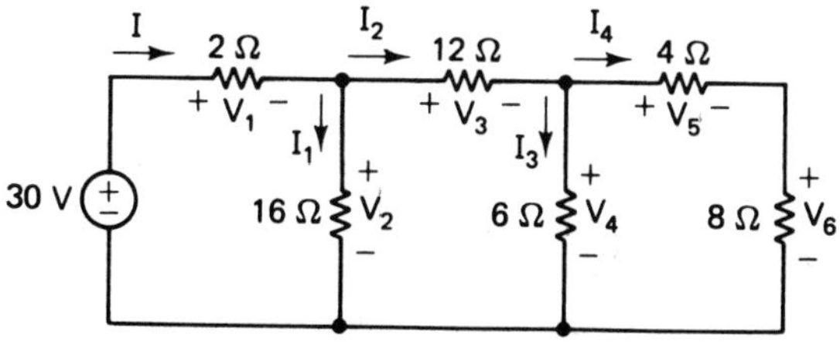

FIGURE 5.16 *Ladder network.*

and

$$I = \frac{30}{R_T} = \frac{30}{10} = 3 \text{ A}$$

With reference to Fig. 5.16, we may use Ohm's and Kirchhoff's laws to obtain the following:

$$V_1 = 2I = 6 \text{ V} \qquad \text{(Ohm's law)}$$

$$V_2 = 30 - V_1 = 24 \text{ V} \qquad \text{(KVL)}$$

$$I_1 = \frac{V_2}{16} = 1.5 \text{ A} \qquad \text{(Ohm's law)}$$

$$I_2 = I - I_1 = 3 - 1.5 = 1.5 \text{ A} \qquad \text{(KCL)}$$

$$V_3 = 12I_2 = 18 \text{ V} \qquad \text{(Ohm's law)}$$

$$V_4 = V_2 - V_3 = 24 - 18 = 6 \text{ V} \qquad \text{(KVL)}$$

$$I_3 = \frac{V_4}{6} = 1 \text{ A} \qquad \text{(Ohm's law)}$$

$$I_4 = I_2 - I_3 = 1.5 - 1 = 0.5 \text{ A} \qquad \text{(KCL)}$$

$$V_5 = 4I_4 = 2 \text{ V} \qquad \text{(Ohm's law)}$$

$$V_6 = V_4 - V_5 = 6 - 2 = 4 \text{ V} \qquad \text{(KVL)}$$

By finding I and then working outward from the source, we may find every current and voltage in the circuit, as we have seen. This is the general case with ladder networks.

PRACTICE EXERCISES

5-3.1   Find I and I_1 if $R = 12 \ \Omega$.   *Ans.* 3 A, 2 A

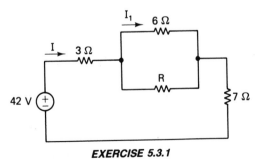

EXERCISE 5.3.1

5-3.2 Find I_1 and R in the circuit of Exercise 5-3.1 if $I = 3.6$ A.   *Ans.* 1 A, $\frac{30}{13}$ Ω

5-3.3 Find V.   *Ans.* 8 V

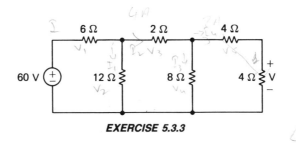

EXERCISE 5.3.3

5.4 OPEN AND SHORT CIRCUITS

An *open* circuit is a break in the circuit, such as between points a and b in the circuit of Fig. 5.17. The open circuit may occur accidentally, as in the case of a power line being struck by a falling tree, or it may occur as the result of the melting of the conducting link of a fuse or the opening of a "circuit breaker." In these latter cases the circuit is opened deliberately to protect it from a large surge of current. Of course, we also may open a circuit deliberately by means of a switch.

Open Circuit like an Infinite Resistance: An open circuit is, in effect, an infinite resistance because the air between points a and b is an insulating material. For example, the resistance between a and b may be many billion ohms and thus the current I is essentially zero. As an example, suppose that the open circuit has a resistance of 24 billion Ω. Then the equivalent resistance seen by the source is 24 billion Ω + 12 Ω, or roughly 24 billion Ω. The current is then 24 V divided by 24 billion Ω, or 1 billionth of an ampere, which for all practical purposes is zero.

Since the current in a series circuit is the same in every element of the circuit, if an open circuit develops there will be no current anywhere. This is the case of the string of Christmas tree lights discussed earlier in Section 5.2. If one light burns out (that is, its filament opens), then all the lights are out.

Voltage across an Open Circuit: An open circuit thus blocks the current, but it may still have a voltage across it. For example, in Fig. 5.17, we have $I = 0$ and thus the IR drops across the two resistors are zero. However, the voltage v_{ab} across the open circuit is 24 V, which is equal to that of the source. This follows from KVL around the circuit, which yields

$$-24 + 0 + v_{ab} + 0 = 0$$

and thus $v_{ab} = 24$ V. (The two zero terms are the IR drops of the resistances.)

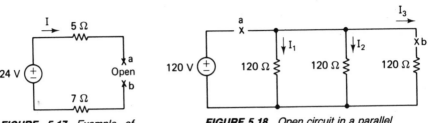

FIGURE 5.17 Example of an open circuit.

FIGURE 5.18 Open circuit in a parallel element or in the main line.

A good example of an open circuit is the 120-V ac voltage across the two terminals of a wall outlet in the home. When an appliance is plugged into the outlet, a current flows. With nothing connected the outlet is, in effect, an open pair of terminals with 120 V across them.

Open Circuit in a Parallel Element: In the case of an open-circuited parallel element, its current is blocked but there still may be current through the other parallel elements. For example, if in Fig. 5.18 the main line is open-circuited at a, no currents will exist anywhere. But if the break at a is restored and the circuit is broken at b, we will have

$$I_1 = I_2 = \frac{120}{120} = 1 \text{ A}$$

and $I_3 = 0$.

Short Circuit: A short circuit is a perfectly conducting wire which may be thought of as a resistor with zero resistance. Of course, an actual wire has a small amount of resistance, but it is negligible by comparison with actual resistors. A short circuit thus has zero voltage across it, because by Ohm's law the voltage is $RI = 0 \times I = 0$.

An example of a short circuit is the path placed from a to b across the 5-Ω resistor of Fig. 5.19(a). Since the short circuit is like a resistance of zero, its IR drop is zero, and since it is in parallel with the resistor, there can be no voltage on the resistor. Thus by Ohm's law the resistor draws no current and may be replaced by the short circuit, as in the equivalent circuit of Fig. 5.19(b).

Short-circuited Parallel Resistors: In the general case, short-circuited parallel resistors have no current. This is because the short placed across one parallel resistor is across all the parallel resistors, and therefore the voltage across each resistor is zero. Consequently, the current through each resistor is zero, and all the current is through the short circuit, bypassing the resistors.

As an example, in Fig. 5.20 the short circuit across the parallel 6-Ω and 4-Ω resistors "shorts out" both resistors. That is, it produces 0 V across their terminals, with the consequence that both their currents are zero. The 10-V source sees only the IR drop of 10 V across the 5-Ω resistor, so that $10/5 = 2$ A flows through the 5-Ω resistor, the short, and the source.

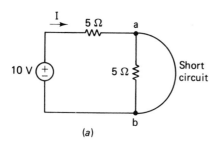

(a)

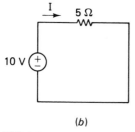

(b)

FIGURE 5.19 *(a) Short-circuited resistor and (b) its equivalent.*

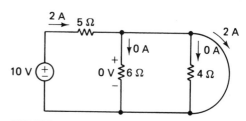

FIGURE 5.20 *Circuit with short-circuited parallel resistors.*

Example 5.7: Find I in Fig. 5.21 if (a) points a and b are shorted together and (b) if points a and c are shorted together.

Solution: In part (a) the short circuit across a-b shorts out the resistor, so that the source sees 9 Ω (the 6-Ω and 3-Ω resistances in series) in parallel with 3 Ω, or $3(9)/(3 + 9) = 9/4$ Ω. Thus we have

$$I = \frac{6}{\frac{9}{4}} = \frac{8}{3} \text{ A}$$

In (b) shorting terminals a and c together produces the circuit of Fig. 5.22, where nodes

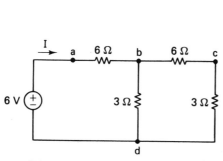

FIGURE 5.21 *Series-parallel circuit*

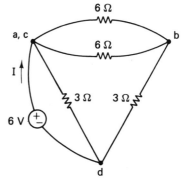

FIGURE 5.22 *Circuit of Fig. 5.21 with points a and c shorted together.*

a and *c* are shown as a common node *a, c*. As before, the source is between *a* and *d*, and the 3-Ω resistors are between *b* and *d* and between *c* and *d*. The resistance seen by the source may be found, as in the other examples, to be $R_T = 2\ \Omega$. Therefore, we have

$$I = \frac{6}{R_T} = 3 \text{ A}$$

PRACTICE EXERCISES

5-4.1 Suppose that element *R*, shown opened, represents a burned-out light in a string of lights. Find V_1, V_2, V_3, and V_4. Note that the faulty light can be found by measuring the terminal voltages, since the full source voltage is across it.   *Ans.* 0, 0, 0, 40 V

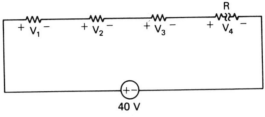

EXERCISE 5.4.1

5-4.2 Find *I* if the circuit is open-circuited at (a) point *a*, (b) point *b*, and (c) point *c*.
                                                              Ans. (a) 0, (b) 6 A, (c) 7 A

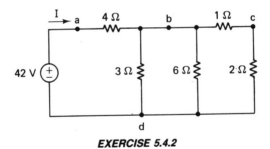

EXERCISE 5.4.2

5-4.3 Find *I* in the circuit of Exercise 5-4.2 if a short circuit is placed between points (a) *a* and *b*, (b) *a* and *c*, and (c) *b* and *d*.   *Ans.* (a) 35 A, (b) 36 A, (c) 10.5 A

5.5 SUMMARY

Series–parallel circuits are combinations of series and parallel circuits, containing strings, or series connections, of resistors and banks, or parallel connections, of resistors. As in the case of series or parallel circuits, an equivalent resistance can be found and used to simplify the analysis of the circuit.

Ladder networks, which resemble an actual ladder in appearance, are special and very important cases of series–parallel circuits. Their analysis can be performed by obtaining the current and voltage at the input terminals, using the equivalent resistance method, and working outward from the source.

An open circuit is a break in the circuit and, since it is equivalent to a resistor with an infinite resistance, it prevents the flow of current. A short circuit is a perfect conductor, or a resistor with zero resistance. Thus it has no voltage across it.

PROBLEMS

5.1 Find I in Fig. 5.1(b) if $R_1 = 4 \ \Omega$, $R_2 = 8 \ \Omega$, $R_3 = 24 \ \Omega$, and $V = 20 \ V$.

5.2 Find V in Fig. 5.1(b) if $R_1 = 3 \ \Omega$, $R_2 = 6 \ \Omega$, $R_3 = 30 \ \Omega$, and $I = 4 \ A$.

5.3 Find R_3 in Fig. 5.1(b) if $R_1 = 2 \ \Omega$, $R_2 = 7 \ \Omega$, $V = 24 \ V$, and $I = 3 \ A$.

5.4 Find R_T.

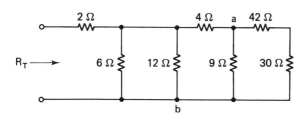

PROBLEM 5.4

5.5 Find I.

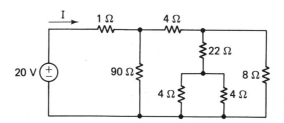

PROBLEM 5.5

5.6 In the circuit of Problem 5.5, find the power absorbed by the 90-Ω resistor.

5.7 Find V_1.

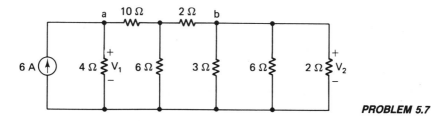

PROBLEM 5.7

5.8 In the circuit of Practice Exercise 5-1.3, find the power absorbed by the 4-Ω resistor.

5.9 If a 10-V source is applied across R_T in the circuit of Problem 5.4, find the power absorbed by the 4-Ω resistor.

5.10 Find the power absorbed by the 2-Ω resistor.

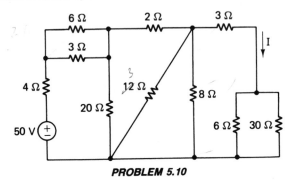

PROBLEM 5.10

5.11 Find I in Problem 5.10.

5.12 Find V_2 in Problem 5.7.

5.13 Find I_1, I_2, and V.

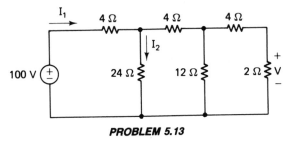

PROBLEM 5.13

5.14 If the 100-V source is replaced by a 200-V source in Problem 5.13, find I_1, I_2, and V.

5.15 A string of twenty 2-Ω resistors is connected in series with a bank of ten 40-Ω resistors. (a) Find the equivalent resistance R_T of the combination. (b) If a voltage source of 88 V is connected across R_T, find the current in each of the 40-Ω resistors.

5.16 Find I_1 and I_2.

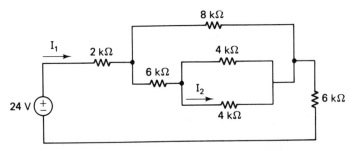

PROBLEM 5.16

5.17 Find R_T in Problem 5.4 if the 4-Ω resistor is open-circuited.

5.18 Find R_T in Problem 5.4 if a short circuit is placed between points *a* and *b*.

5.19 Find V_1 in Problem 5.7 if points *a* and *b* are connected by a short circuit.

5.20 Place a short circuit between points *a* and *b*, open the circuit at point *c*, and find *V* in the resulting circuit.

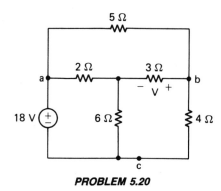

PROBLEM 5.20

6

VOLTAGE AND CURRENT DIVISION

Series circuits, parallel circuits, and series–parallel circuits generally can be analyzed much more easily by the use of the concepts of *voltage division* and *current division,* which we consider in this chapter. Circuits that employ those concepts are often called *voltage dividers* and *current dividers,* respectively, and are useful in tapping voltages or currents from a power supply or source. The tapped voltages may be calculated without the need for knowing the currents involved, as we shall see. The same idea applies to current division, where it is not necessary to know the voltages involved in order to find the individual currents.

The *Wheatstone bridge,* which we consider also, is a well-known circuit that may be used to measure unknown resistances. When the bridge is *balanced,* it is essentially a series–parallel circuit to which the concepts of voltage and current division are directly applicable.

6.1 VOLTAGE DIVISION

In a series circuit the current I is the same in every element, as we have seen in previous chapters. Therefore, each IR drop across a resistor is proportional to the resistance R. For example, in the circuit of Fig. 6.1, the current is I and the IR drops are V_1 and V_2, given respectively by

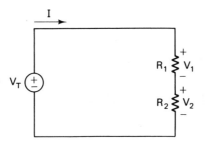

FIGURE 6.1 *Voltage divider with two resistors.*

$$V_1 = IR_1 \tag{6.1}$$

and

$$V_2 = IR_2 \tag{6.2}$$

The current, by Ohm's law, is

$$I = \frac{V_T}{R_T} = \frac{V_T}{R_1 + R_2} \tag{6.3}$$

where R_T is the equivalent resistance

$$R_T = R_1 + R_2 \tag{6.4}$$

Replacing I in (6.1) and (6.2) by its value in (6.3), we have

$$V_1 = \frac{R_1}{R_T} V_T$$

$$V_2 = \frac{R_2}{R_T} V_T \tag{6.5}$$

or equivalently,

$$\frac{V_1}{V_T} = \frac{R_1}{R_T}$$

$$\frac{V_2}{V_T} = \frac{R_2}{R_T} \tag{6.6}$$

That is, the voltages are in the same ratio as the resistances. Substituting (6.4) into (6.5) results in

$$V_1 = \frac{R_1}{R_1 + R_2} V_T \qquad (6.7)$$

and

$$V_2 = \frac{R_2}{R_1 + R_2} V_T \qquad (6.8)$$

Thus each voltage across a series resistor is a fraction of the total voltage, and the fraction is the ratio of the resistance of the resistor to the total resistance. Putting it another way, the voltage of the source V_T *divides* between resistances R_1 and R_2 in direct proportion to their resistances. This is the principle of *voltage division,* and the circuit of Fig. 6.1 is called a *voltage divider.*

From (6.7) and (6.8) we see that the larger resistance has the higher voltage and the smaller resistance has the smaller voltage. This is to be expected since both resistors carry the same current.

Example 6.1: Find the voltages V_1 and V_2 in Fig. 6.1 if $R_1 = 4 \ \Omega$, $R_2 = 6 \ \Omega$, and $V_T = 20$ V.

Solution: By (6.7) and (6.8) the voltages are

$$V_1 = \frac{4}{4+6} V_T = \frac{4}{10} V_T = \frac{4}{10} (20) = 8 \text{ V}$$

and

$$V_2 = \frac{6}{4+6} V_T = \frac{6}{10} V_T = \frac{6}{10} (20) = 12 \text{ V}$$

General Case: Any string of resistances such as R_1 and R_2 in Fig. 6.1 may constitute a voltage divider. For example, if there are N series resistors in the string, denoted by $R_1, R_2, \ldots, R_N$, with voltages $V_1, V_2, \ldots, V_N$, we have

$$R_T = R_1 + R_2 + \cdots + R_N$$

and the *IR* drops are

$$
\begin{aligned}
V_1 &= IR_1 \\
V_2 &= IR_2 \\
&\;\;\vdots \\
V_N &= IR_N
\end{aligned}
\qquad (6.9)
$$

Again we have

$$I = \frac{V_T}{R_T}$$

where V_T is the total voltage across the string. Substituting this value of I into (6.9) results in

$$V_1 = \frac{R_1}{R_T} V_T$$

$$V_2 = \frac{R_2}{R_T} V_T \qquad\qquad (6.10)$$

$$\vdots$$

$$V_N = \frac{R_N}{R_T} V_T$$

Again, the voltage V_T divides among the various resistors in direct proportion to their resistances.

An advantage of the voltage-division procedure is that we may find the IR drops without calculating the current I. We will illustrate this with an example.

Example 6.2: Find V in Fig. 6.2.

Solution: The circuit is a voltage divider with

$$R_T = 2 + 4 + 1 + 5 = 12 \text{ k}\Omega$$

Therefore, by voltage division we have

$$V = \frac{5}{R_T} V_T = \frac{5}{12}(24) = 10 \text{ V}$$

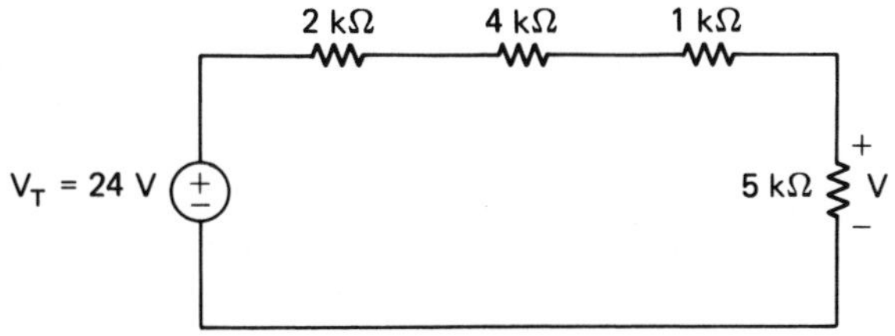

FIGURE 6.2 *Voltage divider with four resistors.*

Case of Equal Resistances: If the resistances in the divider are equal, the voltage divides equally. For example, in the case of two resistances R_1 and R_2 in Fig. 6.1, if $R_1 = R_2$, (6.7) and (6.8) become

$$V_1 = \frac{V_T}{2} \quad \text{and} \quad V_2 = \frac{V_T}{2}$$

The case of a string of N equal resistances is considered in Practice Exercise 6-1.4.

PRACTICE EXERCISES

6-1.1 Find the IR drops in the circuit of Fig. 6.2 across the 2-kΩ, the 4-kΩ, and 1-kΩ resistors. Note that KVL holds by observing that the sum of these IR drops and V equals V_T.   *Ans.* 4 V, 8 V, 2 V

6-1.2 If $V_1 = 12$ V, $V_T = 30$ V, and $R_2 = 6$ Ω in Fig. 6.1, find (a) R_1 using voltage division and (b) I.   *Ans.* (a) 4 Ω, (b) 3 A

6-1.3 Find V using voltage division. (*Suggestion:* First find V_T.)   *Ans.* 14 V

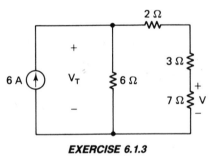

EXERCISE 6.1.3

6-1.4 A voltage divider has N resistances, all of which are equal to R. If V_T is the total voltage, find the voltage across each resistor.   *Ans.* V_T/N

6-1.5 Find the ratio V_1/V_2 in the voltage-divider circuit of Fig. 6.1 in terms of the resistances. [*Suggestion:* Divide (6.7) by (6.8).]   *Ans.* R_1/R_2

6.2 CURRENT DIVISION

The current into a bank of resistors divides in much the same way as the voltage across a string of resistors, or a voltage divider. The voltage V across the bank is, of course, the voltage across each of the parallel resistors in the bank. Therefore, the current in each resistor is proportional to the conductance G of the resistor, since by Ohm's law it is given by

$$I = GV$$

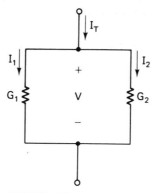

FIGURE 6.3 Current divider with two resistors.

As an example, in the circuit of Fig. 6.3 a total current I_T is flowing into the bank of two resistors having conductances G_1 and G_2 and currents I_1 and I_2. The voltage V is the common voltage across each resistor. The currents I_1 and I_2, by Ohm's law, are given by

$$I_1 = G_1 V \tag{6.11}$$

and

$$I_2 = G_2 V \tag{6.12}$$

The equivalent conductance of the bank is

$$G_T = G_1 + G_2 \tag{6.13}$$

so that the bank voltage is

$$V = \frac{I_T}{G_T} = \frac{I_T}{G_1 + G_2} \tag{6.14}$$

Substituting this value into (6.11) and (6.12), we have

$$I_1 = \frac{G_1}{G_T} I_T \tag{6.15}$$

and

$$I_2 = \frac{G_2}{G_T} I_T \tag{6.16}$$

We see, therefore, that the circuit of Fig. 6.3 is a *current divider,* in which the total current I_T *divides* between the resistors in direct proportion to their conductances

117

G_1 and G_2. In the special case of two resistors, as in Fig. 6.3, we have

$$I_1 = \frac{G_1}{G_1 + G_2} I_T \tag{6.17}$$

and

$$I_2 = \frac{G_2}{G_1 + G_2} I_T \tag{6.18}$$

The larger conductance has the larger current and the smaller conductance has the smaller current.

It is interesting to note that all the results of this section are the duals of corresponding voltage-division results in Section 6.1, and could have been obtained directly by replacing all the quantities by their duals.

Example 6.3: Find the currents I_1 and I_2 in Fig. 6.3 if $I_T = 12$ A, $G_1 = 2$ ℧, and $G_2 = 6$ ℧.

Solution: By (6.17) and (6.18) the currents are

$$I_1 = \frac{2}{2+6} I_T = \frac{2}{8} I_T = \frac{2}{8}(12) = 3 \text{ A}$$

and

$$I_2 = \frac{6}{2+6} I_T = \frac{6}{8} I_T = \frac{6}{8}(12) = 9 \text{ A}$$

General Case: In general, the current divider may be a bank of N resistors with conductances $G_1, G_2, \ldots, G_N$, having currents $I_1, I_2, \ldots, I_N$. In this case the equivalent conductance is

$$G_T = G_1 + G_2 + \cdots + G_N$$

and the currents are

$$I_1 = G_1 V$$
$$I_2 = G_2 V$$
$$\vdots \tag{6.19}$$
$$I_N = G_N V$$

Again we have

$$V = \frac{I_T}{G_T}$$

where I_T is the total current entering the divider. Substituting this value of V into (6.19) yields

$$I_1 = \frac{G_1}{G_T} I_T$$

$$I_2 = \frac{G_2}{G_T} I_T$$

$$\vdots$$

$$I_N = \frac{G_N}{G_T} I_T$$

(6.20)

In the general case, therefore, the total current I_T divides among the various resistors in the bank in direct proportion to their conductances. As we see in (6.20), it is not necessary to know the voltage across the bank to get the individual currents.

Example 6.4: Find I_1, I_2, and I_3 in Fig. 6.4 if $I_T = 36$ mA.

Solution: The circuit is a current divider with

$$G_T = 4 + 6 + 8 = 18 \text{ m}\mho$$

Therefore, by the current-division principle of (6.20), we have

$$I_1 = \frac{4}{18}(36) = 8 \text{ mA}$$

$$I_2 = \frac{6}{18}(36) = 12 \text{ mA}$$

$$I_3 = \frac{8}{18}(36) = 16 \text{ mA}$$

Current-Divider Rule in Terms of Resistances: We have given the current-divider rule in terms of conductances, but as a rule the resistances are specified in the circuit. It is very easy to change the formulas such as (6.15), (6.16), and (6.20) to use resistance values rather than conductance values by replacing G_T, G_1, G_2, and so on, by their equivalent values $1/R_T$, $1/R_1$, $1/R_2$, and so on. In the case of a

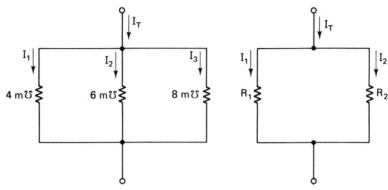

FIGURE 6.4 *Current divider with three resistors.*

FIGURE 6.5 *Current divider using resistance values.*

current divider with two resistors, as shown in Fig. 6.5, the result, obtained from (6.15) and (6.16), is

$$I_1 = \frac{R_T}{R_1} I_T$$

$$I_2 = \frac{R_T}{R_2} I_T$$

(6.21)

Since in the parallel case we have

$$R_T = \frac{R_1 R_2}{R_1 + R_2}$$

we may write (6.21) in the form

$$I_1 = \frac{R_2}{R_1 + R_2} I_T$$

$$I_2 = \frac{R_1}{R_1 + R_2} I_T$$

(6.22)

Therefore, the current divides in inverse proportion to the resistances. The larger current flows through the smaller resistance and the smaller current flows through the larger resistance, as we would expect.

Case of Equal Resistances: In case $R_1 = R_2$ in the current divider of Fig. 6.5, we have by (6.22)

$$I_1 = \frac{I_T}{2}$$

$$I_2 = \frac{I_T}{2}$$

That is, the current divides equally since it sees two identical paths.

Example 6.5: In Fig. 6.5, $I_T = 16$ A. Find I_1 and I_2 if (a) $R_1 = 4\ \Omega$ and $R_2 = 12\ \Omega$ and (b) $R_1 = R_2 = 8\ \Omega$.

Solution: In case (a) we have by (6.22)

$$I_1 = \frac{12}{4+12}(16) = 12 \text{ A}$$

$$I_2 = \frac{4}{4+12}(16) = 4 \text{ A}$$

In case (b), since the resistances are equal, the current divides equally. That is,

$$I_1 = I_2 = \frac{I_T}{2} = 8 \text{ A}$$

The case of more than two resistors can also be derived in terms of resistances, but the result is too cumbersome to make it worthwhile because of the complicated form of the equivalent resistance R_T. We may always convert resistance to conductance and use (6.20), however. In the case of N equal resistances (and therefore conductances) the result is simple, and by (6.20) is given by

$$I_1 = I_2 = \cdots = I_N = \frac{I_T}{N} \qquad (6.23)$$

PRACTICE EXERCISES

6-2.1 Find I_1 and I_2 if $R = 30\ \Omega$. (*Suggestion:* First find I_T and use current division.)

Ans. 5 A, 1 A

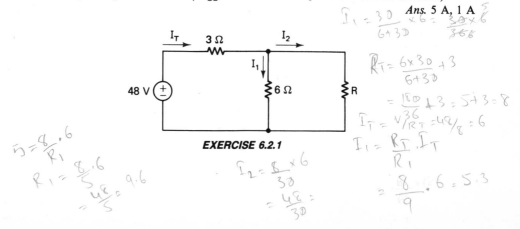

EXERCISE 6.2.1

6-2.2   Solve Exercise 6-2.1 if $R = 6\ \Omega$.   *Ans.* 4 A, 4 A

6-2.3   A current divider consists of a bank of 10 resistors. Nine of them have equal conductances of 20 m℧ and the tenth has a conductance of 70 m℧. If the total current entering the divider is $I_T = 50$ mA, find the current in the tenth resistor.   *Ans.* 14 mA

6.3   EXAMPLES USING VOLTAGE AND CURRENT DIVISION

Voltage and current division often may be used to shorten considerably the analysis of series–parallel circuits. As an example, let us find I_1 and V_1 in the circuit of Fig. 6.6. The resistance R_T seen by the source is given by

$$R_T = 4 + \frac{4(3 + 5)}{4 + 3 + 5} = \frac{20}{3}\ \Omega$$

so that

$$I = \frac{20}{R_T} = 3\ \text{A}$$

The current I divides into I_1 through the 4-Ω path and the current through the $5 + 3 = 8\ \Omega$ path. Therefore, by current division we have

$$I_1 = \frac{8}{8 + 4}\,(3) = 2\ \text{A}$$

The voltage V_2 is therefore

$$V_2 = 4I_1 = 8\ \text{V}$$

and by voltage division we have

$$V_1 = \frac{3}{3 + 5}\,V_2 = 3\ \text{V}$$

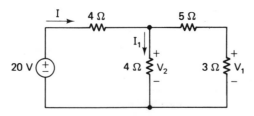

FIGURE 6.6   *Series-parallel circuit.*

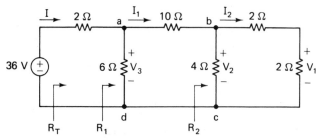

FIGURE 6.7 *Ladder network.*

Application to Ladder Networks: Current and voltage division are especially useful in analyzing ladder networks. To illustrate the procedure, let us find V_1 in the ladder network of Fig. 6.7. We first note that R_2, the resistance seen at terminals *b-c*, is given by

$$R_2 = \frac{4(4)}{4+4} = 2 \ \Omega$$

and that R_1, the resistance seen at *a-d* is given by

$$R_1 = \frac{6(R_2 + 10)}{6 + (R_2 + 10)} = 4 \ \Omega$$

The resistance R_T seen by the source is therefore

$$R_T = R_1 + 2 = 6 \ \Omega$$

and thus the current I is

$$I = \frac{36}{R_T} = 6 \ A$$

The current I divides at node *a* into two paths—a 6-Ω path and a path of $10 + R_2 = 12 \ \Omega$. Therefore, by current division, I_1 through the 12-Ω path is

$$I_1 = \frac{6}{6 + 12} \cdot I = 2 \ A$$

This current divides at node *b* into two 4-Ω paths. Therefore, we have

$$I_2 = \frac{I_1}{2} = 1 \ A$$

The voltage V_1, therefore, by Ohm's law is

$$V_1 = 2I_2 = 2 \ V$$

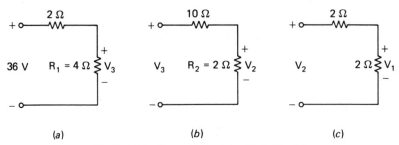

FIGURE 6.8 *Steps in obtaining V_1 in Fig. 6.7.*

We have analyzed the ladder network by repeated use of current division. Now let us analyze it by means of voltage division. The voltage V_3 is across an equivalent resistance of $R_1 = 4\ \Omega$, so that by voltage division we have

$$V_3 = \frac{4}{4+2}(36) = 24\ \text{V} \tag{6.24}$$

The voltage V_2 is across an equivalent resistance of $R_2 = 2\ \Omega$, so that by voltage division we have

$$V_2 = \frac{2}{2+10}V_3 = 4\ \text{V} \tag{6.25}$$

Finally, the voltage V_2 divides equally across the two 2-Ω resistances it sees, and therefore we have, as before,

$$V_1 = \frac{V_2}{2} = 2\ \text{V} \tag{6.26}$$

The steps shown in obtaining (6.24) to (6.26) may be seen more clearly in Fig. 6.8(a) to (c).

Example 6.6: As a last example, let us find I, I_1, V_1, V_2, and V_3 in the ladder network of Fig. 6.9.

Solution: The resistance R_T seen by the source may be found, by combining parallel and series resistances, to be

$$R_T = \frac{6}{3} + \frac{12(4)}{12+4} = 5\ \Omega$$

Therefore, we have

$$I = \frac{45}{R_T} = 9\ \text{A}$$

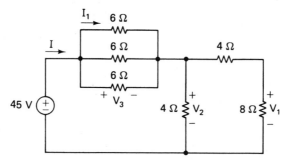

FIGURE 6.9 *Ladder network with a bank of three resistors.*

By current division I_1 is given by

$$I_1 = \frac{I}{3} = 3 \text{ A}$$

and therefore we have

$$V_3 = 6I_1 = 18 \text{ V}$$

By KVL we have

$$V_2 = 45 - V_3 = 27 \text{ V}$$

and by voltage division we have

$$V_1 = \frac{8}{8+4} V_2 = 18 \text{ V}$$

PRACTICE EXERCISES

6-3.1   Find I using current division.   *Ans.* 12.5 A

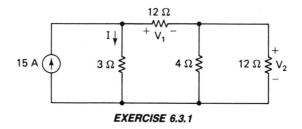

EXERCISE 6.3.1

6-3.2 Find V_1 and V_2 in Exercise 6-3.1 using voltage division.   *Ans.* 30 V, 7.5 V

6-3.3 Solve Problem 5.12 using voltage division.   *Ans.* 1 V

6.4 WHEATSTONE BRIDGE

In Chapter 5 we considered an example of a *bridge* circuit in Fig. 5.7 and pointed out that it was not of the series, parallel, or series–parallel types. However, if the resistances are adjusted so that there is no current in the bridged element (R_6 in the case of Fig. 5.7), then the bridge circuit is, in effect, a series–parallel circuit, and the principles of voltage and current division are applicable.

One of the best known and most useful bridge circuits is the *Wheatstone bridge*, named for the English physicist Sir Charles Wheatstone (1802–1875). It is shown in Fig. 6.10 and consists of four resistances R_1, R_2, R_x, and R_s and a *galvanometer*, symbolized by the circle labeled G between nodes a and b. The galvanometer is a very sensitive meter which measures the current through the bridged element between a and b. (We shall discuss meters in detail in Chapter 9.) The element S is a switch that is closed to connect the bridge to the voltage source V.

Use in Measuring Resistance: The Wheatstone bridge of Fig. 6.10 may be used to measure an unknown resistance R_x by adjusting the variable (known) *standard* resistance R_s, keeping R_1 and R_2 at fixed, known values. To see how this may be done, let us consider what happens as R_s is varied. By KCL the current from point a through R_s is $I_1 - I_G$, and since it changes with R_s, there is a value of R_s for which $I_G = 0$. For this value of R_s, the galvanometer indicates zero current and the bridge is said to be *balanced*.

When the bridge is balanced, the current I_1 is through both R_x and R_s and the

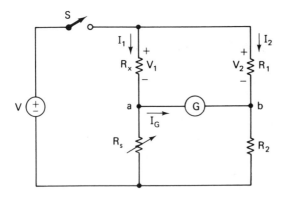

FIGURE 6.10 *Wheatstone bridge circuit.*

current I_2 is through both R_1 and R_2. Therefore, the bridge behaves like a series–parallel circuit with the two strings of two resistors in parallel with each other and the source. Also, the galvanometer, as is typical of current-measuring meters, has almost no voltage across it, and in the ideal case its terminal voltage is zero. This is true even if it draws current, because the ideal galvanometer behaves like a short circuit. Thus by KVL we see in Fig. 6.10 that

$$V_1 = V_2 \tag{6.27}$$

Since with $I_G = 0$ the bridge appears to be a series–parallel circuit, voltage division applies and we may write

$$V_1 = \frac{R_x}{R_x + R_s} V$$

$$V_2 = \frac{R_1}{R_1 + R_2} V$$

Substituting these results into (6.27) we have, after canceling V,

$$\frac{R_x}{R_x + R_s} = \frac{R_1}{R_1 + R_2}$$

or

$$R_1 R_x + R_2 R_x = R_1 R_x + R_1 R_s$$

Subtracting out the common term $R_1 R_x$ we have

$$R_2 R_x = R_1 R_s$$

so that solving for R_x results in

$$R_x = \frac{R_1}{R_2} R_s \tag{6.28}$$

By (6.28) we see that if R_1 and R_2 are known and we may vary R_s in a known way, we may find the value of the unknown resistance R_x. We simply adjust R_s until the galvanometer reads zero current and find R_x by (6.28). In fact, we do not need to know R_1 and R_2. Only their ratio is needed.

Example 6.7: In the Wheatstone bridge circuit of Fig. 6.10, $R_1 = 12 \ \Omega$, $R_2 = 8 \ \Omega$, and the necessary value of R_s to make $I_G = 0$ is 22 Ω. Find R_x.

Solution: By (6.28) we have

$$R_x = \frac{R_1}{R_2} \, R_s = \frac{12}{8} \, (22) = 33 \ \Omega$$

A photograph of an actual Wheatstone bridge circuit is shown in Fig. 6.11. It includes a slide-wire form with a sliding contact key, which provides R_1 and R_2; a dial-type resistance box for R_s; mounted resistors, from which R_x is chosen; a galvanometer; a dry-cell battery; and a switch. The resistance box is used to set R_s to make I_G zero, or almost zero, and the sliding contact is for fine tuning. That is, if the current I_G is not quite zero after setting R_s, it may be made precisely zero by a slight movement of the contact key. At this point R_1 and R_2 may be read from the scale on the slide-wire form. The unknown resistance R_x may be any one of several available on the mounted resistor assembly.

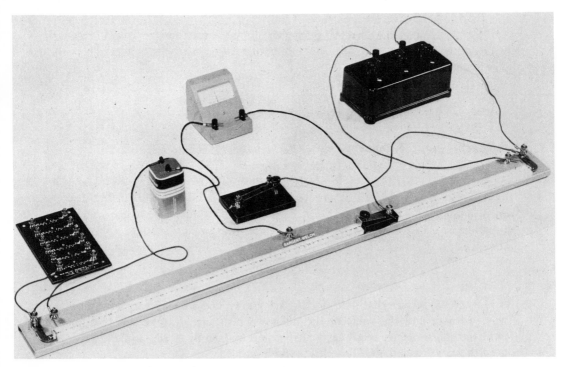

FIGURE 6.11 Photograph of an actual Wheatstone bridge circuit (Courtesy, Sargent-Welch Scientific Co.).

PRACTICE EXERCISES

6-4.1 If in the Wheatstone bridge of Fig. 6.10, $R_1/R_2 = 4$ and $R_s = 6$ Ω is required to balance the bridge, find R_x. *Ans. 24 Ω*

6-4.2 Find V_1, V_2, and I_1 in Fig. 6.10 for the resistance value of Exercise 6-4.1 if $V = 30$ V. *Ans. 24 V, 24 V, 1 A*

6-4.3 If $V = 30$ V and $I_2 = 2$ A in Exercise 6-4.1, find R_1 and R_2. *Ans. 12 Ω, 3 Ω*

6.5 SUMMARY

A voltage across a series connection of resistors divides, with the voltage across each resistor directly proportional to its resistance. For example, if the total voltage across the string of resistances is V, then the voltage V_1 across any resistance, say R_1, is

$$V_1 = \frac{R_1}{R_T} V$$

where R_T is the equivalent resistance of the string.

In a like manner the current I through a parallel connection, or bank, of resistors divides, with the current through each resistor directly proportional to its conductance. That is, the current I_1 through G_1 is

$$I_1 = \frac{G_1}{G_T} I$$

where G_T is the equivalent conductance of the bank of resistors.

Voltage and current division are very useful in analyzing series–parallel circuits because they can be used in shortcut procedures where only a single current or voltage need be found. The concepts are especially useful in analyzing ladder networks and in finding an unknown resistance by means of a Wheatstone bridge circuit. In the latter case, when the bridge is balanced (no current in the bridged element), it is essentially a series–parallel circuit to which voltage and current division are directly applicable.

PROBLEMS

6.1 Find V_1 and V_2 in Fig. 6.1 if in $R_1 = 2$ kΩ, $R_2 = 6$ kΩ, and $V_1 = 16$ V.

6.2 Find the ratio R_1/R_2 in Fig. 6.1 if $V_1 = 8$ V and $V_2 = 2$ V. (*Suggestion:* See Practice Exercise 6-1.5.)

6.3 If in Fig. 6.1 we have $V_T = 30$ V, $I = 5$ A, and $V_1/V_2 = 2$, find R_1 and R_2.

6.4 A string of N equal resistances is connected across a 20-V source. If the voltage across each resistance is 4 V, find N.

6.5 Find V_1 and V_2 using voltage division.

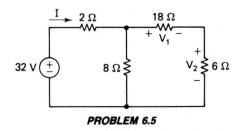

PROBLEM 6.5

6.6 Solve Problem 6.5 using current division. (*Suggestion:* Find the division of I in the 18-Ω and 6-Ω path.)

6.7 Find I_1 and I_2 if $R = 5$ Ω.

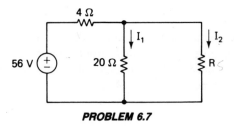

PROBLEM 6.7

6.8 Find I_1 and I_2 in Problem 6.7 if $R = 20$ Ω.

6.9 A current divider consists of a bank of four resistors with conductances of 2, 4, 6, and 8 m$\mho$. If the total current entering the divider is 10 mA, find the current in each resistor.

6.10 Solve Problem 5.16 using voltage and current division.

6.11 Use voltage and current division to find I_1, I_2, V_1, and V_2.

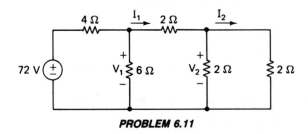

PROBLEM 6.11

6.12 Find R_1 and R_2 for a voltage divider if the voltage across the divider is 100 V, the current entering the divider is 5 A, and there is 25 V across R_1.

⋆ **6.13** Find I if $R = 6 \ \Omega$.

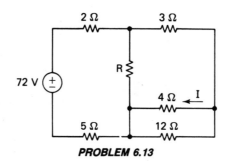

PROBLEM 6.13

6.14 Find I in Problem 6.13 if $R = 3 \ \Omega$.

⋆ **6.15** If the Wheatstone bridge of Fig. 6.10 is balanced and $R_1 = 2$ kΩ, $R_2 = 500 \ \Omega$, and $R_s = 80 \ \Omega$, find R_x.

6.16 Find V_1, V_2, I_1, and I_2 in the circuit of Problem 6.15 if $V = 20$ V.

7

GENERAL RESISTIVE CIRCUITS

As we have noted earlier, many circuits are not of the series, parallel, or series–parallel types, and must be analyzed by more general methods. An example of such a general circuit is the bridge circuit previously considered in Fig. 5.7, in which no two resistors are in series or in parallel. Therefore, we cannot combine resistances to get a single equivalent resistance, and consequently we cannot apply voltage or current division. An even simpler circuit which is not of the series–parallel type is one with two loops, each of which contains a voltage source.

There are two general methods of analysis that may be applied to any resistive circuit. These methods, which we will consider in detail in this chapter, are known as *mesh,* or *loop,* analysis and *nodal* analysis. In the mesh or loop analysis we apply KVL around certain closed paths, obtaining a set of equations in the unknown currents in the circuit. Nodal analysis, in which the unknowns are voltages, is the application of KCL at nodes in the circuit.

In either mesh or nodal analysis we have a set of simultaneous equations which must be solved for the unknowns (currents in mesh analysis and voltages in nodal analysis). In this chapter we will systematically select the set of currents or voltages that must be found for a complete analysis of the circuit, obtain the set of equations they satisfy, and consider methods of solving the equations. As we shall see, the procedures apply to any resistive circuit, from the simple circuits of the series or parallel types to the bridge circuit of Fig. 5.7, or to circuits with any number of nodes or closed paths.

The first general analysis method that we will consider is that of *loop* analysis, in which KVL is applied around certain loops, or closed paths, in the circuit. The currents we will consider at first are element currents, but we will make the method more systematic by later considering a set of currents called *mesh* currents.

We will limit our discussion to *planar* circuits, which are those that may be drawn on a plane surface in a way such that no element crosses any other element. That is, no wires or elements touch except at nodes. The method of loop analysis applies as well to circuits that are not planar (and thus are called *nonplanar*), but it is more difficult to find an appropriate set of loops to use in the analysis. An example of a planar circuit is that of Fig. 7.1, which has two loops identified by the arrows marked 1 and 2. (There is also a loop around the outside of the circuit but, as we shall see, it is not necessary to consider it in the analysis.)

An example of a nonplanar network is shown in Fig. 7.2. The reader may wish to try redrawing it so that no wires cross except at nodes, but the task will doubtless soon be abandoned as hopeless.

Meshes: In the case of a planar circuit, such as that of Fig. 7.1, the plane is divided into distinct areas in the same way that the wooden or metal partitions in a window distinguish the window panes. The closed loop that forms the boundary

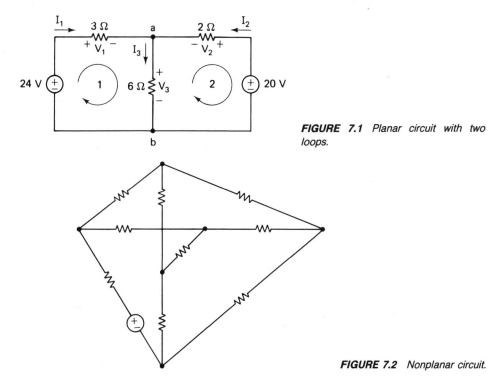

FIGURE 7.1 Planar circuit with two loops.

FIGURE 7.2 Nonplanar circuit.

133

of each of these areas or "windows" is called a *mesh* of the circuit. Thus a mesh is the special type of loop that contains no elements within it. Examples of meshes are loops 1 and 2 of Fig. 7.1. The outside loop is not a mesh because it contains the 6-Ω resistor within it.

The meshes are the easiest loops to use because they are the easiest to find, and, as we shall see, there are exactly the right number of them to analyze the circuit. In the case of Fig. 7.1, there are two meshes and every element in the circuit is in one mesh or the other.

Loop Equations: Loop analysis is begun by writing the loop equations, which are obtained by simply applying Kirchhoff's voltage law around the loops. Since the voltages across the resistors are *IR* drops, the unknowns in the loop equations will be the currents. It does not matter in which direction the loops are traversed as long as Kirchhoff's and Ohm's laws are applied properly.

To illustrate the procedure, let us write the loop equations for the circuit of Fig. 7.1. Around loop 1 in the direction of the arrow we have by KVL

$$V_1 + V_3 - 24 = 0 \tag{7.1}$$

In like manner, around loop 2 in the direction of the arrow we have

$$-V_3 - V_2 + 20 = 0 \tag{7.2}$$

In terms of the element currents I_1, I_2, and I_3, the *IR* drops by Ohm's law are

$$\begin{aligned} V_1 &= 3I_1 \\ V_2 &= 2I_2 \\ V_3 &= 6I_3 \end{aligned} \tag{7.3}$$

Substituting these values into (7.1) and (7.2), we have the loop equations

$$\begin{aligned} 3I_1 + 6I_3 &= 24 \\ 2I_2 + 6I_3 &= 20 \end{aligned} \tag{7.4}$$

Thus we have two equations in the three element currents. We may obtain another equation by applying Kirchhoff's current law at node *a*, resulting in

$$I_3 = I_1 + I_2 \tag{7.5}$$

Substituting this value of I_3 into (7.4), we have

$$3I_1 + 6(I_1 + I_2) = 24$$
$$2I_2 + 6(I_1 + I_2) = 20$$

which simplifies to

$$9I_1 + 6I_2 = 24$$
$$6I_1 + 8I_2 = 20$$

(7.6)

These are the loop equations that we must solve to complete the analysis of the circuit. Their solution yields I_1 and I_2, which may be substituted into (7.5) to yield I_3. Once the element currents are known, we may find any current or voltage in the circuit. For example, the IR drops follow from (7.3).

Solving the Equations: There are several methods of solving a set of simultaneous equations such as (7.6). One method we shall use is that of *eliminating* an unknown by multiplying one or more of the equations by an appropriate constant and adding or subtracting the resulting equations. To illustrate the method, let us eliminate I_2 in (7.6). To accomplish this, we multiply the first equation of (7.6) through by 4 and the second equation through by 3, resulting in

$$36I_1 + 24I_2 = 96$$
$$18I_1 + 24I_2 = 60$$

Now the coefficients of I_2 are the same in both equations, so that if we subtract the second equation term by term from the first equation, I_2 is eliminated. The result of the subtraction is

$$18I_1 = 36$$

from which we have

$$I_1 = 2 \text{ A}$$

(7.7)

Now that we know I_1, we may substitute its value into either of the equations of (7.6) to obtain an equation in I_2. For example, substituting $I_1 = 2$ into the first of (7.6), we have

$$9(2) + 6I_2 = 24$$

or

$$6I_2 = 6$$

and consequently

$$I_2 = 1 \text{ A}$$

(7.8)

Finally, from (7.5) the last element current is

$$I_3 = 2 + 1 = 3 \text{ A} \tag{7.9}$$

and from (7.3) the *IR* drops are

$$V_1 = 3(2) = 6 \text{ V}$$
$$V_2 = 2(1) = 2 \text{ V} \tag{7.10}$$
$$V_3 = 6(3) = 18 \text{ V}$$

We will consider another method of solving a set of simultaneous equations in the following section. This method uses *determinants* and allows us to write down the answer directly, after which it may be simplified by the rules of determinant theory.

As a final note in this section, we observe that the analysis of the circuit of Fig. 7.1 has been carried out by means of only Ohm's and Kirchhoff's laws. No current or voltage division has been applied and no equivalent resistances have been found. Moreover, the method is completely general. If the circuit contains more meshes, more equations are required and the mathematics is more complicated, but the method still applies.

PRACTICE EXERCISES

7-1.1 Find the element currents I_1, I_2, and I_3 if $V_g = 14$ V.   *Ans.* 5 A, 1 A, 4 A

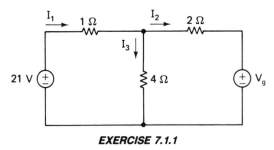

EXERCISE 7.1.1

7-1.2 Solve Exercise 7-1.1 if $V_g = 0$ (the voltage source is replaced by a short circuit). Check by using voltage division.   *Ans.* 9 A, 6 A, 3 A

7-1.3 Solve for I_1 and I_2:

$$2I_1 + 5I_2 = 13$$
$$3I_1 + 4I_2 = 9$$

Ans. −1, 3

7.2 DETERMINANTS

Another method of solving simultaneous equations, such as (7.6) of Section 7.1, is that of *Cramer's rule,* which uses *determinants.* This method is given in most introductory algebra books, but for the reader who is not familiar with determinants, a brief discussion is given in this section.

Coefficient Determinant: In the case of two equations in two unknowns x_1 and x_2, we have

$$
\begin{aligned}
ax_1 + bx_2 &= k_1 \\
cx_1 + dx_2 &= k_2
\end{aligned}
\tag{7.11}
$$

where a, b, c, d, k_1, and k_2 are known constants. The *coefficient determinant* Δ (the capital Greek letter *delta*) is written as the 2×2 array (two *rows* and two *columns*),

$$
\Delta = \begin{vmatrix} a & b \\ c & d \end{vmatrix}
\tag{7.12}
$$

The first row contains the coefficients a and b of the unknowns x_1 and x_2 in the first equation and the second row contains the coefficients c and d in the second equation.

The value of Δ is defined to be

$$
\Delta = ad - bc
$$

which may be obtained by a *diagonal rule* defined by

$$
\Delta = \begin{vmatrix} a & b \\ c & d \end{vmatrix} = ad - bc
\tag{7.13}
$$

That is, Δ is the difference of the product ad of numbers down the diagonal to the right and the product bc of numbers down the diagonal to the left. For example, in the case of the loop equations (7.6) of Section 7.1, the unknowns are $x_1 = I_1$ and $x_2 = I_2$, and the coefficient determinant is

$$
\Delta = \begin{vmatrix} 9 & 6 \\ 6 & 8 \end{vmatrix} = 9(8) - 6(6) = 36
\tag{7.14}
$$

Cramer's Rule: We define determinant Δ_1 as the coefficient determinant Δ of (7.12) with its first column, consisting of the coefficients a and c of x_1, replaced by the constants k_1 and k_2 in (7.11). That is,

$$\Delta_1 = \begin{vmatrix} k_1 & b \\ k_2 & d \end{vmatrix}$$

In a similar manner, the determinant Δ_2 is defined as Δ with the second column, consisting of the coefficients b and d of x_2, replaced by k_1 and k_2. That is,

$$\Delta_2 = \begin{vmatrix} a & k_1 \\ c & k_2 \end{vmatrix}$$

By Cramer's rule the solution of (7.11) is given by

$$x_1 = \frac{\Delta_1}{\Delta} \qquad x_2 = \frac{\Delta_2}{\Delta}$$

As an example, let us solve the loop equations (7.6) of Section 7.1, which we repeat as

$$9I_1 + 6I_2 = 24$$
$$6I_1 + 8I_2 = 20$$

The coefficient determinant Δ has been found already in (7.14), and the determinants Δ_1 and Δ_2 are

$$\Delta_1 = \begin{vmatrix} 24 & 6 \\ 20 & 8 \end{vmatrix} = 24(8) - 6(20) = 72$$

and

$$\Delta_2 = \begin{vmatrix} 9 & 24 \\ 6 & 20 \end{vmatrix} = 9(20) - 24(6) = 36$$

Therefore, by Cramer's rule the currents are

$$I_1 = \frac{\Delta_1}{\Delta} = \frac{72}{36} = 2 \text{ A}$$

$$I_2 = \frac{\Delta_2}{\Delta} = \frac{36}{36} = 1 \text{ A}$$

In the case of three equations in three unknowns,

$$a_1x_1 + b_1x_2 + c_1x_3 = k_1$$
$$a_2x_1 + b_2x_2 + c_2x_3 = k_2$$
$$a_3x_1 + b_3x_2 + c_3x_3 = k_3$$

Cramer's rule yields

$$x_1 = \frac{\Delta_1}{\Delta} \qquad x_2 = \frac{\Delta_2}{\Delta} \qquad x_3 = \frac{\Delta_3}{\Delta}$$

where Δ is the 3×3 coefficient determinant

$$\Delta = \begin{vmatrix} a_1 & b_1 & c_1 \\ a_2 & b_2 & c_2 \\ a_3 & b_3 & c_3 \end{vmatrix}$$

Δ_1 is Δ with its first column replaced by the constants k_1, k_2, and k_3; Δ_2 is Δ with its second column so replaced; and Δ_3 is Δ with its third column so replaced.

There is also a diagonal rule that applies to 3×3 determinants, given by

$$\Delta = \begin{vmatrix} a_1 & b_1 & c_1 & a_1 & b_1 \\ a_2 & b_2 & c_2 & a_2 & b_2 \\ a_3 & b_3 & c_3 & a_3 & b_3 \end{vmatrix}$$

$$= (a_1 b_2 c_3 + b_1 c_2 a_3 + c_1 a_2 b_3)$$
$$- (c_1 b_2 a_3 + a_1 c_2 b_3 + b_1 a_2 c_3)$$

This is easier to see when the first two columns are repeated, as shown. The value is then the difference of products of the numbers down the diagonals to the right and products of the numbers down the diagonals to the left.

As an example, consider the equations

$$\begin{aligned} x_1 + x_2 + x_3 &= 6 \\ 2x_1 - x_2 + x_3 &= 3 \\ -x_1 + x_2 + 2x_3 &= 7 \end{aligned}$$

The coefficient determinant is

$$\Delta = \begin{vmatrix} 1 & 1 & 1 \\ 2 & -1 & 1 \\ -1 & 1 & 2 \end{vmatrix} = \begin{vmatrix} 1 & 1 & 1 & 1 & 1 \\ 2 & -1 & 1 & 2 & -1 \\ -1 & 1 & 2 & -1 & 1 \end{vmatrix}$$

$$= [(1)(-1)(2) + (1)(1)(-1) + (1)(2)(1)]$$
$$- [(1)(-1)(-1) + (1)(1)(1) + (1)(2)(2)]$$

$$= -7$$

The unknowns are

$$x_1 = \frac{\Delta_1}{\Delta} = \frac{\begin{vmatrix} 6 & 1 & 1 \\ 3 & -1 & 1 \\ 7 & 1 & 2 \end{vmatrix}}{-7} = \frac{-7}{-7} = 1$$

$$x_2 = \frac{\Delta_2}{\Delta} = \frac{\begin{vmatrix} 1 & 6 & 1 \\ 2 & 3 & 1 \\ -1 & 7 & 2 \end{vmatrix}}{-7} = \frac{-14}{-7} = 2$$

$$x_2 = \frac{\Delta_3}{\Delta} = \frac{\begin{vmatrix} 1 & 1 & 6 \\ 2 & -1 & 3 \\ -1 & 1 & 7 \end{vmatrix}}{-7} = \frac{-21}{-7} = 3$$

Methods of handling determinants larger than 3×3 may be developed, but we will not need them here. For the interested reader they are easily found in most algebra textbooks.

PRACTICE EXERCISES

7-2.1 Solve Practice Exercise 7-1.1 by using determinants.   *Ans.* 5 A, 1 A, 4 A

7-2.2 Solve for I_1 and I_2 by using determinants:

$$3I_1 + I_2 = 5$$
$$4I_1 + 5I_2 = 3$$

Ans. 2, −1

7.3 MESH CURRENTS

The loop analysis method is sometimes simplified by expressing the loop equations in terms of *mesh* currents rather than element currents, as we will see in this section. We begin by defining a mesh current as the current that flows around a mesh. The mesh current may constitute the entire current in an element, in which case the mesh current is an element current, or it may be only a portion of the element current. For example, the mesh currents in Fig. 7.3 are I_a and I_b, whereas the element currents are I_1, I_2, and I_3. To see how the mesh and element currents are related, let us analyze the circuit by the loop method.

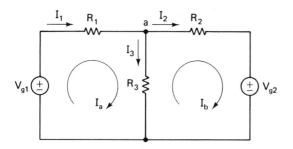

FIGURE 7.3 *Circuit with two mesh currents.*

Applying KVL around the first mesh (containing V_{g1}) and replacing the IR drops by their values in terms of the element currents, we have

$$-V_{g1} + R_1 I_1 + R_3 I_3 = 0$$

Similarly, around the second mesh we have

$$-R_3 I_3 + R_2 I_2 + V_{g2} = 0$$

Rewriting these results, we have the loop equations

$$R_1 I_1 + R_3 I_3 = V_{g1}$$
$$R_2 I_2 - R_3 I_3 = -V_{g2}$$

(7.15)

Also by KCL at node a we may write

$$I_3 = I_1 - I_2 \qquad (7.16)$$

Relationship of the Element Currents to the Mesh Currents: Since I_a flows to the right through R_1, it must be the element current I_1. That is,

$$I_1 = I_a \qquad (7.17)$$

Similarly, I_b flows to the right through R_2, as does I_2, and we must have

$$I_2 = I_b \qquad (7.18)$$

In these two cases, therefore, the mesh current constitutes the entire element current. In the case of element current I_3, however, we have by (7.16) to (7.18)

$$I_3 = I_a - I_b \qquad (7.19)$$

Thus the element current in R_3 is a composite of the mesh currents.

General Case: In general, an element current is an algebraic sum of mesh currents if two or more mesh currents pass through the element. This is illustrated in Fig. 7.4(a) and (b), where in (a) the element current I_1 is given in terms of mesh currents I_a and I_b by

$$I_1 = I_a - I_b \qquad\qquad (7.20)$$

and in (b) the element current I_2 is given by

$$I_2 = I_a + I_b \qquad\qquad (7.21)$$

In Fig. 7.4(a), I_1 is the total current downward in the element and thus is the mesh current I_a downward minus the mesh current I_b upward. In Fig. 7.4(b), the total downward current I_2 is the sum of the downward mesh currents I_a and I_b.

Returning to the solution of the circuit of Fig. 7.3, the mesh equations (7.15) may be written in terms of the mesh currents I_a and I_b by substituting (7.17) to (7.19) for the element currents. The result is

$$\begin{aligned} R_1 I_a + R_3(I_a - I_b) &= V_{g1} \\ R_2 I_b - R_3(I_a - I_b) &= -V_{g2} \end{aligned} \qquad\qquad (7.22)$$

These results may be found directly from Fig. 7.3 by writing the IR drops in terms of the mesh currents.

Collecting terms in (7.22) results in the simpler form

$$(R_1 + R_3)I_a - R_3 I_b = V_{g1}$$

$$-R_3 I_a + (R_2 + R_3)I_b = -V_{g2}$$

which may be solved for I_a and I_b. Then all the element currents and voltages may be found from a knowledge of the mesh currents.

Example 7.1: Find the mesh currents, the element currents, and the IR drops in the circuit of Fig. 7.5

Solution: In terms of the mesh currents I_a and I_b the IR drops are

$$V_1 = 3I_a$$

$$V_2 = 12I_b$$

$$V_3 = 6(I_a - I_b)$$

Therefore, the mesh equations are, around the first mesh,

$$3I_a + 6(I_a - I_b) = 51$$

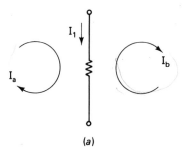

(a)

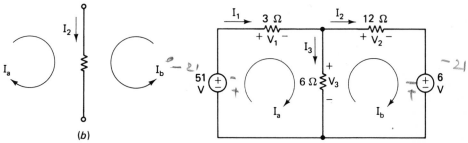

(b)

FIGURE 7.4 Relationship of mesh currents to element currents.

FIGURE 7.5 Two-mesh circuit example.

$$21 - 9 I_a + 6 I_b = 0$$
$$9 I_2 - 6 I_b = 21$$

and around the second mesh,

$$-6(I_a - I_b) + 12I_b = -6$$

Collecting terms, we have

$$9I_a - 6I_b = 51$$
$$-6I_a + 18I_b = -6$$

(7.23)

Leaving the first equation as it is and dividing the second equation through by 3, we have

$$9I_a - 6I_b = 51$$
$$-2I_a + 6I_b = -2$$

Adding the two equations results in

$$7I_a = 49$$

or

$$I_a = 7\text{A}$$

Substituting this value into the first of (7.23), we have

$$9(7) - 6I_b = 51$$

from which

$$I_b = 2A$$

The element currents are

$$I_1 = I_a = 7 \text{ A}$$
$$I_2 = I_b = 2 \text{ A}$$
$$I_3 = I_a - I_b = 5 \text{ A}$$

and the IR drops are

$$V_1 = 3I_a = 21 \text{ V}$$
$$V_2 = 12I_b = 24 \text{ V}$$
$$V_3 = 6(I_a - I_b) = 30 \text{ V}$$

Shortcut Procedure: If all the mesh currents are assumed in the same direction, say clockwise, as in Fig. 7.5, then there is a shortcut method of writing the mesh equations. In going around a given mesh in the direction of its mesh current, say I_a, the coefficient of I_a in the left member of the equation is the sum of the resistances in the mesh. This is because for every resistance in the mesh the IR drop due to I_a is positive and there is one for each resistance. The coefficients of any other mesh current, say I_b, in the left member of the equation, is the negative of the resistance common to its mesh and that of I_a. This follows from the fact that the other mesh current, say I_b, flows in the opposite direction in the common resistance to that of I_a, and therefore its contribution to the IR drop in the mesh is negative. To complete the equation, the right member is the algebraic sum of the voltage rises in the mesh due to the voltage sources.

As an example, let us apply the shortcut procedure to the circuit of Fig. 7.5. In the mesh equation for mesh current I_a, the coefficient of I_a is $3 + 6 = 9$, the sum of the resistances in the mesh. The coefficient of I_b is -6, the negative of the resistance common to the two meshes. The right member of the equation is 51, which is the voltage rise of the source encountered in going clockwise with I_a around the mesh.

In the mesh equation for mesh current I_b, the coefficient of I_b is $12 + 6 = 18$, the sum of the resistances in the mesh. The coefficient of I_a is -6, the negative of the resistance common to the two meshes. The right member of the equation is -6. (Since $+6$ encountered in going clockwise through the source is a drop, the rise is -6.) The reader may check these results by comparing them with (7.23).

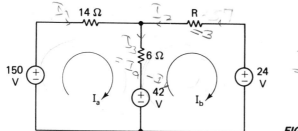

FIGURE 7.6   *Circuit with three sources.*

Example 7.2: Using the shortcut procedure, write the mesh equations for the circuit of Fig. 7.6 with $R = 3\ \Omega$. Complete the solution for the mesh currents I_a and I_b.

Solution: Using the shortcut procedure, the equations are

$$20I_a - 6I_b = 150 - 42 = 108$$
$$-6I_a + 9I_b = 42 - 24 = 18$$

(7.24)

To solve the equations we may multiply the first equation through by 3 and the second through by 2 to obtain

$$60I_a - 18I_b = 324$$
$$-12I_a + 18I_b = 36$$

Adding the equations, we have

$$48I_a = 360$$

or $I_a = 360/48 = 7.5$ A. Substituting this value into the second equation of (7.24) results in

$$-6(7.5) + 9I_b = 18$$

or

$$9I_b = 18 + 45 = 63$$

Therefore, $I_b = 7$ A.

Negative Mesh Currents: As a last example let us replace both voltage sources in Fig. 7.5 by 21-V sources and find the mesh currents. The mesh equations are

$$9I_a - 6I_b = 21$$
$$-6I_a + 18I_b = -21$$

Solving these equations, we have the mesh currents $I_a = 2$ A and $I_b = -0.5$ A.

One of the mesh currents, I_b, is negative, which simply means that a positive mesh current of 0.5 A is flowing counterclockwise around the second loop. The current I_3 through the 6-Ω resistor is therefore

$$I_3 = 2 + 0.5 = 2.5 \text{ A}$$

PRACTICE EXERCISES

7-3.1 Find I_a and I_b in Fig. 7.6 if $R = 7 \ \Omega$.   *Ans.* 6.75 A, 4.5 A

7-3.2 Find I_1 and I_2 using mesh analysis.   *Ans.* 6 A, 11 A

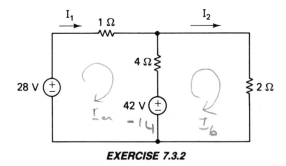

EXERCISE 7.3.2

7-3.3 Solve Exercise 7-3.2 if the 42-V source is replaced by a 14-V source.

Ans. 10 A, 9 A

7.4 NODE-VOLTAGE ANALYSIS

In the mesh current analysis, as we have seen, Kirchhoff's voltage law is applied around the meshes, resulting in a set of equations in the mesh currents. From the mesh currents we may find the element currents and voltages throughout the circuit. The other popular general circuit analysis method is *node-voltage analysis,* or *nodal analysis,* in which Kirchhoff's current law is applied at nodes resulting in equations in the so-called *node* voltages. As we shall see in this section, the node voltages may also be used to find all the element currents and voltages, and thus nodal analysis is also a general analysis method.

Node Voltages: Before performing nodal analysis, we first select arbitrarily one node to be the *reference,* or *ground,* node, in which case all the other nodes are *nonreference* nodes. The voltage of each of the nonreference nodes with respect to the reference node is then defined to be a *node* voltage. For example, the reference node in Fig. 7.7 is node *c,* identified by the attached symbol for *ground.* The nonreference nodes *a* and *b* are labeled with their node voltages V_a and V_b, meaning that these nodes are, respectively, V_a and V_b volts above the ground potential, as indicated.

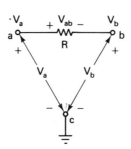

FIGURE 7.7 Reference
and nonreference nodes.

Ground Potential: Frequently, the analysis is easier, as we shall see, when the reference node is chosen to be the node to which the largest number of elements are connected. Many practical circuits are built on a metallic chassis, which is the logical choice for the ground node. This is the case in most electronic circuits and in the electrical circuitry in an automobile. In other cases, such as in electric power systems, the ground is the earth itself. In every case, ground potential is usually taken as zero volts, so that the nonreference nodes are at potentials above zero.

Element Voltages: Once the nodes are labeled with their node voltages, it is a simple matter to find all the element voltages. For example, in Fig. 7.7 the element voltage V_{ab} across the resistor R is simply

$$V_{ab} = V_a - V_b \qquad (7.25)$$

This may be seen by applying KVL around loop *abca* (real or imagined). This results in

$$V_{ab} + V_b - V_a = 0$$

from which (7.25) follows.

> *Example 7.3:* To illustrate nodal analysis, let us find the currents in the resistors of the circuit of Fig. 7.8(a), with the reference node as shown. The nonreference node voltages are labeled V_a, V_b, and V_c, but some of these may be determined by inspection of the circuit. For example, the node labeled V_a is 16 V above the ground (there is a rise of 16 V in going from the reference node to V_a), and thus the node voltage is
>
> $$V_a = 16 \text{ V}$$
>
> By similar reasoning the node labeled V_c is 22 V above the ground, so that
>
> $$V_c = 22 \text{ V}$$
>
> These node voltages are labeled in Fig. 7.8(b), where the ground node has been redrawn to identify it more clearly as a point.

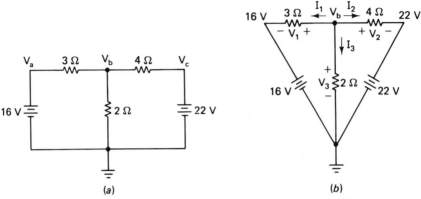

FIGURE 7.8 *Two versions of a circuit with labeled node voltages.*

Solution: There is now only one unknown node voltage V_b, so that only one application of KCL is needed. Applying KCL at node V_b, we have the *node* equation

$$I_1 + I_2 + I_3 = 0 \tag{7.26}$$

By Ohm's law we have

$$I_1 = \frac{V_1}{3}$$

$$I_2 = \frac{V_2}{4} \tag{7.27}$$

$$I_3 = \frac{V_3}{2}$$

Also, the element voltages are related to the node voltages by the relations

$$V_1 = V_b - 16$$
$$V_2 = V_b - 22 \tag{7.28}$$
$$V_3 = V_b$$

which with (7.27) may be substituted into (7.26) to yield the node equation

$$\frac{V_b - 16}{3} + \frac{V_b - 22}{4} + \frac{V_b}{2} = 0 \tag{7.29}$$

This equation, of course, may be obtained without (7.26) to (7.28), by direct inspection of Fig. 7.8(b). It simply says that the sum of the currents leaving node V_b is zero. The currents are expressed in terms of the node voltages by means of Ohm's law. For example, the current from V_b to $V_a = 16$ is $(V_b - 16)/3$.

To complete the analysis of Fig. 7.8, we may solve (7.29) for V_b, resulting in

$$\left(\frac{1}{3}+\frac{1}{4}+\frac{1}{2}\right) V_b = \frac{16}{3} + \frac{22}{4}$$

or

$$\frac{13}{12} V_b = \frac{130}{12}$$

which yields $V_b = 10$ V. The resistor currents, by (7.27) and (7.28), are given by

$$I_1 = \frac{V_b - 16}{3} = -2 \text{ A}$$

$$I_2 = \frac{V_b - 22}{4} = -3 \text{ A}$$

$$I_3 = \frac{V_b}{2} = 5 \text{ A}$$

We note in this example that the node equation was written at a node (V_b in this case) where it is not necessary to know the current through a voltage source. This is always possible, since the presence of voltage sources reduces the number of unknown node voltages and thus the number of necessary node equations. Node equations at current source terminals are easily handled because the current through the current source is specified. We will consider circuits of these and other types in Section 7.5.

PRACTICE EXERCISES

7-4.1 Use nodal analysis to find V_b and I_3 in Fig. 7.8(b) if the 16-V source is replaced by a 29-V source.   *Ans.* 14 V, 7 A

7-4.2 Repeat Exercise 7-4.1 if the 16-V source is replaced by a 42-V source.

                                                                Ans. 18 V, 9 A

7-4.3 Find I using nodal analysis. Check by using current division.   *Ans.* 2 A

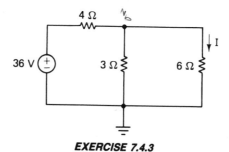

EXERCISE 7.4.3

7.5 OTHER NODAL ANALYSIS EXAMPLES

In this section we will apply nodal analysis to circuits containing current sources and to more complicated circuits with voltage sources. The procedure is, of course, the same as that of the previous section, but the mathematics involved may, in some cases, be slightly more complicated, as we shall see.

Example 7.4: Find I in Fig 7.9 using nodal analysis.

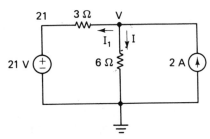

FIGURE 7.9 *Circuit with a voltage and a current source.*

Solution: The ground node is as marked and the nonreference nodes have node voltages of 21 and V, as indicated. The node equation at node V is

$$I_1 + I = 2$$

or

$$\frac{V - 21}{3} + \frac{V}{6} = 2$$

Solving for the node voltage V, we have

$$\frac{3}{6} V = 2 + \frac{21}{3}$$

or $V = 18$ V. The current I therefore is given by

$$I = \frac{V}{6} = 3 \text{ A}$$

Example 7.5: Find the node voltages V_1 and V_2 and the current I in the circuit of Fig. 7.10.

150

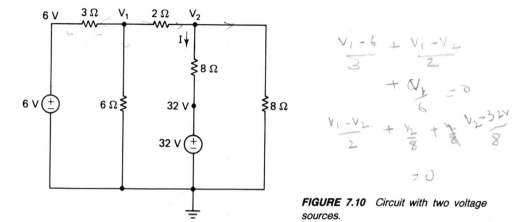

FIGURE 7.10 *Circuit with two voltage sources.*

Solution: With the reference node as shown, there are two known node voltages of 6 V and 32 V, as indicated. The node equation at node V_1 is

$$\frac{V_1 - 6}{3} + \frac{V_1 - V_2}{2} + \frac{V_1}{6} = 0 \tag{7.30}$$

which equates the sum of the three currents leaving the node to zero. In a similar way at node V_2 we have

$$\frac{V_2 - V_1}{2} + \frac{V_2 - 32}{8} + \frac{V_2}{8} = 0 \tag{7.31}$$

Multiplying (7.30) through by 6 and (7.31) through by 8 and collecting terms, we have

$$6V_1 - 3V_2 = 12$$
$$-4V_1 + 6V_2 = 32$$

If we now divide the second equation through by 2, we have

$$6V_1 - 3V_2 = 12$$
$$-2V_1 + 3V_2 = 16 \tag{7.32}$$

Now adding the two equations yields

$$4V_1 = 28$$

or $V_1 = 7$ V. Substituting this value of V_1 into the first of (7.32), we have

$$42 - 3V_2 = 12$$

or

$$-3V_2 = -30$$

from which $V_2 = 10$ V.

Finally, from Fig. 7.10 we see that

$$I = \frac{V_2 - 32}{8}$$

$$= \frac{10 - 32}{8}$$

$$= -2.75 \text{ A}$$

Therefore, a current of 2.75 A flows upward out of the positive terminal of the 32-V source.

Example 7.6: As a last example of nodal analysis let us find the current I_1 in Fig. 7.11. There are four nodes, labeled *a, b, c,* and *d,* and thus there are three nonreference nodes. Choosing node *d* as the reference and labeling the node voltage at *c* as *V,* we note that nodes *a* and *b* have node voltages that may be expressed in terms of *V.* There is a voltage rise of 10 V in going from *c* to *b,* so that the node voltage at *b* must be 10 V more than that at *c.* Therefore, the node voltage at *b* must be $V + 10$, as indicated. Similarly, there is another rise, this one of 6 V, in moving from *b* to *a,* so that the node voltage at *a* must be $V + 10 + 6 = V + 16$, again as indicated.

Solution: We therefore have all the three nonreference node voltages expressed in terms of a single unknown *V,* so that only one node equation is needed. At node *a* we have

$$I_3 + I_4 = 18 \qquad\qquad (7.33)$$

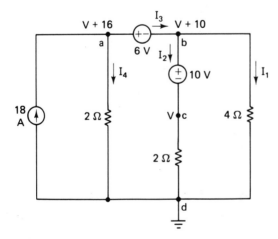

FIGURE 7.11 *Circuit with three sources.*

but since I_3 is the current through a voltage source, it cannot be expressed using Ohm's law. However, at node b we have

$$I_3 = I_2 + I_1$$

so that (7.33) becomes

$$I_4 + I_2 + I_1 = 18 \qquad\qquad (7.34)$$

It is interesting to note that this last result is Kirchhoff's current law applied to the closed curve of the redrawn circuit in Fig. 7.12. That is, the sum of the currents I_4, I_2, and I_1 leaving the curve equals the 18-A current entering the curve. Thus KCL applies not only to a node, or point, but to any closed curve drawn through the circuit.

Returning to (7.34), we may replace I_4, I_2, and I_1 by their equivalent expressions using Ohm's law and Fig. 7.12, resulting in

$$\frac{V+16}{2} + \frac{V}{2} + \frac{V+10}{4} = 18$$

Collecting terms, we have

$$\frac{5}{4}V = \frac{15}{2}$$

or $V = 6$ V. From Fig. 7.11 the current I_1 is given by

$$I_1 = \frac{V+10}{4} = \frac{16}{4} = 4 \text{ A}$$

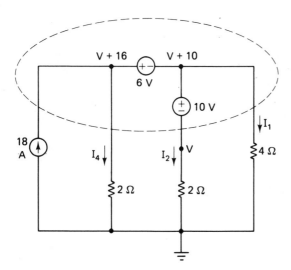

FIGURE 7.12 *Circuit of Fig. 7.11 redrawn.*

PRACTICE EXERCISES

√7-5.1   Using nodal analysis, find V if the element x is a 12-V source with the positive terminal at the top.   *Ans.* 8 V

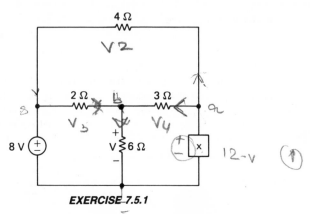

EXERCISE 7.5.1

7-5.2   Repeat Exercise 7-5.1 if the element x is an 8-A current source directed upward.
                                                                                                  Ans. 12 V

7-5.3   Repeat Exercise 7-5.1 if the element x is a 12-Ω resistor.   *Ans.* 6 V

7.6   SUMMARY

There are two general circuit analysis methods, known as loop analysis and nodal analysis, that may be applied to any electric circuit. The methods of the previous chapters apply only to series, parallel, or series–parallel circuits, but these more general methods may be used to analyze any type of circuit.

Loop analysis, or its special case of mesh analysis, consists of writing KVL around certain closed loops in the circuit. The resulting loop equations contain unknowns which are the loop currents. The loop currents may be found from these equations and used to obtain any element current or voltage in the circuit.

In the case of nodal analysis, a reference or ground node is chosen and the node voltages assigned to the remaining nonreference nodes. A node voltage is the voltage of the node with respect to the reference node. Applying KCL at nonreference nodes yields a set of node equations in the node voltages. The node voltages may then be found and used to obtain any element current or voltage in the circuit.

PROBLEMS

7.1   Solve Practice Exercise 6-2.1 using mesh analysis.

7.2   Solve Practice Exercise 6-2.2 using mesh analysis.

7.3 Find V_1 in Fig. 6.6 using mesh analysis.

7.4 Solve Problem 6.5 using mesh analysis.

7.5 Find I using mesh analysis if $V_g = 21$ V.

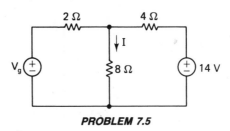

PROBLEM 7.5

7.6 Solve Problem 7.5 if $V_g = 7$ V.

7.7 Find I_3 in Fig. 7.8(b) using mesh analysis.

7.8 Find V_3 in Fig. 7.1 using nodal analysis.

7.9 Find I_2 in Fig. 7.5 using nodal analysis.

7.10 Find I_b in Fig. 7.6 for $R = 3$ Ω using nodal analysis.

7.11 Repeat Problem 7.10 for $R = 7$ Ω.

7.12 Solve Practice Exercise 7-3.2 using nodal analysis.

7.13 Find I_1 and I_2 using mesh analysis. (*Suggestion:* Note that mesh current $I_a = 6$ A and write a mesh equation around the right mesh.)

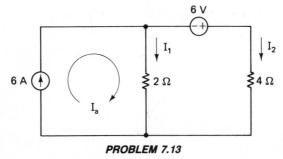

PROBLEM 7.13

7.14 Solve Problem 7.13 using nodal analysis.

7.15 Find V.

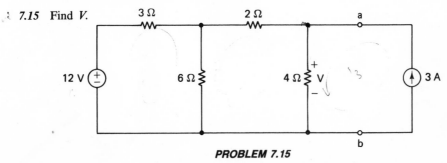

PROBLEM 7.15

7.16   Find *I*.

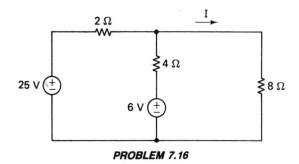

PROBLEM 7.16

7.17   Find the power delivered to the 4-Ω resistor. (*Suggestion:* Write a mesh equation around the center mesh.)

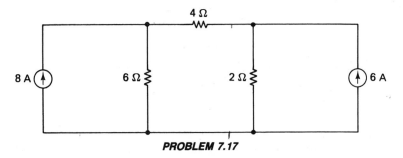

PROBLEM 7.17

7.18   Find *V* in Fig. 7.11 if the 18-A source is changed to a 3-A source.

8

NETWORK THEOREMS

In the previous chapters we have considered straightforward methods of analyzing circuits from the simple series or parallel types to the most complex types. In many cases we may considerably shorten the analysis, however, by means of certain network *theorems*, which we will consider in this chapter. For example, if we are interested only in the voltage or current of one element in a circuit, it may be possible by means of a network theorem to replace the rest of the circuit by an equivalent and simpler circuit.

The network theorems that we will consider apply to *linear* circuits, which are circuits made up of sources and *linear* elements. A linear element is simply one whose voltage is multiplied by a constant K if its current is multiplied by K. For example, a resistor defined by Ohm's law,

$$v = Ri$$

is linear because it follows that

$$Kv = K(Ri)$$
$$= R(Ki)$$

Thus if i is multiplied by K, then v is multiplied by K. The resistive circuits we have considered thus far are all linear because their elements are sources and linear resistors. Other linear elements that we will consider in later chapters are inductors and capacitors, which, with resistors, are important in ac circuits. Thus the network theorems apply not only to the dc circuits of this chapter, but to ac circuits as well.

8.1 SUPERPOSITION

The first network theorem that we consider is that of *superposition,* which says the following:

> *In a network containing two or more sources, an element voltage or current is the algebraic sum of the voltages or currents produced by each source acting alone.*

That is, we may find the voltage or current produced by one source at a time by making all the other sources zero. This procedure is repeated for each source and the resulting voltages or currents are *superimposed,* or algebraically added, to find the total effect of all the sources working together.

Killing the Sources: The operation of making a source zero is sometimes referred to as "killing" the source, or making the source "dead." Thus killing a voltage source V_g means making $V_g = 0$, and since this is the equation of a short circuit, the operation is merely to replace the voltage source by a short circuit. In like manner, making a current source $I_g = 0$ (the equation of an open circuit) is accomplished by replacing the current source by an open circuit.

Example 8.1: As an illustration let us apply the principle of superposition to find the current I in the circuit of Fig. 8.1. We will first use conventional circuit analysis and then check the result by superposition.

Solution: By KCL at node a we have

$$\frac{V_1}{8} + \frac{V_1 - V_g}{4} = I_g$$

from which

$$V_1 = \frac{2V_g}{3} + \frac{8I_g}{3} \tag{8.1}$$

Since $I = V_1/8$, we have from (8.1)

$$I = \frac{V_g}{12} + \frac{I_g}{3} \tag{8.2}$$

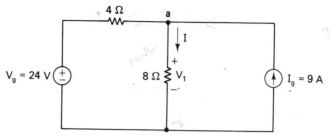

FIGURE 8.1 *Circuit with two sources.*

Since $V_g = 24$ V and $I_g = 9$ A, we have

$$I = \frac{24}{12} + \frac{9}{3} = 2 + 3 = 5 \text{ A} \tag{8.3}$$

Working the problem now by superposition, we may write

$$I = I_a + I_b \tag{8.4}$$

where I_a is due to V_g alone (with the current source I_g made zero) and I_b is due to I_g alone (with the voltage source V_g made zero). Evidently, from (8.2) if $I_g = 0$, then $I = I_a$, given by

$$I_a = \frac{V_g}{12} = \frac{24}{12} = 2 \text{ A} \tag{8.5}$$

and if $V_g = 0$, then $I = I_b$, given by

$$I_b = \frac{I_g}{3} = \frac{9}{3} = 3 \text{ A} \tag{8.6}$$

Then by (8.4) we have, as before,

$$I = 2 + 3 = 5 \text{ A}$$

Of course, we may apply the operations of killing the sources to the circuit itself and avoid the process of analyzing the original circuit. This is the great advantage of superposition, since it allows us to perform the analysis by analyzing separately circuits with only one source. Such *single-input* circuits are often easily analyzed since we can use network reduction properties, such as equivalent resistance and voltage and current division.

As an example, to find I_a, the component of I in Fig. 8.1 due to V_g alone, we kill the source I_g (replacing it by an open circuit). This results in the circuit of Fig. 8.2(a), from which we have, by Ohm's law,

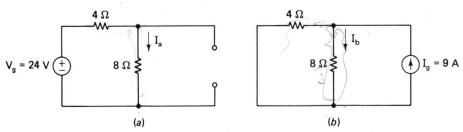

FIGURE 8.2 *Circuit of Fig. 8.1 with (a) the current source killed and (b) the voltage source killed.*

$$I_a = \frac{V_g}{12} = \frac{24}{12} = 2 \text{ A}$$

In like manner, to find I_b, the component of I in Fig. 8.1 due to I_g alone, we kill the source V_g (replacing it by a short circuit), resulting in Fig. 8.2(b). In this circuit we have, by current division,

$$I_b = \frac{4}{12}I_g = \frac{4(9)}{12} = 3 \text{ A}$$

These are, of course, the values obtained before in (8.5) and (8.6).

Example 8.2: As a last example, let us find the power delivered to the 6-Ω resistor of Fig. 7.1, which for convenience is repeated as Fig. 8.3.

Solution: Since power is proportional to the square of the current $(P = RI^2)$, it is not a linear quantity like voltage $(V = RI)$ or current $(I = V/R)$. Therefore, we cannot superimpose powers to get the total power, as in superposition of voltages or currents. However, we can get the current I in Fig. 8.3 by superposition and use the result to obtain the power,

$$P = 6I^2 \tag{8.7}$$

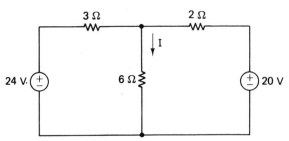

FIGURE 8.3 *Circuit with two voltage sources.*

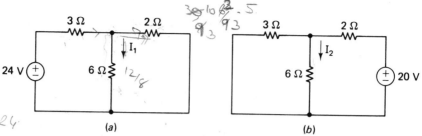

FIGURE 8.4   Circuit of Fig. 8.3 with (a) the 20-V source killed and (b) the 24-V source killed.

If I_1 is the component of I due to the 24-V source alone (the 20-V source killed) and I_2 is the component of I due to the 20-V source alone (the 24-V source killed), then

$$I = I_1 + I_2$$

where I_1 and I_2 may be found from Fig. 8.4(a) and (b), respectively. In Fig. 8.4(a) we have, by Ohm's law and current division,

$$I_1 = \frac{24}{3 + [6(2)/8]} \cdot \frac{2}{6+2} = \frac{4}{3} \text{ A}$$

Similarly, in Fig. 8.4(b) we have

$$I_2 = \frac{20}{2 + [3(6)/9]} \cdot \frac{3}{6+3} = \frac{5}{3} \text{ A}$$

Therefore, the current I is

$$I = I_1 + I_2 = \frac{4}{3} + \frac{5}{3} = 3 \text{ A}$$

and by (8.7) the power is

$$P = 6(3)^2 = 54 \text{ W}$$

PRACTICE EXERCISES

8-1.1   Find I_3 in Practice Exercise 7-1.1 using superposition.   *Ans.* $3 + 1 = 4$ A

8-1.2   Find I_1 in Practice Exercise 7-1.1 using superposition.   *Ans.* $9 - 4 = 5$ A

8-1.3   Find V using superposition. (*Suggestion:* Note that $V = V_1 + V_2 + V_3$, where V_1 is due to the 5-A source alone, V_2 is due to the 2-A source alone, and V_3 is due to the 10-V source alone.) $Ans. \dfrac{25}{3} + \dfrac{5}{2} + \dfrac{25}{6} = 15$ V

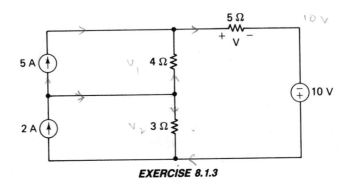

EXERCISE 8.1.3

8.2 THÉVENIN'S THEOREM

There are two very important network theorems, known as *Thévenin's theorem* and *Norton's theorem*, that allow us to replace an entire network with two available terminals by a network equivalent at the terminals, which contains a single source and a single resistor. Thus if we are interested in the voltage or current of a single element, we may replace the entire circuit other than the single element by its equivalent circuit and find the voltage or current using a much simpler circuit. We shall consider Thévenin's theorem in this section and use the results to develop Norton's theorem in Section 8.3.

Historically, the first of these network theorems was Thévenin's theorem, named in honor of the French telegraph engineer Charles Léon Thévenin (1857–1926), who published his results in 1883. Thévenin's theorem states that a circuit such as that of Fig. 8.5(a) is equivalent at terminals *a-b* to the circuit of Fig. 8.5(b), containing a voltage source V_{oc} in series with a resistance R_{th}. The quantity V_{oc} is the *open-circuit* voltage that appears across terminals *a-b* when they are opened, as shown in Fig. 8.6(a). The resistance R_{th}, called the *Thévenin* resistance, is that seen looking

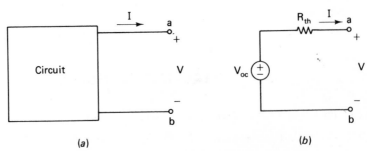

FIGURE 8.5   (a) Circuit and (b) its Thévenin equivalent.

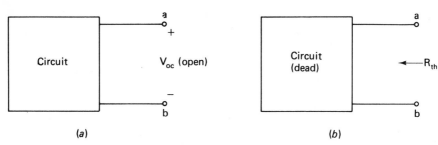

FIGURE 8.6 (a) Open-circuited network for V_{oc} and (b) dead circuit for R_{th}.

in terminals *a-b* of the *dead* circuit (that is, with all the internal sources killed). The Thévenin resistance is symbolized by Fig. 8.6(b).

Thévenin Equivalent Circuit: The circuit of Fig. 8.5(b) is called the *Thévenin equivalent circuit* of Fig. 8.5(a). The polarity of V_{oc} is such that it will produce a current from *a* to *b* in the same direction as in the original circuit of Fig. 8.5(a). Also, by the definition of equivalent circuits, if *V* is the same in Fig. 8.5(a) and (b), then *I* must be the same.

Example 8.3: To illustrate the use of Thévenin's theorem, let us replace the network to the left of terminals *a-b* in Fig. 8.7 by its Thévenin equivalent and use the result to find the voltage *V*.

Solution: The open-circuit voltage V_{oc} may be found from Fig. 8.8(a). Since there is no current in the 8-Ω resistor because of the open circuit, the voltage V_{oc} is across the 6-Ω resistor. Therefore, by voltage division we have

$$V_{oc} = \frac{6}{6+3} \cdot 18 = 12 \text{ V}$$

The Thévenin resistance R_{th} is that seen at the terminals of the dead circuit (the voltage source is replaced by a short circuit) shown in Fig. 8.8(b). Evidently, we have

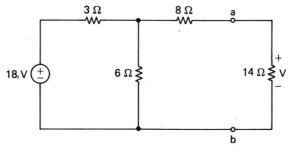

FIGURE 8.7 Circuit to be analyzed by Thévenin's theorem.

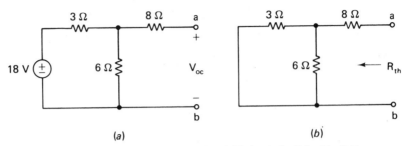

(a)   (b)

FIGURE 8.8 (a) Open-circuited and (b) dead circuit for Fig. 8.7.

$$R_{th} = 8 + \frac{3(6)}{3+6} = 10 \; \Omega$$

The Thévenin equivalent circuit is thus a voltage source of $V_{oc} = 12$ V in series with a resistance $R_{th} = 10 \; \Omega$. This is shown in Fig. 8.9 with the 14-Ω load connected to terminals a-b. It is now easily seen by voltage division that V, the voltage across the 14-Ω resistor, is given by

$$V = \frac{14}{14+10} \cdot 12 = 7 \text{ V}$$

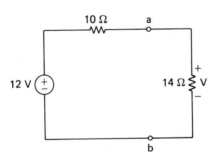

FIGURE 8.9 Loaded Thévenin equivalent of the circuit of Fig. 8.7.

Example 8.4: Replace everything in the circuit of Fig. 8.10 except the 2-Ω resistor by its Thévenin equivalent and find I.

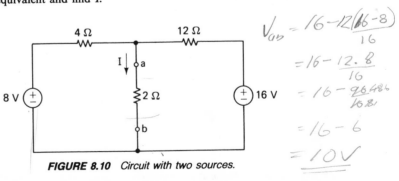

FIGURE 8.10 Circuit with two sources.

$V_{ab} = 16 - 12 \frac{(16-8)}{16}$

$= 16 - 12 \cdot \frac{8}{16}$

$= 16 - \frac{96.486}{168}$

$= 16 - 6$

$= 10 V$ ✓

$10 V \cdot \overset{3\Omega}{\underset{2\Omega}{\rule{0pt}{0pt}}}$

$I_2 = \frac{10}{3} \approx 2A.$

$R_{TH} \; I_{SC} = \frac{8}{4} + \frac{16}{12} = 2 + 1\frac{1}{3} = 3\frac{1}{3} A$

$R_{TH} = 10 \frac{3}{10} = 3 \Omega$

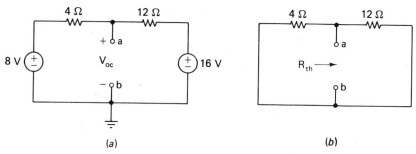

(a)   (b)

FIGURE 8.11 *Circuits for obtaining the Thévenin equivalent of Fig. 8.10.*

Solution: The circuit has two sources, but its Thévenin equivalent is found in the same way as that of Fig. 8.7. We find V_{oc} by open-circuiting terminals *a-b*, as shown in Fig. 8.11(a), and R_{th} by killing both sources, as shown in Fig. 8.11(b).

Taking the reference node in Fig. 8.11(a) as node *b*, we see that the node voltage at node *a* is V_{oc}. Thus a nodal equation at node *a* yields

$$\frac{V_{oc}-8}{4}+\frac{V_{oc}-16}{12}=0$$

from which we have

$$V_{oc}=10 \text{ V}$$

In Fig. 8.11(b) R_{th} is the parallel combination of the 4-Ω and 12-Ω resistances, so that

$$R_{th}=\frac{4(12)}{4+12}=3 \ \Omega$$

The Thévenin equivalent circuit with the 2-Ω resistance attached to terminals *a-b* is therefore that of Fig. 8.12, from which we see that

$$I=\frac{10}{5}=2 \text{ A}$$

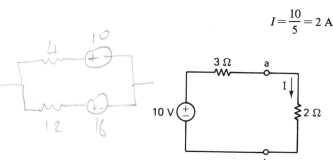

FIGURE 8.12 *Loaded Thévenin equivalent of the circuit of Fig. 8.10.*

PRACTICE EXERCISES

8-2.1 Replace the circuit to the left of terminals a-b by its Thévenin equivalent and find I.

Ans. $V_{oc} = 14$ V, $R_{th} = 5$ Ω, $I = 2$ A

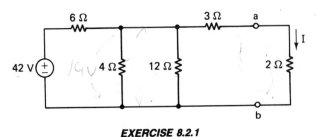

EXERCISE 8.2.1

8-2.2 Replace the circuit to the left of terminals a-b in Problem 7.15 by its Thévenin equivalent and find V.

Ans. $V_{oc} = 4$ V, $R_{th} = 2$ Ω, $V = 10$ V

8-2.3 Replace everything except the 4-Ω resistor in the circuit of Problem 7.13 by its Thévenin equivalent and find I_2.

Ans. $V_{oc} = 18$ V, $R_{th} = 2$ Ω, $I_2 = 3$ A

8.3 NORTON'S THEOREM

As we have seen in Section 8.2, the Thévenin equivalent circuit is equivalent at the terminals a-b to any circuit for which the open-circuit voltage across a-b is V_{oc} and whose resistance seen at a-b with the sources killed is R_{th}. Since the two circuits are equivalent at the terminals a-b, if a short circuit is placed across a-b in either case, the current through the short circuit from a to b will be the same for each circuit. We illustrate this with the circuit of Fig. 8.5(a) and its Thévenin equivalent of Fig. 8.5(b) with terminals a-b connected by a short circuit. The resulting circuits with the *short-circuit* current I_{sc} are shown in Fig. 8.13(a) and (b). Because of the equivalence, I_{sc} is the same for both circuits.

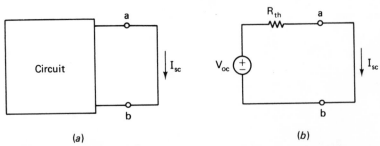

(a)   (b)

FIGURE 8.13 (a) Circuit and (b) its Thévenin equivalent with terminals a-b short circuited.

166

Relationship between the Open-Circuit Voltage and the Short-Circuit Current: From Fig. 8.13(b) and Ohm's law we have

$$I_{sc} = \frac{V_{oc}}{R_{th}} \tag{8.8}$$

which relates the short-circuit current I_{sc} to the open-circuit voltage V_{oc}. We may also write (8.8) in the form

$$V_{oc} = R_{th} I_{sc} \tag{8.9}$$

which yields V_{oc} in terms of I_{sc}.

Norton Equivalent Circuit: Let us now consider the circuit of Fig. 8.14, which consists of a current source I_{sc} in parallel with the Thévenin resistance R_{th}. If the terminals a-b are open, the voltage V_{ab} is given, by Ohm's law, as

$$V_{ab} = R_{th} I_{sc}$$

Since by (8.9) this means that $V_{ab} = V_{oc}$, we see that Fig. 8.14 is also an equivalent circuit at a-b of the circuit of Fig. 8.5(a). That is, it produces the same V_{oc} and R_{th} and therefore the same Thévenin equivalent. The circuit of Fig. 8.14 is called the *Norton equivalent circuit* of Fig. 8.5(a), in honor of the American scientist E. L. Norton (1898–), whose work was published some 50 years after Thévenin's. The statement that Fig. 8.14 is equivalent at the terminals a-b to Fig. 8.5(a) is known as *Norton's theorem*.

To obtain the Norton equivalent circuit we find R_{th} as before, as the resistance seen at terminals a-b of the dead circuit. The short-circuit current I_{sc} may be found by finding V_{oc} of the Thévenin circuit and using (8.8), or it may be found directly from the given circuit by placing a short circuit across terminals a-b. In any case,

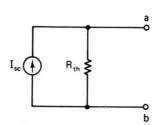

FIGURE 8.14 *Norton equivalent circuit of the circuit of Fig. 8.5(a).*

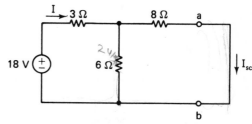

FIGURE 8.15 *Circuit for finding I_{sc}.*

if we know any two of the quantities R_{th}, V_{oc}, or I_{sc}, we may find the other from (8.8) or (8.9).

Example 8.5: To illustrate Norton's theorem, let us find the Norton equivalent of the circuit to the left of terminals *a-b* in Fig. 8.7, and use the result to find V.

Solution: The Thévenin resistance was found earlier from Fig. 8.8(b) as

$$R_{th} = 10 \ \Omega$$

To find I_{sc} we short-circuit terminals *a-b* in Fig. 8.7, resulting in the circuit of Fig. 8.15. The current I is given by

$$I = \frac{18}{3 + [6(8)/(6+8)]} = 2.8 \ A$$

and by current division the short-circuit current is

$$I_{sc} = \frac{6}{6+8} I = 1.2 \ A \tag{8.10}$$

The Norton equivalent of the circuit of Fig. 8.7 is therefore as shown in Fig. 8.16, from which we see that

$$V = \frac{10(14)}{10+14} (1.2) = 7 \ V$$

This, of course, is the result obtained earlier using Thévenin's theorem.

Alternatively, since we already know $V_{oc} = 12$ V from the Thévenin equivalent circuit of Section 8.2, we may use (8.9) to find I_{sc} in Example 8.5. The result is

$$I_{sc} = \frac{V_{oc}}{R_{th}} = \frac{12}{10} = 1.2 \ A$$

which checks (8.10).

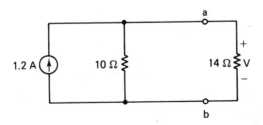

FIGURE 8.16 Norton equivalent of the circuit of Fig. 8.7.

PRACTICE EXERCISES

8-3.1 Find the Norton equivalent of the circuit to the left of terminals *a-b* in the figure for Practice Exercise 8-2.1.   *Ans.* $I_{sc} = 2.8$ A, $R_{th} = 5$ Ω

8-3.2 Replace the circuit to the left of terminals *a-b* by its Norton equivalent, and use the result to find *I*.   *Ans.* $I_{sc} = 3.5$ A, $R_{th} = 4$ Ω, $I = 2$ A

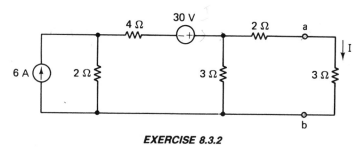

EXERCISE 8.3.2

8-3.3 Solve Practice Exercise 8-2.3 using the Norton equivalent circuit instead of the Thévenin equivalent circuit.   *Ans.* $I_{sc} = 9$ A, $R_{th} = 2$ Ω, $I_2 = 3$ A

8.4 VOLTAGE AND CURRENT SOURCE CONVERSIONS

As we have seen, the Thévenin equivalent of a circuit is a voltage source V_{oc} in series with a resistor R_{th}, and the Norton equivalent is a current source I_{sc} in parallel with the same resistor R_{th}, as shown in Fig. 8.17(a) and (b). Since the Thévenin and Norton circuits are general equivalents, it follows that we may convert a Thévenin equivalent to a Norton, and vice versa. Indeed, as we know, the resistance R_{th} is the same for both circuits and V_{oc} and I_{sc} are related by

$$V_{oc} = R_{th}I_{sc} \tag{8.11}$$

As an example, the circuit of Fig. 8.18(a), which may be thought of as the Thévenin equivalent of some other circuit, has a Norton equivalent described by

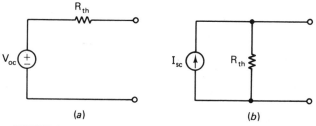

(a)   (b)

FIGURE 8.17 (a) Thévenin and (b) Norton equivalent circuits.

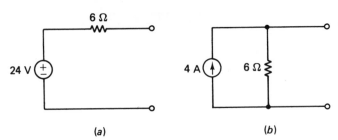

FIGURE 8.18   (a) Thévenin and (b) Norton equivalents of each other.

$R_{th} = 6 \ \Omega$ and

$$I_{sc} = \frac{V_{oc}}{R_{th}} = \frac{24}{6} = 4 \text{ A} \tag{8.12}$$

The Norton equivalent circuit is shown in Fig. 8.18(b).

Source Conversions: The equivalence at the terminals of the Thévenin and Norton circuits of Fig. 8.18 suggests a method of *converting* one type of source, such as a voltage source, to the other type source, such as a current source. Such source *conversions* are accomplished by simply replacing a Thévenin equivalent circuit by its Norton equivalent circuit, and vice versa. In many instances such conversions may be repeatedly performed (a voltage to a current source, a current to a voltage source, and so on) to reduce a complicated circuit to a relatively simple circuit.

As an example, let us apply source conversions to the circuit of Fig. 8.19 and obtain I by reducing the circuit to an equivalent circuit (as far as I is concerned) with only one loop. We will first replace the 32-A source and its parallel 2-Ω resistance by its equivalent Thévenin circuit. The result is a resistance of 2 Ω in series with a voltage source of 32(2) = 64 V, as shown in Fig. 8.20(a).

Next we replace the series combination of 3 Ω and 24 V by its Norton equivalent of 3 Ω in parallel with a source of 24/3 = 8 A, as shown in Fig. 8.20(b). Combining the parallel 6-Ω and 3-Ω resistors into their equivalent of 2 Ω, we have Fig. 8.20(c). Finally, the Thévenin equivalent of the 2-Ω resistor in parallel with the 8-A source is 2 Ω in series with a source of 8(2) = 16 V, as shown in Fig. 8.20(d). From this last circuit we may write

$$I = \frac{64 - 16}{2 + 4 + 2} = 6 \text{ A}$$

Ideal and Practical Sources: The 24-V source of Fig. 8.19 is an *ideal* voltage source. That is, it maintains 24 V between its terminals regardless of any external circuitry that may be connected to it and regardless of any current that may flow out of its terminals. This is an *ideal* rather than a *practical* situation because such

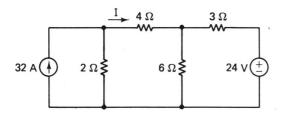

FIGURE 8.19 *Circuit with two sources.*

a source may supply *any* amount of power. A *practical,* or real, source, on the other hand, is limited by the amount of current it can deliver. For example, a 12-V automobile battery supplies 12 V when its terminals are open-circuited, but it supplies less than 12 V as current is drawn through its terminals. This is illustrated by the lights becoming dim when the starter is actuated.

A practical voltage source thus appears to have an *internal* drop in voltage when current is drawn through its terminals. This can be accounted for by representing the practical voltage source as an ideal voltage source in series with an *internal* resistance. An example is the series combination of the ideal 24-V source and the 3-Ω resistance of Fig. 8.19. A more general example is the Thévenin equivalent circuit with the ideal source V_{oc} in series with the internal resistance R_{th}.

In an analogous manner, the Norton equivalent circuit represents a *practical current source,* consisting as it does of an *ideal* current source I_{sc} in parallel with the internal resistance R_{th}. Thus the source conversions we are considering in this section may be thought of as conversions of practical voltage sources to equivalent practical current sources, and vice versa.

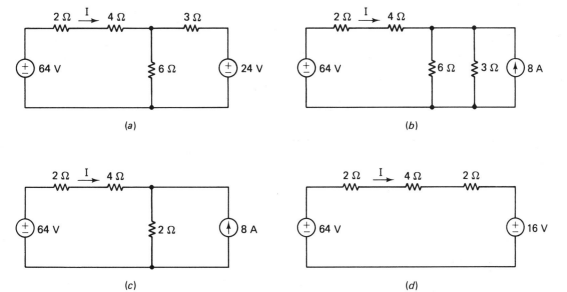

FIGURE 8.20 *Steps in obtaining I in Fig. 8.19.*

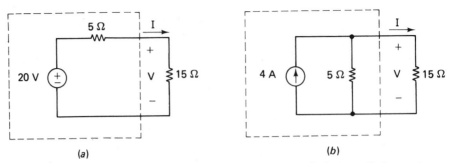

(a)   (b)

FIGURE 8.21 *(a) Practical voltage source and (b) its equivalent practical current source loaded with 15 Ω.*

Example 8.6: Replace the practical voltage source in the dashed box of Fig. 8.21(a) by its equivalent practical current source, and show that V and I are the same in both cases.

Solution: Applying Norton's theorem to the practical source of Fig. 8.21(a), we see that it is equivalent to a practical current source of $20/5 = 4$ A in parallel with a 5-Ω resistor. This result is shown in Fig. 8.21(b) with the 15-Ω load connected. From Fig. 8.21(a) we have

$$I = \frac{20}{5+15} = 1 \text{ A} \qquad \text{and} \qquad V = \frac{15}{5+15}(20) = 15 \text{ V}$$

From Fig. 8.21(b) we have

$$I = \frac{5}{5+15}(4) = 1 \text{ A} \qquad \text{and} \qquad V = 4\left[\frac{5(15)}{5+15}\right] = 15 \text{ V}$$

which illustrates the equivalence of the practical sources.

PRACTICE EXERCISES

8-4.1 By repeatedly applying source conversions, replace the circuit by an equivalent circuit with a single voltage source V_g in series with an internal resistance R_g.

Ans. $V_g = 6$ V, $R_g = 5$ Ω

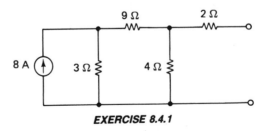

EXERCISE 8.4.1

8-4.2 By repeated source conversions change the circuit of Exercise 8-4.1 to an equivalent practical current source.   *Ans.* $I_g = 1.2$ A, $R_g = 5 \, \Omega$

8-4.3 A practical source delivers the greatest amount of power to a resistive load R when R is equal to the internal resistance R_g of the source. Illustrate this by finding the power delivered to the load R when (a) $R = R_g = 4 \, \Omega$, (b) $R = 3 \, \Omega$, (c) $R = 5 \, \Omega$, and (d) $R = 10 \, \Omega$.   *Ans.* (a) 49, (b) 48, (c) 48.4, (d) 40 W

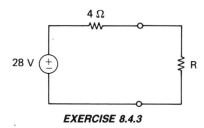

EXERCISE 8.4.3

8.5 MILLMAN'S THEOREM

If the circuit to be replaced by its Thévenin equivalent is a parallel connection of practical voltage sources, then the Thévenin voltage V_{oc} and the Thévenin resistance R_{th} may be expressed in a very special and useful way. The result is *Millman's theorem*, which we will first illustrate for the case of three practical sources in Fig. 8.22.

Special Case: In this special case there are three parallel practical sources with voltage sources of V_1, V_2, and V_3, having internal resistances of R_1, R_2, and R_3, respectively. To find V_{oc} we take the bottom node (node b) as reference and write a nodal equation at the top node (node a). The node voltage at node a is V_{oc} and the node voltages at the nodes between the sources and the internal resistances are V_1, V_2, and V_3. Thus the nodal equation is

$$\frac{V_{oc} - V_1}{R_1} + \frac{V_{oc} - V_2}{R_2} + \frac{V_{oc} - V_3}{R_3} = 0$$

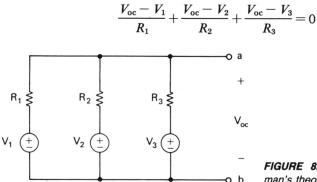

FIGURE 8.22 *Circuit illustrating Millman's theorem.*

or

$$V_{oc}\left[\frac{1}{R_1}+\frac{1}{R_2}+\frac{1}{R_3}\right]=\frac{V_1}{R_1}+\frac{V_2}{R_2}+\frac{V_3}{R_3}$$

Therefore, we have

$$V_{oc}=\frac{V_1/R_1+V_2/R_2+V_3/R_3}{1/R_1+1/R_2+1/R_3} \tag{8.13}$$

The Thévenin resistance R_{th} is the resistance seen looking in terminals a-b with the sources replaced by short circuits. Therefore, R_{th} is simply the equivalent resistance of the resistances R_1, R_2, and R_3 connected in parallel. This is, of course,

$$R_{th}=\frac{1}{1/R_1+1/R_2+1/R_3} \tag{8.14}$$

These results constitute Thévenin's theorem for the circuit of Fig. 8.22. Specifically, (8.13) is known as Millman's theorem and may be used to find the voltage, denoted here by V_{oc}, across a parallel combination of practical sources.

Example 8.7: Find V in the circuit of Fig. 8.23.

Solution: By Millman's theorem (8.13) we have

$$V=\frac{16/4+12/6+6/12}{1/4+1/6+1/12}$$

$$=\frac{13/2}{1/2}=13\text{ V}$$

If we want the Thévenin equivalent of the circuit of Fig. 8.23, then $V_{oc}=V=13$ V and by (8.14) the Thévenin resistance is

$$R_{th}=\frac{1}{1/4+1/6+1/12}=2\ \Omega$$

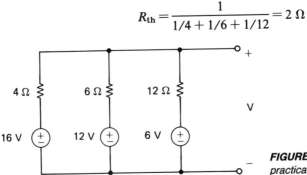

FIGURE 8.23 *Circuit of three parallel practical voltage sources.*

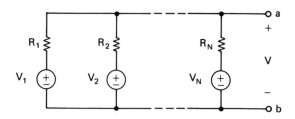

FIGURE 8.24 Circuit with N parallel practical voltage sources.

The General Case: In the general case of N parallel practical voltage sources, shown in Fig. 8.24, we may write a nodal equation at node a, as we did for Fig. 8.22. The result is

$$\frac{V - V_1}{R_1} + \frac{V - V_2}{R_2} + \cdots + \frac{V - V_N}{R_N} = 0$$

Solving for V, we have the general case of Millman's theorem,

$$V = \frac{V_1/R_1 + V_2/R_2 + \cdots + V_N/R_N}{1/R_1 + 1/R_2 + \cdots + 1/R_N} \tag{8.15}$$

In the case of Thévenin's theorem, V is V_{oc} and the Thévenin resistance is

$$R_{\text{th}} = \frac{1}{1/R_1 + 1/R_2 + \cdots + 1/R_N} \tag{8.16}$$

Example 8.8: Use Millman's theorem to find V in Fig. 8.25.

Solution: By (8.15) we have

$$V = \frac{20/2 + 32/4 + 0/16 + (-24)/8}{1/2 + 1/4 + 1/16 + 1/8} \tag{8.17}$$

$$= \frac{10 + 8 - 3}{15/16} = 16 \text{ V}$$

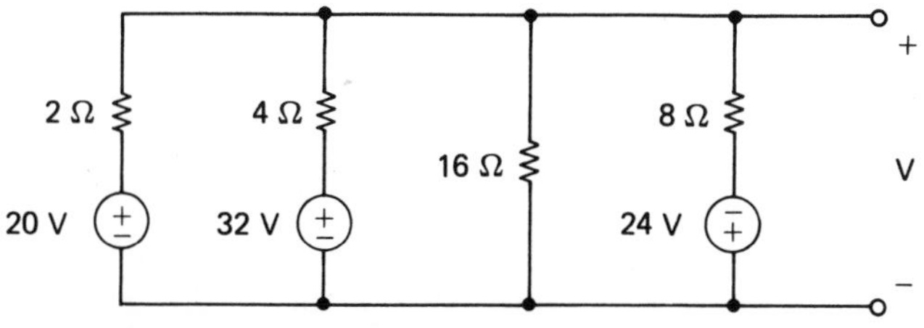

FIGURE 8.25 Circuit with four parallel components.

We note in (8.17) that the voltage "source" in series with the 16-Ω resistor is zero (a short circuit), and the source in series with the 8-Ω resistor has polarity opposite to those of V and the other sources. Therefore, its voltage is negative.

PRACTICE EXERCISES

8-5.1 Use Millman's theorem to find V.   *Ans.* 4 V

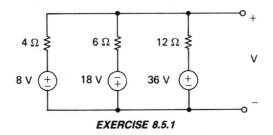

EXERCISE 8.5.1

8-5.2 Find the Thévenin equivalent of the circuit of Exercise 8-5.1.

Ans. $V_{oc} = 4$ V, $R_{th} = 2$ Ω

8-5.3 Use Millman's theorem to find V.   *Ans.* 3 V

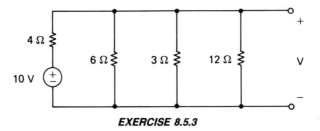

EXERCISE 8.5.3

8.6 Y AND Δ NETWORKS

The two equivalent networks of Fig. 8.26 are called T (tee) or Y (wye) networks because of their shape. They are, of course, the same network drawn in different ways.

The networks of Fig. 8.27 are called Π (pi) or Δ (delta) networks because of their resemblance to these Greek letters. They are, of course, the same network drawn in different ways, because the bottom node c of the Π network may be drawn as a single point, as in the Δ network.

If the Y or Δ networks are embedded in a circuit, there may not be any two resistors in series or in parallel. An example is the bridge circuit noted earlier in Fig. 5.7, in which R_2, R_3, and R_6 form a Y, and R_2, R_4, and R_6 form a Δ network.

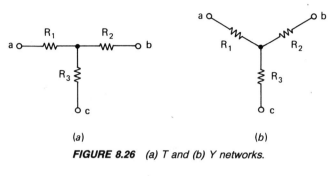

FIGURE 8.26 *(a) T and (b) Y networks.*

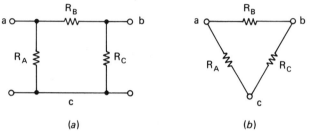

FIGURE 8.27 *(a) π and (b) Δ networks.*

In these cases, if the Δ (or Y) can be replaced by an equivalent Y (or Δ) network, there will usually be resistors in series or in parallel which can be combined to produce a simpler circuit. (By equivalent, of course, we mean equivalent at the terminals *a*, *b*, and *c* of Figs. 8.26 and 8.27.)

Y-Δ Conversions: It may be shown by means of Kirchhoff's laws that a Y network may be replaced by an equivalent Δ network, and vice versa. With reference to Fig. 8.28, we may replace the Y network of R_1, R_2, and R_3, with terminals *a*, *b*, and *c*, by the Δ network of R_A, R_B, and R_C. The conversion formulas, or Y-Δ conversion, are given by

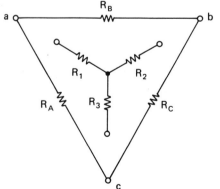

FIGURE 8.28 *Circuit for Y-Δ, Δ-Y conversions.*

$$R_A = \frac{R_1 R_2 + R_2 R_3 + R_3 R_1}{R_2}$$

$$R_B = \frac{R_1 R_2 + R_2 R_3 + R_3 R_1}{R_3} \qquad (8.18)$$

$$R_C = \frac{R_1 R_2 + R_2 R_3 + R_3 R_1}{R_1}$$

We may note from these equations and Fig. 8.28 that in each case the numerator is the sum of the products of the resistances of the Y network taken two at a time and the denominator is the resistance in the Y which is *opposite* the resistance being computed in the Δ. That is,

$$R_\Delta = \frac{\text{sum of products in Y}}{\text{opposite } R \text{ in Y}} \qquad (8.19)$$

Example 8.9: Find the equivalent Δ network of the Y network of Fig. 8.29(a).

Solution: The sum of the products of the resistances taken two at a time in the Y network is

$$6(2) + 2(3) + 3(6) = 36$$

This is the numerator in each of the right members of (8.18) and of (8.19). Referring to the Δ network of Fig. 8.29(b) and imagining it superimposed on Fig. 8.29(a) as was done in Fig. 8.28, we see that the resistance in the Y which is opposite R_A in the Δ is $R_2 = 2$ Ω. Therefore, by (8.19) we have

$$R_A = \frac{36}{2} = 18 \ \Omega$$

In like manner we have, by (8.19) and Fig. 8.29,

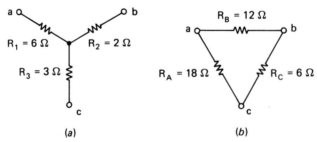

(a)   *(b)*

FIGURE 8.29 *(a) Y network and (b) its equivalent Δ network.*

$$R_B = \frac{36}{R_3} = \frac{36}{3} = 12 \ \Omega$$

and

$$R_C = \frac{36}{R_1} = \frac{36}{6} = 6 \ \Omega$$

Therefore, the equivalent Δ network is that of Fig. 8.29(b).

Δ-Y Conversions: Changing from a given Δ network to an equivalent Y, as in Fig. 8.28, may be done with the Δ-Y conversion, given by

$$R_1 = \frac{R_A R_B}{R_A + R_B + R_C}$$

$$R_2 = \frac{R_B R_C}{R_A + R_B + R_C} \qquad (8.20)$$

$$R_3 = \frac{R_A R_C}{R_A + R_B + R_C}$$

From Fig. 8.28 we may summarize these results by noting that in each case the denominator is the sum of the resistances of the Δ and the numerator is the product of the two Δ resistances that are *adjacent* to the Y resistance (on each side of the Y resistance). That is,

$$R_Y = \frac{\text{product of two adjacent } Rs \text{ in } \Delta}{\text{sum of } Rs \text{ in } \Delta} \qquad (8.21)$$

Example 8.10:   Find the equivalent Y of the Δ network of Fig. 8.29(b).

Solution: The equivalent Y is the circuit of Fig. 8.29(a), since the sum of the Rs in the Δ is

$$18 + 12 + 6 = 36$$

and by (8.20) or (8.21) we have

$$R_1 = \frac{18(12)}{36} = 6 \ \Omega$$

$$R_2 = \frac{12(6)}{36} = 2 \ \Omega$$

$$R_3 = \frac{18(6)}{36} = 3 \ \Omega$$

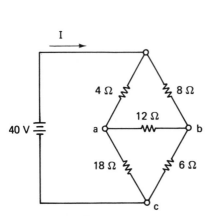

FIGURE 8.30 Bridge circuit to be converted.

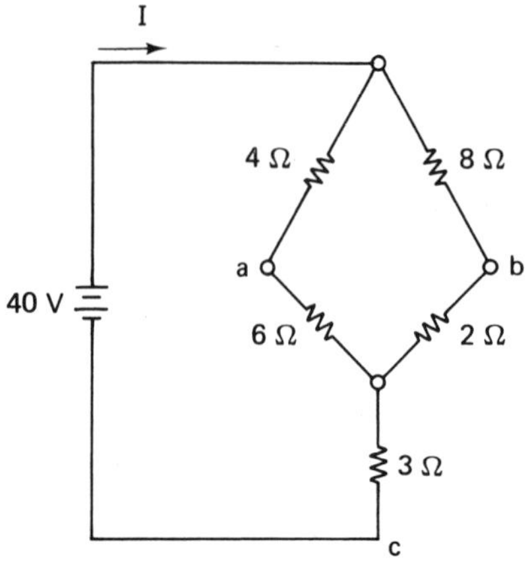

FIGURE 8.31 Equivalent of the bridge circuit of Fig. 8.30.

Example 8.11: Replace the Δ consisting of the 12-Ω, 18-Ω, and 6-Ω resistors of the bridge network of Fig. 8.30 by its equivalent Y and use series–parallel circuit theory to find *I*.

Solution: The Δ network is the same as that of Fig. 8.29(b), which has been shown to be equivalent to the Y network of Fig. 8.29(a). Replacing the Δ by the Y we have the equivalent circuit of Fig. 8.31, which is a series–parallel circuit. The resistance seen by the source is

$$R = \frac{(4+6)(8+2)}{4+6+8+2} + 3 = 8 \ \Omega$$

and thus we have

$$I = \frac{40}{8} = 5 \text{ A}$$

PRACTICE EXERCISES

8-6.1 In Fig. 8.28, if the Y network has $R_1 = 12 \ \Omega$, $R_2 = 4 \ \Omega$, and $R_3 = 3 \ \Omega$, find the equivalent Δ.   *Ans.* $R_A = 24 \ \Omega$, $R_B = 32 \ \Omega$, $R_C = 8 \ \Omega$

8-6.2 In Fig. 8.28, if the Δ network has $R_A = 12 \ \Omega$, $R_B = 16 \ \Omega$, and $R_C = 4 \ \Omega$, find the equivalent Y.   *Ans.* $R_1 = 6 \ \Omega$, $R_2 = 2 \ \Omega$, $R_3 = 1.5 \ \Omega$

8-6.3 Convert the bridge circuit to an equivalent series–parallel circuit by a Δ-Y conversion and find *I*.   *Ans.* 4 A

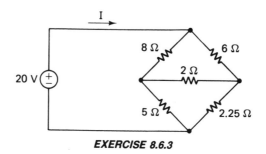

EXERCISE 8.6.3

8.7 SUMMARY

In many cases network theorems may be used to shorten considerably the work of analyzing a circuit. One such theorem is the principle of superposition, which says that a voltage or current in a circuit containing two or more sources is the algebraic sum of the voltages or currents produced by each source acting alone (with the others made zero). This allows us to analyze any circuit by solving circuits containing only one source.

Two other theorems—Thévenin's theorem and Norton's theorem—allow us to replace all of a circuit with two terminals by a single source and a resistor. In the case of Thévenin's theorem, the source is a voltage source V_{oc} in series with a resistance R_{th}. The voltage V_{oc} is the voltage across the terminals if they are open-circuited, and R_{th} is the resistance seen at the terminals when all the sources are killed, or made zero. In the case of Norton's theorem the source is a current source I_{sc} in parallel with the same R_{th}. The value of I_{sc} is the current that would flow through the terminals if they are short-circuited. Thévenin's and Norton's theorems allow us to find the current or voltage associated with a single element by replacing the rest of the circuit by a much simpler circuit, having only one source and one resistor.

A practical voltage source is a voltage source in series with an internal resistor. Without the resistor the source is called an ideal source and is, of course, the type we have considered in previous chapters. An ideal source cannot exist by itself in practice, however, because it cannot account for the change in voltage across its terminals when current is drawn from it. A practical current source is a current source in parallel with an internal resistance, as contrasted with an ideal current source having no parallel resistor connected across it. In practice, of course, current sources, like voltage sources, are practical sources.

Thévenin's and Norton's theorems may be used to change a practical voltage source to an equivalent practical current source, and vice versa. The internal resistance R is the same in both cases and the voltage V of the ideal source in the practical voltage source and the current I of the ideal source in the practical current source are related by

$$V = RI$$

Millman's theorem is a formula for finding the voltage across a parallel connection of practical voltage sources. It holds in the general case as well as the case where one or more of the parallel connections is a resistor only (the ideal voltage source in the connection is zero).

Finally, Y-Δ and Δ-Y conversions are theorems that allow us to change a Y network (one with three resistors connected in a shape resembling a Y) to an equivalent Δ network (one with three resistors in the shape of a Δ), and vice versa. Such conversions are useful in changing a non-series–parallel circuit, such as a bridge, to a series–parallel circuit.

PROBLEMS

8.1 Find V using superposition. Check by combining the sources into an equivalent source.

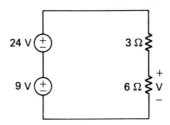

PROBLEM 8.1

8.2 Find V_1 in Fig. 8.1 using superposition.

8.3 Using superposition find the current to the right in the 2-Ω resistor of Fig. 8.3.

8.4 Find V_2 in Fig. 7.10 using superposition.

8.5 Find I_1 in Fig. 7.11 using superposition.

8.6 Solve Problem 7.17 using superposition.

8.7 Replace everything except the 3-Ω resistor in Practice Exercise 8-2.1 by its Thévenin equivalent circuit and use the result to find the power delivered to the 3-Ω resistor.

8.8 Replace the circuit to the left of terminals a-b by its Thévenin equivalent and find V_1.

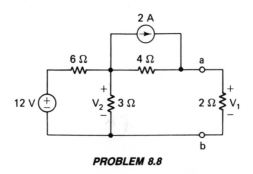

PROBLEM 8.8

8.9 Replace everything in the circuit of Problem 8.8 except the 3-Ω resistor by its Norton equivalent and find V_2.

8.10 Solve Problem 7.16 by replacing everything in the circuit except the 8-Ω resistor by its Thévenin equivalent.

8.11 Solve Problem 8.10 by using Norton's theorem instead of Thévenin's theorem.

8.12 Find the Norton equivalent of the circuit external to the 6-Ω resistor and use the results to obtain I.

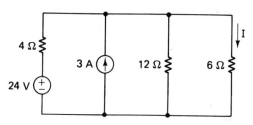

PROBLEM 8.12

8.13 Solve Problem 8.12 using the Thévenin equivalent circuit instead of the Norton equivalent circuit.

8.14 Find the Thévenin and the Norton equivalent circuits of the circuit external to the 2-Ω resistor and use the results to find I.

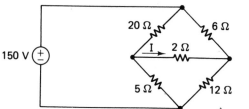

PROBLEM 8.14

8.15 By successive source conversions in Problem 8.8 obtain an equivalent one-loop circuit containing the 2-Ω resistor, one other resistor R, and a single voltage source V_g. Using this result, find V_1.

8.16 Using successive source conversions, replace the circuit to the left of terminals a-b by its Thévenin equivalent and find V.

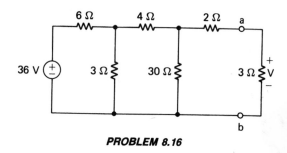

PROBLEM 8.16

8.17 Convert all the practical voltage sources to practical current sources, and combine the sources and resistors to form the Norton equivalent circuit.

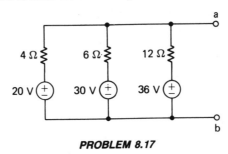

PROBLEM 8.17

8.18 Use Millman's theorem to find the voltage V_{ab} across terminals a-b in Problem 8.17.

8.19 Replace the practical current source consisting of the 3-A source and the 12-Ω resistor in Problem 8.12 by its equivalent voltage source and use Millman's theorem to find the Thévenin equivalent of the circuit external to the 6-Ω resistor.

8.20 In Fig. 8.28, if the Y network has $R_1 = 12$ Ω, $R_2 = 4$ Ω, and $R_3 = 6$ Ω, find the equivalent Δ.

8.21 Repeat Problem 8.20 if $R_1 = R_2 = R_3 = R$.

8.22 Convert the Δ of the 12-Ω, 4-Ω, and 8-Ω resistors to an equivalent Y and find R_T and I.

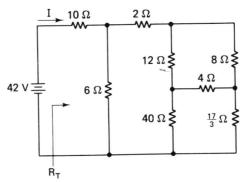

PROBLEM 8.22

9

DIRECT-CURRENT METERS

Everyone who uses electric circuits, from the engineer and scientist to the hobbyist, has a need for making measurements. To design and check a circuit, we must be able to measure the currents through certain elements, or the voltages across the elements. We may want to know what size the resistances are, how close their actual values are to their nominal values, or how much power is being supplied to a given element.

Measurement devices, or *meters,* come in a wide variety of shapes and sizes. *Voltmeters* and *ammeters* (ampere meters) measure voltage and current, *wattmeters* measure power, and *ohmmeters* measure resistance. A very versatile example of a meter is the *volt-ohm-milliammeter* (VOM) of Fig. 9.1, which may be used to measure voltage in volts, resistance in ohms, or current in milliamperes. The quantity to be measured (voltage, resistance, or current) is selected by the dials and its value is indicated by the position of the pointer on the meter scale.

Regardless of their appearances or purposes, most measurement devices use a common current- or voltage-sensing mechanism, called a *meter movement,* which deflects a needle or pointer to indicate the magnitude of the quantity being measured. This, of course, is the case with the VOM of Fig. 9.1. In this chapter we shall consider in detail one such device, called the *d'Arsonval movement,* and see how it may be used in the construction of various types of meters. We shall restrict ourselves, for the most part to direct current meters, but as we shall see, the d'Arsonval movement may also be used in the construction of certain ac meters.

185

FIGURE 9.1 Volt-ohm-milliammeter (VOM) (Courtesy, Simpson Electric Company, Elgin, Illinois).

9.1 D'ARSONVAL MOVEMENT

The simplest and most commonly used meter movement is the *d'Arsonval*, or *permanent-magnet-moving-coil*, movement, which was developed in 1881 by the French physicist Jacques Arsène d'Arsonval. This movement consists of a coil of wire wound on a drum, which is pivoted on a shaft between the poles of a permanent magnet, as shown in Fig. 9.2. When the current *i* flows through the external terminals into the coil, the reaction with the magnet produces a force that is exerted on the drum, causing it to rotate. Thus, a pointer attached to the drum, as in Fig. 9.2, will turn with the drum, indicating a deflection on the scale, which is proportional to the current in the coil.

The rotation of the drum is opposed by two restraining springs, one of which is visible on the front in Fig. 9.2. The other is located on the other end of the shaft at the back of the meter. In the case shown, the springs also provide the conducting path for the current flow. The springs are calibrated so that a known current causes

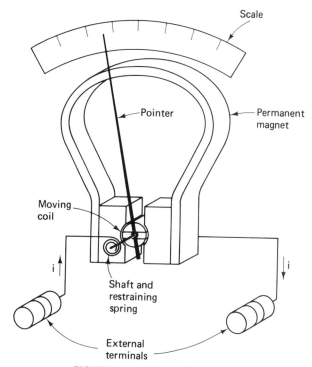

FIGURE 9.2 *d'Arsonval movement.*

a specific rotation of the pointer, which corresponds to an appropriate number on the scale.

Principle of the d'Arsonval Movement: As we shall see in Chapter 13, a *magnetic field* exists between the poles of a magnet, and current flowing in a wire produces a magnetic field of its own. The interaction of the magnetic field of the permanent magnet in Fig. 9.2 and the magnetic field produced by the current in the coil produces a force that causes the drum to rotate. This is also the principle of the electric motor discovered by Michael Faraday in the nineteenth century.

Meter Construction: In order for the meter reading to be as accurate as possible, the meter movement itself should not influence the amount of current that flows through its rotating coil. Therefore, any friction that would oppose the rotation of the coil should be minimized. One way that this is done is to use jewel bearings to support the pivot shaft. The jewels, which are usually synthetic sapphire, provide very low friction at the contact points with the steel shaft, and also serve to keep the shaft properly centered. An exploded view of a movement assembly with pivot and jewel construction is shown in Fig. 9.3.

Another method of construction is to use a *taut-band* support rather than the

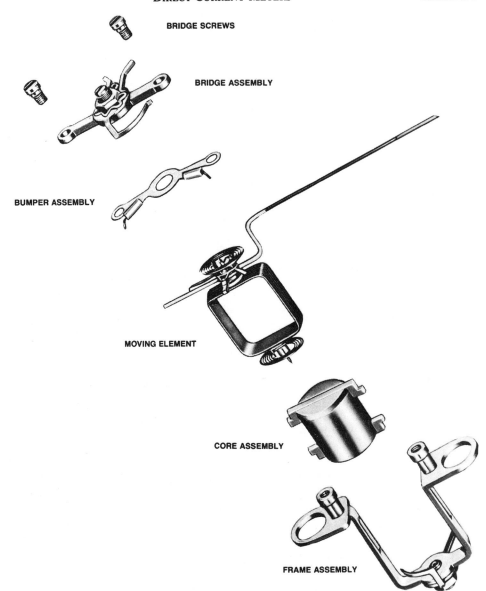

FIGURE 9.3 *Movement assembly, pivot and jewel exploded view (Courtesy, Simpson Electric Company, Elgin, Illinois).*

shaft, jewel, and spring arrangement. In this case, the drum is supported by two thin metal ribbons or bands, which take the place of the shaft and provide the electrical connection and the restoring force of the restraining springs. The chief advantage of using taut bands is the absence of friction between moving parts. A taut-band movement is shown in Fig. 9.4.

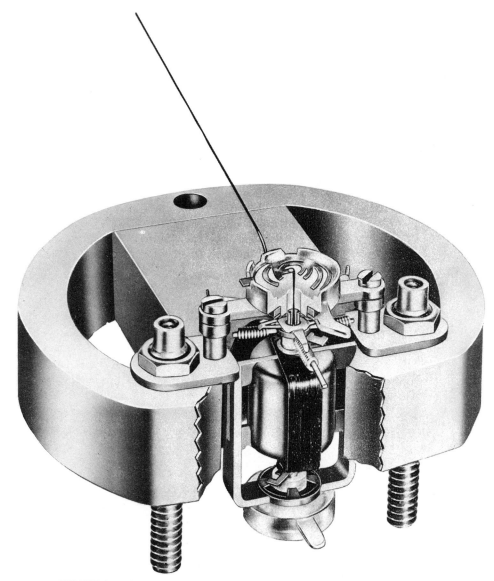

FIGURE 9.4 *Annular taut-band movement (Courtesy, Simpson Electric Company, Elgin, Illinois).*

Types of Scales: Two types of scales are generally used with the d'Arsonval movement. One, such as the galvanometer discussed in Chapter 6, has a zero at the center of the scale. When a positive current (from the positively marked external terminal to the negatively marked external terminal) flows through the coil, the needle is deflected *upscale* (to the right). When a current flows in the opposite direction,

FIGURE 9.5 *Zero-center-scale meter (Courtesy, Simpson Electric Company, Elgin, Illinois).*

the deflection is *downscale* (to the left). A galvanometer with a zero-center scale is shown in Fig. 9.5.

The other type of scale usually has the zero on the left end. Such a meter will only deflect upscale. If the current is negative, the pointer will rest on a stop pin at the zero reading. To obtain the magnitude of the current in this case, we must reverse the connections of the leads to the meter. A meter that is calibrated to read voltage (a voltmeter), with a zero on the left end of the scale, is shown in Fig. 9.6.

FIGURE 9.6 *Meter with zero on the left end of the scale (Courtesy, Simpson Electric Company, Elgin, Illinois).*

FIGURE 9.7 d'Arsonval movement notations.

(a)   (b)

Rating of D'Arsonval Movements: D'Arsonval movements are usually rated by current I_M and resistance R_M. The value of I_M is the *full-scale current*, which is the amount of current needed to deflect the pointer all the way to the last mark on the right of the meter scale. The value of full-scale current typically may range from about 10 μA to 30 mA. However, the current range may be increased to almost any value by means of *shunt*, or *parallel*, resistances, as we shall see in Section 9.2.

The value of R_M is the resistance of the wire of the moving coil, and may range from about 1 Ω to several thousand ohms. The smaller the I_M rating, however, the larger the R_M rating, because many turns of fine wire must be used. As an example, one meter may have a rating of 1 mA, 50 Ω, whereas another's rating may be 50 μA, 2000 Ω. Common notations for d'Arsonval movements are shown in Fig. 9.7. In Fig. 9.7(a), the internal resistance R_M is indicated in series with an *ideal* movement (one with no voltage drop across its terminals). In Fig. 9.7(b), both specifications are shown. The movement has an *IR* drop across it because of the resistance R_M assumed to be incorporated in it.

Example 9.1: A d'Arsonval movement is rated 1 mA, 100 Ω. Find the voltage across the movement if the current flowing is one-half the full-scale current I_M.

Solution: The full-scale current is $I_M = 1$ mA, and thus the current flowing in the meter is $I = I_M/2 = 0.5$ mA. Since in Fig. 9.7(a) the voltage across the ideal movement is zero, the voltage V across the d'Arsonval movement is $V = R_M I$, where $R_M = 100$ Ω. Therefore, we have

$$V = (100)(0.5)(10^{-3})$$

$$= 0.05 \text{ V}$$

PRACTICE EXERCISES

9-1.1 Find the voltage across a d'Arsonval movement rated 50 μA, 1000 Ω if the current through the movement is one-fifth of the full-scale current.   *Ans.* 0.01 V

9-1.2 Find the maximum voltage that would ever appear across the d'Arsonval movement of Exercise 9-1.1 if the current is never to exceed the full-scale current.

Ans. 0.05 V

9.2 AMMETERS

The d'Arsonval movement described in Section 9.1 is a current-detecting device, and thus is an ammeter, because the deflection of its pointer is proportional to the current flowing through the movement. However, the full-scale current I_M, which is the maximum current the device can safely carry, is limited to a relatively small value. Thus, if the d'Arsonval movement is to function as a practical ammeter, we must provide a way for higher currents to be measured while the current through the movement is limited to I_M.

Meter Shunt: A resistive path connected in parallel with a d'Arsonval movement is called a *meter shunt,* because it can be used as a bypass to *shunt* a specific fraction of current by the meter movement. For example, Fig. 9.8 illustrates how an ammeter can be constructed with a d'Arsonval movement shunted by a parallel resistance R_p. The value of R_p may be calculated so that the current i_m through the meter movement may be a given fraction of the total current i entering the ammeter. Since there is no voltage across the ideal movement, we have by current division

$$i_m = \left(\frac{R_p}{R_M + R_p}\right) i \tag{9.1}$$

Thus the meter scale may be calibrated so that a given current i yields the correct needle deflection caused by the meter movement current i_m.

The value of R_p in (9.1) is determined so that the meter movement current i_m never exceeds the full-scale current I_M, as long as the total current i does not exceed some maximum allowable value, say $i = I_{max}$. From (9.1) we may write

$$(R_M + R_p)i_m = R_p i$$

or

$$R_p(i - i_m) = R_M i_m$$

Therefore, we have

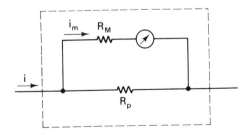

FIGURE 9.8 *Ammeter using d'Arsonval movement and shunt.*

192

$$R_p = \frac{R_M i_m}{i - i_m}$$

Therefore, when $i_m = I_M$ and $i = I_{max}$, we have

$$R_p = \frac{R_M I_M}{I_{max} - I_M} \qquad (9.2)$$

Thus for this value of R_p, the maximum allowable current that can be measured by the ammeter of Fig. 9.8 is $i = I_{max}$, which causes the full-scale current $i_m = I_M$ to flow in the meter movement. The ammeter dial, of course, is calibrated to read i rather than i_m.

The maximum current $i = I_{max}$ that can be read for a given I_M, R_M, and R_p may be found by solving (9.1) for i and replacing i by I_{max} and i_m by I_M. The result is

$$I_{max} = \left(\frac{R_M + R_p}{R_p}\right) I_M \qquad (9.3)$$

Example 9.2: A d'Arsonval movement rated 1 mA, 50 Ω is to be used as in Fig. 9.8 to construct an ammeter that can measure up to 1 A of current. Find the required shunt resistance R_p.

Solution: We have $R_M = 50$ Ω, $I_M = 1$ mA $= 10^{-3}$ A, and $I_{max} = 1$ A, so that by (9.2) the shunt resistance is

$$R_p = \frac{50(10^{-3})}{1 - 10^{-3}} = \frac{50}{999} = 0.05005 \text{ Ω}$$

Thus a shunt resistance of approximately 0.05 Ω is to be used.

Polarity: A dc ammeter has polarity markings on its terminals, as shown in Fig. 9.9(a). If a positive current i flows as shown from the positive to the negative terminal, the deflection of the needle will be upscale, indicating a positive reading. Also, instead

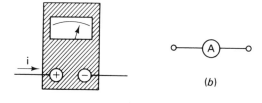

(a)

(b)

FIGURE 9.9   (a) Ammeter and (b) its circuit symbol.

of having $+$ and $-$ signs, the terminals may be color-coded, with red for plus and black for minus.

The circuit symbol often used for an ammeter is that of Fig. 9.9(b). Ideally, there is no voltage across its terminals, but in the practical case there is a small voltage, because of the resistors R_p and R_M of Fig. 9.8. This voltage is usually negligible because R_p is normally very small compared to the resistance of the rest of the circuit.

Multirange Ammeter: By varying the value of the shunt resistance R_p in Fig. 9.8, we may, as we have seen by (9.3), change the range of current that the ammeter can safely measure. For instance, in Example 9.2 for a 1-mA, 50-Ω meter movement, a value of $R_p = 0.05$ Ω allowed us to measure currents up to 1 A. If $R_p = 0.005$ Ω, the maximum current that may be measured is by (9.3)

$$I_{\max} = \left(\frac{R_M + R_p}{R_p}\right)I_M$$

$$= \left(\frac{50 + 0.005}{0.005}\right)(10^{-3})$$

$$= 10 \text{ A}$$

A *multirange* ammeter is one that can measure several ranges of current. It may be constructed by building in several different-valued shunt resistors, each of which may be used as R_p. A current lead may be connected manually to one of several available external terminals, or, as in Fig. 9.10, a rotary switch may be used with two external connections to select the appropriate shunt resistor. For example, if in Fig. 9.10, the d'Arsonval movement is rated 1 mA, 50 Ω, and $R_1 = 0.05$ Ω, $R_2 = 0.005$ Ω, and $R_3 = 0.0005$ Ω, then by (9.3) the range of currents that can be measured is 0 to 1 A (using R_1), 0 to 10 A (using R_2), and 0 to 100 A (using R_3).

As an example, the meter of Fig. 9.11 has the capability of measuring both ac and dc current and voltage. In the case where it functions as a dc or ac ammeter, the shunt resistance may be selected, by making connections to the appropriate terminals, to measure current ranges of 0 to 0.6 A, 0 to 3 A, and 0 to 15 A. The switching arrangement shown allows us to measure voltages up to 3, 15, 30, and 150 V, as will be discussed in Section 9.3.

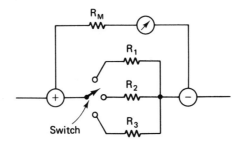

FIGURE 9.10 *Multirange ammeter.*

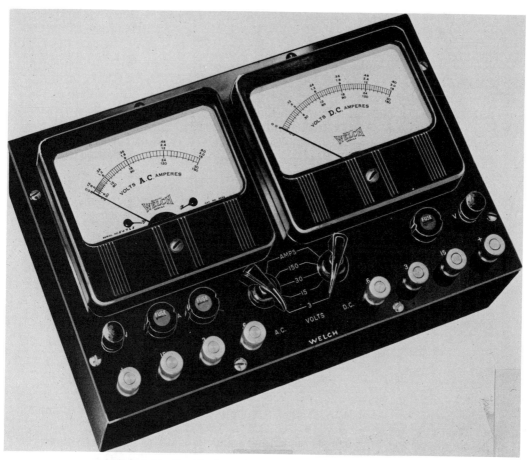

FIGURE 9.11 *Combination ammeter and voltmeter (Courtesy, Sargent-Welch Scientific Company).*

An additional advantage of the multirange meter is that if more than one range is applicable, we may select the smallest for the greatest accuracy. For example, a current of 1.5 A is more accurately read on a range of 0–3 A than on one of 0–15 A.

PRACTICE EXERCISES

9-2.1   A d'Arsonval movement rated 50 μA, 1000 Ω is to be used as in Fig. 9.8 to construct an ammeter that can measure up to 1 mA of current. Find the shunt resistance R_p.
                                                                 Ans. 52.63 Ω

9-2.2   Repeat Exercise 9-2.1 if the maximum current to be measured is 1 A.   *Ans.* 0.05 Ω

9-2.3 When the ammeter is measuring 0.5 A in Exercise 9-2.2, find the current through the meter movement and through the shunt resistor.　　　*Ans.* 25 μA, 499,975 μA

9.3 VOLTMETERS

The d'Arsonval movement may also function as a voltmeter since the scale could be calibrated to read the IR drop across the resistor R_M. For example, a 1-mA, 50-Ω movement has a voltage of $50I$ volts across it when it carries a current I (restricted to $I \le 1$ mA $= 0.001$ A). Thus if the scale is marked millivolts rather than milliamps, the device is acting as a voltmeter with a full-scale voltage of 50 mV.

Multiplier Resistance: As it was in the case of the ammeter, the d'Arsonval movement is impractical acting alone as a voltmeter. However, we may greatly increase the voltage that can be read by inserting a resistance R_s in series with the movement, as shown in Fig. 9.12. This resistance, called a *multiplier,* is much higher than the coil resistance R_M, in order to limit the current through the movement. The resulting device is a voltmeter with a full-scale voltage determined by R_s, as we shall see.

Applying KVL around the loop of Fig. 9.12 and noting that the voltage is zero across the ideal movement, we have

$$-v + (R_s + R_M)i = 0$$

or

$$v = (R_s + R_M)i \tag{9.4}$$

Since the current i through the meter movement must not be allowed to exceed the full-scale value $i = I_M$, the maximum, or full-scale, voltage that the meter can read is $v = V_{max}$, given by (9.4) to be

$$V_{max} = (R_s + R_M)I_M \tag{9.5}$$

Solving this equation for the multiplier resistance R_s results in

$$R_s = \frac{V_{max} - R_M I_M}{I_M} \tag{9.6}$$

Thus for a given I_M, R_M rating and a given full-scale voltage V_{max} that is to be allowed, we may calculate the required R_s.

Example 9.3: A 1-mA, 50-Ω d'Arsonval movement is to be used with a multiplier resistance R_s to construct a voltmeter, as in Fig. 9.12. If the full-scale voltage is to be 10 V, find R_s.

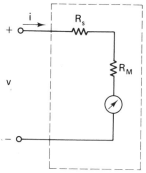

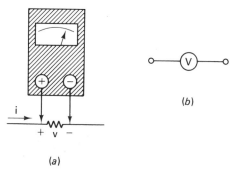

FIGURE 9.12 Voltmeter using d'Arsonval movement and multiplier R_s.

FIGURE 9.13 (a) Voltmeter connected across a resistor, and (b) voltmeter circuit symbol.

Solution: In (9.6) we have $V_{max} = 10$ V, $R_M = 50$ Ω, and $I_M = 1$ mA $= 0.001$ A. Thus the multiplier resistance is

$$R_s = \frac{10 - (50)(0.001)}{0.001} = 9950 \ \Omega$$

As we saw in Section 9.2, to measure a current we have to break the circuit connection and insert the ammeter in series with the element whose current is to be found. Thus it is important that the ammeter have as small a voltage as possible across its external terminals so that its presence does not influence the current it is measuring. (The short circuit it replaces should remain very nearly a short circuit.) A voltmeter, on the other hand, is to measure a voltage *across* an element. Thus it has two external terminals with polarity markings, which are connected as shown in Fig. 9.13(a) across the element whose voltage is to be found. If the current i is flowing as shown in Fig. 9.13(a), the voltage v across the resistor is positive and the needle will register an upscale reading.

It is easier to make voltage readings than current readings because the circuits do not need to be broken to insert a voltmeter. The leads are simply connected across the element of interest. However, since the presence of the voltmeter should not influence the voltage it is measuring, there should be very little current drawn through its leads. (The open circuit it replaces should remain very nearly an open circuit.) Thus the multiplier resistance, as stated earlier, must be relatively large, and ideally is infinite. The circuit symbol usually used for a voltmeter is that of Fig. 9.13(b). In the ideal case, of course, it acts as an open circuit.

Multirange Voltmeter: As in the case of the ammeter, a voltmeter may be designed to have a variety of ranges of measurable voltages. Such a *multirange* voltmeter is constructed by providing a number of possible multiplier resistances, one to be selected for each range. A multirange voltmeter with three possible values of multiplier resist-

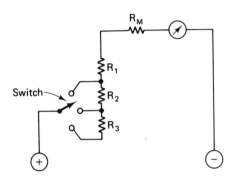

FIGURE 9.14 *Multirange voltmeter.*

ance is shown in Fig. 9.14. If the switch is at the top position, the multiplier resistance is $R_s = R_1$, at the middle position $R_s = R_1 + R_2$, and at the bottom position $R_s = R_1 + R_2 + R_3$. By (9.5) we may see that the higher the value of the multiplier resistance, the greater the range of the meter.

Example 9.4: Find R_1, R_2, and R_3 in Fig. 9.14 so that the voltmeter has ranges up to 10 V, 20 V, and 100 V. The d'Arsonval movement is rated 1 mA, 50 Ω.

Solution: We have $I_M = 1$ mA $= 0.001$ A, $R_M = 50$ Ω, and $V_{max} = 10$ V, 20 V, and 100 V. For the case $V_{max} = 10$ V, we have already found $R_s = 9950$ Ω in Example 9.3. Therefore, $R_1 = 9950$ Ω. For $V_{max} = 20$, we have $R_s = R_1 + R_2$, which by (9.6) is

$$R_1 + R_2 = \frac{20 - (50)(0.001)}{0.001} = 19{,}950 \ \Omega$$

Therefore, we have

$$R_2 = 19{,}950 - R_1 = 10{,}000 \ \Omega$$

Finally, for $V_{max} = 100$ V, we have $R_s = R_1 + R_2 + R_3$, given by

$$R_1 + R_2 + R_3 = \frac{100 - (50)(0.001)}{0.001} = 99{,}950 \ \Omega$$

Therefore, we have

$$R_3 = 99{,}950 - R_1 - R_2 = 80{,}000 \ \Omega$$

Ohms/Volt Rating: Voltmeters are usually rated in *ohms/volt* (Ω/V), which is the ohms of resistance needed for a 1-V deflection. Since the total resistance associated with a voltmeter is $R_s + R_M$ and the maximum deflection is V_{max}, the ohms/volt rating is given by

$$\Omega/V = \frac{R_s + R_M}{V_{max}} \qquad (9.7)$$

This value is a constant tor a given voltmeter and may be simplified by substituting for V_{max} from (9.5). The result is

$$\Omega/V = \frac{1}{I_M} \qquad (9.8)$$

The Ω/V rating is also called the *sensitivity* of the voltmeter, and, as is seen in (9.8), it is the reciprocal of the full-scale meter current. As we will see, it is also a measure of the quality of the voltmeter.

Example 9.5: Find the ohms/volt rating of the voltmeter of Example 9.3, by (a) the method of (9.7) and (b) the method of (9.8).

Solution: We have $I_M = 0.001$ A, $R_M = 50\ \Omega$, $V_{max} = 10$ V, and $R_s = 9950\ \Omega$. Therefore by (9.7) the Ω/V rating is

$$\Omega/V = \frac{9950 + 50}{10} = 1000$$

Similarly, by (9.8) we have

$$\Omega/V = \frac{1}{0.001} = 1000$$

The ohms/volt rating is a measure of how well the voltmeter approaches the ideal voltmeter. In the ideal case the Ω/V rating is infinite, so that the voltmeter is an open circuit and thus draws no current. In general, the higher the Ω/V rating, the better the voltmeter will perform, as illustrated by the following example.

Example 9.6: Find the reading of the voltmeter in Fig. 9.15 if its full-scale voltage is 100 V and its Ω/V rating is (a) 1000 and (b) 20,000. Compare the readings with the true value of the voltage V_{ab}.

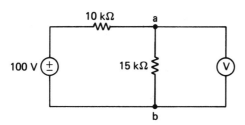

FIGURE 9.15 *Circuit containing a volt-meter.*

Solution: In case (a) we have by (9.7)

$$R_s + R_M = (\Omega/V) V_{\max}$$

$$= (1000)(100) = 100,000 \ \Omega = 100 \ \text{k}\Omega$$

Thus the equivalent resistance R_{ab} seen between points a and b is the parallel combination of the voltmeter resistance $R_s + R_M$ and the 15-kΩ resistance. That is,

$$R_{ab} = \frac{100(15)}{100 + 15} = \frac{300}{23} \ \text{k}\Omega$$

Therefore, by voltage division, the meter reads V_1 given by

$$V_1 = 100\left(\frac{300/23}{300/23 + 10}\right) = 56.6 \ \text{V}$$

In case (b) we have $R_s + R_M = (20,000)(100) \ \Omega = 2000 \ \text{k}\Omega$ and thus

$$R_{ab} = \frac{2000(15)}{2015} = \frac{6000}{403} \ \text{k}\Omega$$

Thus, by voltage division, the meter reads V_2, given by

$$V_2 = 100\left(\frac{6000/403}{6000/403 + 10}\right) = 59.8 \ \text{V}$$

The correct reading, by voltage division, should be

$$V_{ab} = 100\left(\frac{15}{15 + 10}\right) = 60 \ \text{V}$$

Therefore, we see that the voltmeter with the higher Ω/V rating is much more accurate than the one with the lower rating.

PRACTICE EXERCISES

9-3.1 Find the multiplier resistance in the voltmeter of Example 9.3 if the full-scale voltage is to be 1 V.
Ans. 950 Ω

9-3.2 Find R_1, R_2, and R_3 in Figure 9.14 if the d'Arsonval movement is rated 50 μA, 1000 Ω, and the voltmeter has ranges up to 10 V, 20 V, and 100 V.
Ans. 199, 200, 1600 kΩ

9-3.3 In the circuit shown the correct voltmeter reading should be 30 V. Find the actual reading if the voltmeter has a full-scale deflection of 50 V and its Ω/V rating is 1000.

Ans. 28.6 V

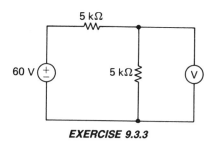

EXERCISE 9.3.3

9-3.4 Repeat Exercise 9-3.3 if the Ω/V rating is 20,000.   *Ans.* 29.9 V

9.4 OHMMETERS

The voltmeter circuit of Fig. 9.12, consisting of a d'Arsonval movement in series with a multiplier resistance, may be converted to an ohmmeter by the addition of a battery, as shown in Fig. 9.16(a). When a resistance to be measured is connected across the input terminals, as in the case of the resistance R_x in Fig. 9.16(b), a current I flows, which causes a deflection of the needle. If the battery voltage V and the resistances R_s and R_M are fixed, the current I will depend on the resistance R_x. Therefore, the scale can be calibrated in ohms to read R_x, and thus the device is an ohmmeter.

The meter of Fig. 9.16 is called a *series* ohmmeter because of the manner in which its constituent elements are connected. The unknown resistor to be measured is inserted in series with the meter movement.

Applying KVL to the circuit of Fig. 9.16(b), we have

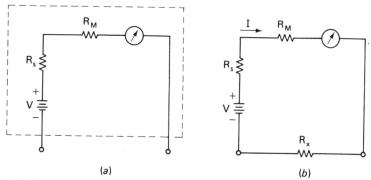

(a)   (b)

FIGURE 9.16 (a) Series ohmmeter and (b) its connection to measure a resistance R_x.

$$V = (R_s + R_M + R_x)I \tag{9.9}$$

or

$$I = \frac{V}{R_s + R_M + R_x} \tag{9.10}$$

Also, from (9.9) we may find R_x, given by

$$R_x = \frac{V}{I} - R_s - R_M \tag{9.11}$$

The amount of deflection on the scale of the ohmmeter indicates the value of the resistance R_x in ohms. The reading is always upscale because the battery is connected internally to give the correct direction of current flow.

Calibration of the Scale: To calibrate the scale to read the resistance being measured, we note that the full-scale current $I = I_M$ will occur when $R_x = 0$ (a short circuit across the ohmmeter terminals). Thus the full-scale current is by (9.10)

$$I_M = \frac{V}{R_s + R_M} \tag{9.12}$$

and this full-scale deflection will be marked 0 Ω. With the ohmmeter leads open, R_x is infinite ($R_x = \infty$), and no current will flow. Thus the position on the scale corresponding to zero current (no deflection) will be marked ∞. Values of R_x between 0 and ∞ will be marked accordingly. A typical ohmmeter scale is the top scale of the VOM of Fig. 9.1, which is very similar to the ohmmeter scale shown in Fig. 9.17. The scale is nonlinear because the resistance varies with $1/I$.

Example 9.7: The ohmmeter of Fig. 9.16(a) has a 1-mA, 100-Ω d'Arsonval movement. Determine V and R_s so that half-scale deflection occurs at $R_x = 1500$ Ω. (Half-scale deflection is at the center of the scale, as is 15 Ω in Fig. 9.17.)

Solution: We have $I_M = 1$ mA $= 0.001$ A and $R_M = 100$ Ω. Therefore, for full-scale deflection ($R_x = 0$ and $I = I_M$) we have by (9.9)

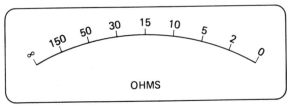

FIGURE 9.17 Ohmmeter scale.

$$V = (R_s + 100)(0.001) \qquad (9.13)$$

For the half-scale deflection ($I = I_M/2 = 0.0005$ A and $R_x = 1500$ Ω) we have

$$V = (R_s + 100 + 1500)(0.0005) \qquad (9.14)$$

Substituting for V from (9.13) into (9.14) results in

$$0.001(R_s + 100) = 0.0005(R_s + 1600)$$

or

$$2(R_s + 100) = R_s + 1600$$

Solving for R_s, we have

$$R_s = 1400 \ \Omega$$

Substituting this result into (9.13) yields

$$V = 1.5 \ V$$

Zero-Adjust Resistor: In practice the effective value of the resistance R_s in the ohmmeter can be adjusted by combining it with a variable resistor R_{za}, as shown in Fig. 9.18. This variable resistor is called the *zero-adjust* resistor and is used to set the pointer to zero (by allowing I_M to flow) when the input leads are connected by a short circuit. In this case R_s is replaced by $R_s + R_{za}$ in (9.9) through (9.12).

Shunt Ohmmeter: The series ohmmeter is most useful for measuring high resistances, because the current through the movement is restricted to very small values. For measuring low resistances a *shunt* ohmmeter, such as that of Fig. 9.19, is usually used. In a shunt ohmmeter, there are three parallel paths containing the movement, the battery and R_s, and the unknown resistance R_x.

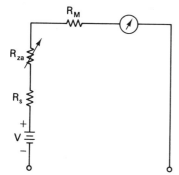

FIGURE 9.18 Series ohmmeter with zero-adjust resistor.

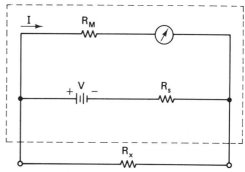

FIGURE 9.19 Shunt ohmmeter with resistance R_x to be measured.

Pointer Deflection: Full-scale current $I = I_M$ will flow in the shunt ohmmeter when $R_x = \infty$ (open circuit), since in this case we have from Fig. 9.19

$$I_M = \frac{V}{R_s + R_M} \tag{9.15}$$

which is the same as (9.12). If $R_x = 0$ (the input leads are connected by a short circuit), all the current will flow through the short, so that $I = 0$. Thus, in contrast to the series ohmmeter, the shunt ohmmeter pointer will deflect from left to right, with zero ohms corresponding to $I = 0$ and infinite ohms corresponding to $I = I_M$.

As in the series ohmmeter, there is usually a zero-adjust resistor R_{za} in series with R_s. Therefore, in this case we must replace R_s by $R_s + R_{za}$ in all the results such as (9.15).

Multiple Ohmmeter Ranges: By providing a variety of values of R_s inside the ohmmeter, we may construct a device with multiple ranges of resistance values. The procedure is similar to that used in the multirange ammeters and voltmeters. A *range switch,* such as that of Fig. 9.20, is used to change the value of R_s and thus change the range of resistances that can be measured. If the needle points to 15 and the range switch is at $R \times 1$, the resistance being measured is $15 \times 1 = 15\ \Omega$. If the switch is at $R \times 100$, the resistance is $15 \times 100 = 1500\ \Omega$, and if the switch is at $R \times 10,000$, the resistance is $15 \times 10,000 = 150,000\ \Omega$.

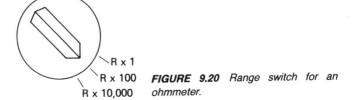

FIGURE 9.20 *Range switch for an ohmmeter.*

A range switch is shown on the VOM of Fig. 9.1. It serves to provide a multirange of values for the voltmeter, the ammeter, and the ohmmeter, with $R \times 1$, $R \times 100$, and $R \times 10,000$ positions in the latter case.

PRACTICE EXERCISES

9-4.1 A 1-mA, 50-Ω d'Arsonval movement is used in the series ohmmeter of Fig. 9.16(a). If $V = 4.5$ V and $R_s = 4450\ \Omega$, find the resistance R_x for which (a) half-scale deflection occurs, (b) one-quarter-scale deflection occurs, and (c) one-tenth-scale deflection occurs. (*Suggestion:* $I = I_M/2$, $I = I_M/4$, and $I = I_M/10$ in these cases.)

Ans. (a) 4.5 kΩ, (b) 13.5 kΩ, (c) 40.5 kΩ

9-4.2 If the ohmmeter of Fig. 9.16(a) has a 1-mA, 50-Ω movement, find V and R_s so that half-scale deflection occurs at $R_x = 2$ kΩ.   *Ans.* 2 V, 1950 Ω

9-4.3 For the shunt ohmmeter of Fig. 9.19 the movement is rated 1 mA, 50 Ω, and $V = 4.5$ V. Find R_s. [*Suggestion:* Note that when $I = I_M$, (9.15) must hold.]
Ans. 4450 Ω

9.5 OTHER METERS

Since ammeters, voltmeters, and ohmmeters may all be designed with the same d'Arsonval movement, they are often combined into one meter. An example of such a meter is the *volt-ohm-milliammeter* (VOM) mentioned earlier and illustrated in Fig. 9.1. The VOM is a very useful device which is capable of measuring dc and ac voltages, dc current, and resistance. For dc and resistance measurements the d'Arsonval movement is used as described in the previous sections. For ac measurements the signal of interest (the voltage, in this case) is converted to a *pulsating* dc signal by a rectifier and the scale is calibrated to read an *average* value.

Rectified AC Voltage: The rectifier produces a pulsating dc signal by blocking the negative portions of the ac voltage or by changing their sign to positive. In the first case the signal is a *half-wave-rectified* voltage which consists only of the positive portions of the original ac voltage. In the second case, the signal is a *full-wave-rectified* voltage. An example of this case is shown in Fig. 9.21, where the ac signal in Fig. 9.21(a) is fully rectified to the pulsating dc signal in Fig. 9.21(b).

Multimeters: The VOM is a special case of a general-purpose meter called a *multimeter,* which uses a single movement to measure both dc and ac quantities and resistance. The VOM of Fig. 9.1 has standard dial settings for dc and ac voltages from 2.5 to 500 V, dc currents from 1 to 500 mA, and resistances from 0 to ∞ Ω.

Other examples of multimeters are the 8020A multimeter of the John Fluke Manufacturing Company, shown in Fig. 9.22, and the VOM *clamp-on* model of the Simpson Electric Company, shown in Fig. 9.23. The 8020A is a highly versatile, portable meter, which can be used to measure seven different functions with 26 ranges. With the Simpson clamp-on model, the problem of opening a circuit to measure an

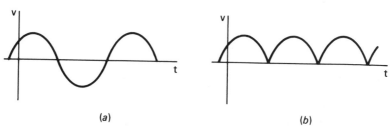

(a)   (b)

FIGURE 9.21   *(a) ac voltage and (b) its corresponding rectified voltage.*

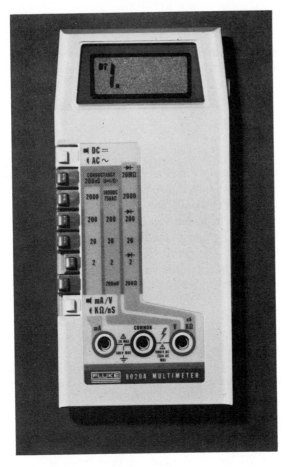

FIGURE 9.22 Multimeter (Courtesy, John Fluke Manufacturing Company).

ac current is eliminated. The probe is clamped around the wire carrying the current, and the magnetic field produced by the current causes the deflection of the pointer. The leads attached are for measuring voltage and resistance. The Fluke 8020A also has a clamp-on accessory available.

Electronic Multimeters: Other types of multimeters are electronic multimeters, which use electronic elements such as vacuum tubes or transistors in their construction. These may be classified as *analog* meters, which employ electromechanical movements and pointers to display the quantity being measured, and *digital* meters, which indicate their measurements in the form of a numerical display.

An example of an analog multimeter is the *electronic voltmeter* (EVM) or *transistor voltmeter* (TVM). Older meters used vacuum tubes rather than transistors and were called *vacuum-tube voltmeters* (VTVM). In every case, the electronic circuitry is used to rectify the signal (if it is ac) and amplify it before it causes the meter movement to deflect.

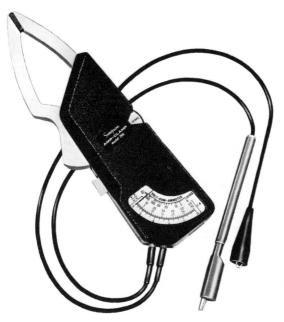

FIGURE 9.23 Clamp-on ac meter (Courtesy, Simpson Electric Company, Elgin, Illinois).

FIGURE 9.24 Digital multimeter (Courtesy, John Fluke Manufacturing Company).

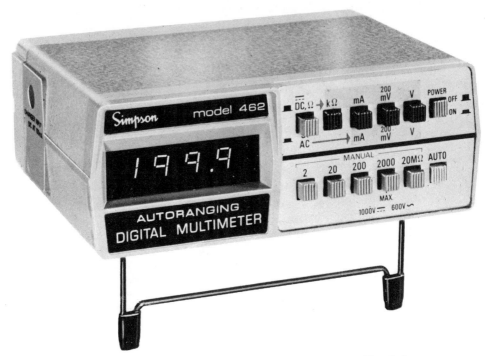

FIGURE 9.25   *Autoranging digital multimeter (Courtesy, Simpson Electric Company, Elgin, Illinois).*

Digital multimeters, such as the *digital voltmeter* (DVM), are much more accurate at low frequencies than are analog multimeters because they display a definite number, which eliminates errors in reading the scale. The number is produced electronically without the need for an electromechanical movement. A great advantage is that the current drawn by the meter is very small. The Fluke 8020A multimeter of Fig. 9.22 is an example of a digital device.

Another example of a digital multimeter is the 8600A model of the John Fluke Manufacturing Company, shown in Fig. 9.24. It is capable of displaying ac and dc voltage and current, as well as resistance, in a number of ranges. It may also be set to adjust automatically the decimal point in the readings of voltage and resistance.

Another example is the autoranging digital multimeter of the Simpson Electric Company, shown in Fig. 9.25. It, too, may be set to adjust automatically the decimal point in the display number.

Other Movements: In this chapter we have discussed in some detail only one electromechanical movement—that of d'Arsonval. There are others, such as the *iron-vane movement* and the *electrodynamometer movement,* that are also widely used. These are both current-detecting devices and are mostly used in ac meters.

The iron-vane movement consists of two bars of iron, one stationary and one movable, each of which is magnetized in the same way by the current being detected.

This causes a force of repulsion between the bars which is proportional to the current. Thus the movable bar can be made to rotate and move an attached pointer, as in the case of the d'Arsonval movement.

The electrodynamometer movement uses two coils, a stationary coil and a movable coil with an attached pointer. A fraction of the current being detected flows through each coil, producing a magnetic field around each coil. The interaction of these fields causes a rotation of the movable coil and pointer, which is proportional to the square of the current. Thus the scale may be calibrated to read dc or an average of the square of an alternating current. (The square of an alternating current is pulsating dc.) The dynamometer movement is often used in the construction of wattmeters, which we will consider later in a chapter on ac circuits.

9.6 SUMMARY

Meters for measuring electrical quantities, such as current, voltage, resistance, and power, may employ electromechanical movements with pointers or analog-to-digital converters with numerical displays. The electromechanical movements include the d'Arsonval movement, the iron-vane movement, and the electrodynamometer movement, which deflect a pointer by the rotation of a movable coil of wire or a movable bar of iron.

The d'Arsonval movement and an additional resistor may be used to construct an ammeter (to measure current), a voltmeter (to measure voltage), and an ohmmeter (to measure resistance). In the case of the ohmmeter, a battery is also required.

More versatile meters, called multimeters, may be constructed to measure, with a single movement, current, voltage, and resistance. Common types are volt-ohm-milliammeters (VOM); electronic multimeters, such as the transistor voltmeter (TVM) and the vacuum-tube voltmeter (VTVM); and digital multimeters, such as the digital voltmeter (DVM). The digital meters provide a numerical display of the measurement and use an analog-to-digital converter rather than an electromechanical movement.

PROBLEMS

9.1 A d'Arsonval movement rated 1 mA, 100 Ω is to be used to construct an ammeter by connecting a shunt resistor R_p, as shown in Fig. 9.8. Find R_p if the full-scale current of the ammeter is to be (a) 2 mA, (b) 10 mA, and (c) 1 A.

9.2 Repeat Problem 9.1 if the d'Arsonval movement is rated 500 μA, 500 Ω.

9.3 A 1-mA, 50-Ω movement is to be used, as in Fig. 9.8, to construct an ammeter capable of measuring up to 150 mA. Find R_p and the current flowing through the shunt when the meter reads half-scale current.

9.4 Find R_1, R_2, and R_3 in the multirange ammeter of Fig. 9.10 if the meter movement is rated 1 mA, 50 Ω, and the maximum currents that may be measured if the three ranges are 2 mA, 10 mA, and 100 mA.

9.5 Repeat Problem 9.4 if the movement is rated 500 μA, 1000 Ω.

9.6 The ammeter in the circuit shown has a 1-mA, 54-Ω movement, with $R_p = 6$ Ω. Find (a) its full-scale deflection, (b) its actual reading in the circuit, and (c) the percent error in its reading based on the 2 mA that would flow with an ideal ammeter. (*Suggestion:* Find the actual resistance seen by the source.)

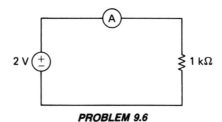

PROBLEM 9.6

9.7 Repeat Problem 9.6 if the meter movement is rated 1 mA, 49 Ω, and $R_p = 1$ Ω.

9.8 A 1-mA, 100-Ω movement is used as in Fig. 9.12 to construct a voltmeter. Find the multiplier resistance R_s if the full-scale voltage is (a) 10 V, (b) 50 V, and (c) 200 V.

9.9 Repeat Problem 9.8 if the movement is rated 10 mA, 4 Ω.

9.10 The multirange voltmeter of Fig. 9.14 has a 1-mA, 100-Ω movement and resistances of $R_1 = 9.9$ kΩ, $R_2 = 40$ kΩ, and $R_3 = 150$ kΩ. Find the three full-scale voltages the meter can measure.

9.11 The multirange voltmeter of Fig. 9.14 has a 1-mA, 50-Ω movement. Find R_1, R_2, and R_3 if the voltmeter has ranges up to 10 V, 50 V, and 100 V.

9.12 Find the ohms/volt rating of a voltmeter with $R_M = 40$ Ω, $R_s = 4960$ Ω, and (a) $V_{max} = 2$ V, (b) $V_{max} = 5$ V, and (c) $V_{max} = 10$ V.

9.13 Repeat Problem 9.12 if $R_s = 9960$ Ω.

9.14 The voltmeter in the circuit shown has a sensitivity of 20,000 Ω/V and a full-scale deflection of 50 V. Find its actual reading and the percent error in the reading based on the 40-V reading of an ideal voltmeter.

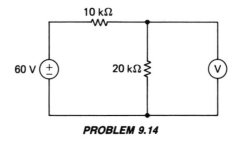

PROBLEM 9.14

9.15 Repeat Problem 9.14 if the sensitivity is 2000 Ω/V.

9.16 The series ohmmeter of Fig. 9.16(a) has a 1-mA, 100-Ω movement. Find V and R_s so that half-scale deflection occurs at $R_x = 3$ kΩ.

9.17 Repeat Problem 9.16 if half-scale deflection occurs at $R_x = 6$ kΩ.

9.18 Find R_s for a shunt ohmmeter with a movement rated 1 mA, 50 Ω and a battery voltage of 1.5 V.

9.19 Show that in the shunt ohmmeter of Fig. 9.19 the current through the meter movement is

$$I = \frac{R_x V}{R_s R_M + R_M R_x + R_x R_s}$$

Using this result find the value of R_x required for the ohmmeter of Problem 9.18 for a deflection of (a) one-fourth scale, (b) one-half scale, and (c) three-fourths scale.

9.20 The shunt ohmmeter of Fig. 9.19 uses a 1-mA, 100-Ω movement and is to have a half-scale deflection for $R_x = 95$ Ω. Find V and R_s.

10

CONDUCTORS AND INSULATORS

As we have noted in Chapter 1, a *conductor* is a material in which it is very easy to have electric current. In other words, a conductor easily *conducts* electricity because its resistance, in the practical case, is negligible compared to that of a resistor. In the ideal case a conductor has zero resistance.

At the other extreme, an *insulator* is a material in which it is very difficult, if not impossible, to have current. By comparison, a conductor may have a resistance of a fraction of an ohm, whereas that of an insulator may be millions of ohms.

The best conducting material is silver, as was also noted previously, but for economic reasons the most common conductor is copper wire. Aluminum wire is also often used, but aluminum is not as good a conductor as copper and therefore aluminum conductors must be larger to carry the same current as copper conductors.

Common examples of insulating materials are air, mica, glass, rubber, and ceramics. They have the common property of presenting an extremely high resistance to current.

Semiconductors are materials such as carbon, silicon, and germanium whose conducting abilities lie between those of conductors and insulators. That is, they have a high resistance compared to conductors and a low resistance compared to insulators. Solid-state electronic elements, such as diodes and transistors, contain silicon and germanium, and, of course, carbon is an important element in the construction of resistors.

In this chapter we will consider in some detail the properties of conductors

212

and insulators, with special emphasis on wire conductors and resistors. In doing so, we will introduce the concepts of *resistivity* and *temperature coefficient,* which may be used to illustrate the dependence of resistance on the kind of material used and the temperature in which the resistance is being used.

We will also briefly consider switches and fuses, which serve dual roles of conductors at some times and insulators at other times.

10.1 RESISTANCE

Every material, whether it be a conductor, a semiconductor, or an insulator, has the property of resistance. Its resistance may be very low, as in the case of a conductor, or it may be high, as in the case of an insulator. But in every case, electric charges passing through the material experience collisions, to some degree, with the atoms of the material, and thus encounter "electrical friction," or resistance.

Factors That Affect Resistance: If a conductor (or an insulator) has a uniform cross section, such as that shown in Fig. 10.1, then its resistance depends on several factors. The four principal ones are (1) the kind of material, (2) the length, (3) the cross-sectional area, and (4) the temperature.

Evidently, the resistance is affected by the material, since materials such as copper have many free electrons and thus can conduct more easily than materials such as carbon or mica, which have relatively few or almost no free electrons. In the case of length, it should be clear that the longer the conductor, the more collisions a charge moving through it would experience and thus the higher the resistance. Therefore, *resistance is directly proportional to length,* identified as l in Fig. 10.1. On the other hand, it is easier to drive charges through a large cross-sectional area than through a small one. Thus *resistance is inversely proportional to cross-sectional area,* shown as A in Fig. 10.1. (The smaller the area, the larger the resistance, and vice versa.)

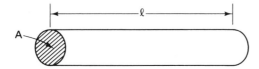

FIGURE 10.1 *Conductor with a uniform cross-sectional area.*

Resistance Formula: If the fourth factor, the temperature, is kept fixed, the foregoing discussion allows us to write a formula for the resistance of a conductor, such as that of Fig. 10.1. Since the resistance R is directly proportional to l and inversely proportional to A, we may write

$$R = \frac{\rho l}{A} \qquad (10.1)$$

where ρ (the lowercase Greek letter *rho*) is a constant of proportionality, called the *resistivity* of the material. Resistivity varies with the material and is relatively high for poor conductors (those with high resistance) and low for good conductors (those having low resistance).

Circular Mils: In the use of (10.1) we will depart from our tradition of using SI units exclusively, because conducting wire is still sold by the foot in the United States. Also, since solid wire comes in circular cross sections, we shall find it convenient to define a new unit called a *circular mil*. A *mil* is defined simply as $\frac{1}{1000}$ in. That is,

$$1 \text{ mil} = \frac{1}{1000} \text{ in.} = 0.001 \text{ in.} = 10^{-3} \text{ in.}$$

or

$$1000 \text{ mils} = 1 \text{ in.}$$

A square mil is an area of 1 mil $\times$ 1 mil, as shown in Fig. 10.2(a). By definition, a circle of diameter 1 mil has an area of 1 *circular mil* (abbreviated 1 cmil), as shown by the shaded area of Fig. 10.2(b). Thus the area in square mils of the circle of Fig. 10.2(b), given by

$$A = \frac{\pi d^2}{4} = \frac{\pi (1)^2}{4} = \frac{\pi}{4} \text{ sq mils}$$

is by definition 1 cmil. That is,

$$\frac{\pi}{4} \text{ sq mils} = 1 \text{ cmil} \tag{10.2}$$

or

$$1 \text{ sq mil} = \frac{4}{\pi} \text{ cmil} \tag{10.3}$$

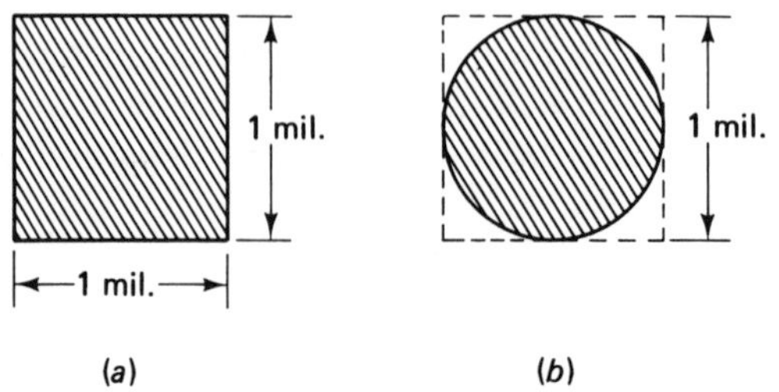

FIGURE 10.2   (a) A square mil, and (b) a circular mil.

(a)   (b)

A circle of diameter d mils has an area of $\pi d^2/4$ sq mils, which by (10.2) is d^2 cmil. Thus the area of a circle in cmil is simply its diameter (in mils) squared. Therefore, to find the area in cmil, given the diameter in inches, change the diameter to mils, by moving the decimal point three places to the right and square the result.

Example 10.1: Find the area in cmils of a circle with diameter 0.025 in.

Solution: The diameter is

$$d = 0.025 \text{ in.} = 25 \text{ mils}$$

Therefore, the area is

$$A = d^2 = (25)^2 = 625 \text{ cmil}$$

Units of Resistivity: If the conductor has a cross-sectional area A measured in cmil, a length l measured in feet, and a resistance R measured in ohms, the units of resistivity ρ may be found from (10.1). Solving for ρ, we have

$$\rho = \frac{AR}{l} \tag{10.4}$$

so that its units are cmil–Ω/ft.

If the conductor is not round, it is more convenient to use units other than cmil. In the SI, A is measured in square meters, l is in meters, and R is in ohms. Therefore, by (10.4) the SI units for ρ are Ω–m²/m, or ohm-meters (Ω–m). If A is in square centimeters, l is in centimeters, and R in ohms, the units of ρ are ohm-centimeters (Ω–cm).

To illustrate the difference in the resistivities of conductors, semiconductors, and insulators, a number of resistivities are compiled in Table 10.1. The first three entries (silver, copper, and aluminum) are conductors, the next three (carbon, germa-

TABLE 10.1 *Selected resistivities.*

Material	Resistivity (Ω–cm at 20° C)
Silver	1.6×10^{-6}
Copper	1.7×10^{-6}
Aluminum	2.8×10^{-6}
Carbon	4×10^{-3}
Germanium	65
Silicon	55×10^3
Glass	17×10^{12}
Rubber	10^{18}

nium, and silicon) are semiconductors, and the last two (glass and rubber) are insulators. The resistivities are given at the room temperature of 20°C (68°F), and increase, of course, as we proceed through the table from conductors to insulators.

Example 10.2: Find the resistance of a piece of copper that has a cross-sectional area of 5 sq cm and a length of 200 cm.

Solution: By Table 10.1 the resistivity is $\rho = 1.7 \times 10^{-6}$ Ω–cm, so that by (10.1) we have

$$R = \frac{\rho l}{A} = \frac{(1.7 \times 10^{-6} \text{ } \Omega\text{–cm})(200 \text{ cm})}{5 \text{ cm}^2} = 6.8 \times 10^{-5} \text{ } \Omega$$

Note that all the units cancel except ohms.

PRACTICE EXERCISES

10-1.1 Find the area in cmil of a circle of diameter (a) 0.002 in., (b) 0.15 in., and (c) 0.1 ft.   *Ans.* (a) 4, (b) 22,500, (c) 1.44×10^6

10-1.2 Convert to cmil the area (a) 20 sq mils, (b) 20 sq in., and (c) 20 sq ft. (*Suggestion:* 1 sq in. $= 10^3 \times 10^3 = 10^6$ sq mils.)
                                  Ans. (a) $80/\pi = 25.46$, (b) 2.546×10^7, (c) 3.67×10^9

10-1.3 A material has a cross-sectional area of 2 sq cm and a length of 1000 cm. Find its resistance in ohms if the material is (a) copper, (b) silicon, and (c) glass.
                                  Ans. (a) 8.5×10^{-4}, (b) 2.75×10^7, (c) 8.5×10^{15}

10.2 WIRE CONDUCTORS

The principal use of conductors is for transmitting electricity in the form of current from one electrical element to another in a circuit. The most common conductor for this purpose is round metal wire, and the most common material used is copper. As an example, the circuit of Fig. 10.3 consists of a 240-Ω resistor connected to a 120-V source by means of two conductors, which may be two pieces of round copper wire.

    Depending on their lengths and cross-sectional areas, the wire conductors of Fig. 10.3 may have a total resistance of, say, 0.5 Ω. Thus the conductors are not ideal (that is, their resistance is not zero), but in most cases they may be approximated

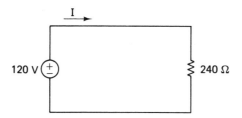

FIGURE 10.3 *Circuit consisting of two elements connected by conductors.*

by ideal conductors, since 0.5 Ω is almost negligible compared to the 240 Ω of the resistor. In the actual case the source sees a resistance R equal to that of the resistor plus that of the conductors, since they are all in series. Therefore, we have

$$R = 240 + 0.5 = 240.5 \ \Omega$$

and the current in the circuit is

$$I = \frac{120}{240.5} = 0.499 \ A$$

This compares to the current of $120/240 = 0.5$ A that would exist with ideal conductors.

From the foregoing example we see that often the resistance of a conductor is small enough to be neglected. There are cases, however, where the conductor's resistance must be taken into account. For this reason we need a relatively easy method of calculating the resistance. This is readily done by means of (10.1), and the work is simplified if the cross-sectional area of the conductor is given in cmil and we have the resistivity ρ in cmil-Ω/ft.

Some common conducting materials are listed with their resistivities in cmil–Ω/ft in Table 10.2. They are ranked in the order of increasing resistivity. Also included for comparison purposes is the semiconductor material carbon.

TABLE 10.2 *Resistivities ρ (cmil-Ω/ft at 20° C).*

Material	ρ
Silver	9.9
Copper	10.4
Aluminum	17.0
Tungsten	33.0
Nickel	47.0
Iron	74.0
Carbon	21,000.0

Example 10.3: Find the resistance at 20°C of a round copper wire of diameter 40.3 mils and length 2000 ft.

Solution: The cross-sectional area is $A = (40.3)^2$ cmil, and by Table 10.2 the resistivity is $\rho = 10.4$ cmil–Ω/ft. Therefore, by (10.1) we have

$$R = \frac{\rho l}{A} = \frac{(10.4 \text{ cmil–}\Omega\text{/ft})(2000 \text{ ft})}{(40.3)^2 \text{ cmil}} = 12.8 \ \Omega$$

We note that all the units cancel except ohms.

Standard Wire Gage Sizes: The work of computing the resistance of copper wire may be greatly simplified by means of Table 10.3, which lists the standard wire sizes in the system known as the *American Wire Gage* (AWG). The gage numbers 1, 2, 3, and so on, relate to the diameter of the wire, with higher gage numbers indicating thinner wire. That is, the diameter decreases, and thus the resistance increases, as the gage number increases.

The area in cmil halves approximately for every three gage sizes. Therefore, the resistance doubles approximately for every three gage sizes. For example, No. 14 wire has 2.525 ohms per thousand feet and No. 17 wire has 5.064 ohms per thousand feet.

Example 10.4: To illustrate the use of Table 10.3, let us find the resistance of 2000 ft of No. 18 wire at 20°C.

Solution: From Table 10.3 we see that the resistance per thousand feet is 6.385 Ω. Therefore, the resistance of 2000 ft is

$$R = 2 \times 6.385 = 12.8 \ \Omega$$

We note that this is the same answer as that obtained in Example 10.3. From Table 10.3 we see that No. 18 wire has a diameter of 40.3 mils, so that the conductors considered in Examples 10.3 and 10.4 are the same; the answers should therefore be the same. These two examples illustrate the relative ease of using the wire gage table as opposed to using the formula for resistance.

Example 10.5: Repeat Example 10.4 if the length of the wire is 750 ft.

Solution: Since 750 ft is 750/1000 of 1000 ft, we have

$$R = \frac{750}{1000}(6.385) = 4.8 \ \Omega$$

TABLE 10.3 *American Wire Gage (AWG) sizes (solid round copper wire).*

Gage No.	Diameter (mils)	Area (cmil)	Ohms per 1000 ft at 20° C
1	289.3	83,690	0.1239
2	257.6	66,370	0.1563
3	229.4	52,640	0.1970
4	204.3	41,740	0.2485
5	181.9	33,100	0.3133
6	162.0	26,250	0.3951
7	144.3	20,820	0.4982
8	128.5	16,510	0.6282
9	114.4	13,090	0.7921
10	101.9	10,380	0.9989
11	90.74	8,234	1.260
12	80.81	6,530	1.588
13	71.96	5,178	2.003
14	64.08	4,107	2.525
15	57.07	3,257	3.184
16	50.82	2,583	4.016
17	45.26	2,048	5.064
18	40.30	1,624	6.385
19	35.89	1,288	8.051
20	31.96	1,022	10.15
21	28.46	810.1	12.80
22	25.35	642.4	16.14
23	22.57	509.5	20.36
24	20.10	404.0	25.67
25	17.90	320.4	32.37
26	15.94	254.1	40.81
27	14.20	201.5	51.47
28	12.64	159.8	64.90
29	11.26	126.7	81.83
30	10.03	100.5	103.2
31	8.928	79.70	130.1
32	7.950	63.21	164.1
33	7.080	50.13	206.9
34	6.305	39.75	260.9
35	5.615	31.52	329.0
36	5.000	25.00	414.8
37	4.453	19.83	523.1
38	3.965	15.72	659.6
39	3.531	12.47	831.8
40	3.145	9.88	1049.0

In typical electronic circuits where the current is on the order of milliamperes, suitable conductors are about No. 22 gage. This size may safely carry up to 1 A of current without heating. House wiring, where the currents may be 5 to 15 A, should be done with No. 14 gage (or lower) wire.

PRACTICE EXERCISES

10-2.1 Find the resistance of 4000 ft of No. 10 copper wire at 20°C.   *Ans.* 3.996 Ω

10-2.2 Find the resistance at 20°C of No. 31 copper wire with a length of (a) 6000 ft, (b) 500 ft, and (c) 25 ft.   *Ans.* (a) 780.6 Ω, (b) 65.05 Ω, (c) 3.25 Ω

10-2.3 Find the current I in the circuit of Fig. 10.3 if the two conductors joining the two elements are No. 22 copper wire each with lengths of 600 ft. (The temperature is 20°C.)   *Ans.* 0.46 A

10.3 EFFECT OF TEMPERATURE

For most conductors, an increase in temperature is accompanied by an increase in resistance. This is because higher temperatures cause more molecular movement within the conductor, which in turn results in more collisions with the charges constituting the current. In this section we will see how temperature affects the resistance of certain conductors, and obtain a resistance formula that takes temperature into account.

Resistance–Temperature Sketch: If the resistance R of a conductor is plotted versus temperature in °C, the result will resemble the solid-line graph of Fig. 10.4. The graph is a straight line over most of the temperature range, but it becomes nonlinear for very high and very low temperatures. The dashed line represents the resistance curve if it were a straight line and provides a good approximation to the actual curve for most practical temperature ranges.

    The resistance is zero at absolute zero temperature (-273°C), as shown, because that is the temperature at which all the molecular motion ceases. The dashed-straight-line approximation reaches zero at T_0, called the *inferred absolute zero*. The slope of the straight line at the point where $R = R_1$ and $T = T_1$ is the *rise* R_1 over the

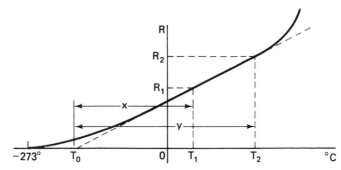

FIGURE 10.4 *Resistance-temperature curve.*

run x. Similarly, at $R = R_2$ and $T = T_2$, the slope is the rise R_2 over the run y. Since the slopes are the same, we have

$$\frac{R_1}{x} = \frac{R_2}{y} \tag{10.5}$$

From Fig. 10.4, we see that $x = T_1 - T_0$ and $y = T_2 - T_0$, so that (10.5) becomes

$$\frac{R_1}{T_1 - T_0} = \frac{R_2}{T_2 - T_0}$$

from which we have

$$R_2 = \frac{T_2 - T_0}{T_1 - T_0} R_1 \tag{10.6}$$

Calculation of Resistance: From (10.6) we see that if the inferred absolute zero T_0 is known, and we have a known value of resistance R_1 at a temperature T_1, we may find the resistance R_2 at a temperature T_2. In the case of copper wire the value of T_0 is known to be −234.5°C, so that for copper (10.6) becomes

$$R_2 = \frac{234.5 + T_2}{234.5 + T_1} R_1 \tag{10.7}$$

Thus if $R = R_1$ is known at $T = T_1$, we may find the resistance R_2 at a temperature T_2.

Example 10.6: Find the resistance of 1000 ft of No. 14 copper wire at $T = 50°C$.

Solution: At the temperature $T_1 = 20°C$ the resistance is $R_1 = 2.525$ Ω, from Table 10.3. Therefore, if R_2 is the resistance at the temperature $T_2 = 50°C$, then by (10.7) we have

$$R_2 = \frac{234.5 + 50}{234.5 + 20}(2.525) = 2.823 \text{ Ω}$$

Temperature Coefficient of Resistance: By adding and subtracting T_1 to the numerator of the right member of (10.6), we obtain

$$R_2 = \frac{T_1 - T_1 + T_2 - T_0}{T_1 - T_0} R_1$$

or

$$R_2 = \left(\frac{T_1 - T_0}{T_1 - T_0} + \frac{T_2 - T_1}{T_1 - T_0}\right) R_1$$

$$= \left(1 + \frac{T_2 - T_1}{T_1 - T_0}\right) R_1 \tag{10.8}$$

Next we define the quantity α (the lowercase Greek letter *alpha*) by

$$\alpha = \frac{1}{T_1 - T_0} \tag{10.9}$$

so that (10.8) becomes

$$R_2 = R_1 \left[1 + \alpha(T_2 - T_1)\right] \tag{10.10}$$

The quantity α is called the *temperature coefficient of resistance*, and is dependent on the conductor material and the temperature T_1 (since T_0 is fixed for a given material). Values of α are listed in Table 10.4 for various materials and for $T_1 = 20°C$.

TABLE 10.4   *Temperature coefficient α for various materials at 20°C.*

Material	α
Silver	0.0038
Copper	0.00393
Aluminum	0.00391
Tungsten	0.005
Nickel	0.006
Iron	0.0055
Carbon	−0.0005

We note that carbon has a negative temperature coefficient. This is typical of semiconductors and indicates that the resistance will drop with increasing temperature.

In the case of copper, if $T_1 = 20°C$, then $\alpha = 0.00393$. Therefore, if we denote R_1, the resistance at 20°C, by R_{20}, and R_2 and T_2 by R and T, (10.10) becomes

$$R = R_{20}[1 + 0.00393(T - 20)] \tag{10.11}$$

Thus the resistance R of copper wire at any temperature T is given in terms of R_{20}, the temperature at 20°C, which is tabulated in Table 10.3.

Example 10.7:   Solve Example 10.6 by using (10.11).

Solution: Since $R_{20} = 2.525 \ \Omega$ and $T = 50°C$, we have, as before,

$$R = 2.525[1 + 0.00393(50 - 20)]$$
$$= 2.823 \ \Omega$$

PRACTICE EXERCISES

10-3.1   Find the resistance of 4000 ft of No. 20 copper wire at (a) 20°C, (b) 100°C, and (c) 0°C.   *Ans.* (a) 40.6 Ω, (b) 53.36 Ω, (c) 37.41 Ω

10-3.2   A material has a resistance of 10 Ω at 20°C. Find its resistance at 70°C if the material is (a) silver, (b) tungsten, and (c) carbon.   *Ans.* (a) 11.9 Ω, (b) 12.5 Ω, (c) 9.75 Ω

10.4   INSULATORS

As we have noted, insulators, or *dielectric materials,* have extremely high resistance, and will not conduct current for ordinary voltages. Insulators have two primary functions. One is to block current flow between conductors by physically isolating them. For example, if there is a possibility that two conductors in a circuit may touch each other at a point other than a node, a solid insulating material is used to separate the conductors physically. The two conductors in a lamp wire, for instance, are wrapped in insulating material to keep them separated. The other principal function of an insulator is to store an electric charge when voltage is applied. We shall see how this is done in Chapter 11 when we consider capacitors, which are simply two conductors separated by a dielectric, or insulator.

Dielectric Strength: Common insulating materials are air, rubber, glass, mica, and ceramics, and, of course, some of these are better than others. No matter how good an insulator is, however, a high-enough voltage can be applied to rupture the physical structure of the material and cause it to conduct. The voltage at which this breakdown in the internal structure occurs is called the *breakdown voltage,* or *dielectric strength,* of the insulator, and may be used as a measure of how well it insulates.

Some common insulators are listed with their average dielectric strengths in Table 10.5. The dielectric strengths are given in kilovolts per centimeter (kV/cm), and may be converted to volts/mil by multiplying them by 2.54.

TABLE 10.5 *Dielectric strengths of common insulators.*

Material	Average Dielectric Strength (kV/cm)
Air	30
Porcelain	70
Bakelite	150
Rubber	270
Teflon	600
Glass	900
Mica	2000

Example 10.8: Find the breakdown voltage of air in volts/mil. Also find the voltage required to break down an air dielectric between two conductors $\frac{1}{8}$ in. apart.

Solution: The breakdown voltage in kV/cm is 30, by Table 10.5. Therefore, in V/mil we have

$$V/mil = 2.54 \times 30 = 76.2$$

Since 1 in. = 1000 mils, then $\frac{1}{8}$ in. = 125 mils. Therefore, the voltage V required to break down $\frac{1}{8}$ in. of dielectric is 125 times that required to break down 1 mil. Thus we have

$$V = 125 \times 76.2 = 9525 \text{ V}$$

That is, a voltage of 9525 V or more will cause an arc across an air gap of length $\frac{1}{8}$ in.

PRACTICE EXERCISES

10-4.1 The insulation between two conductors is $\frac{1}{4}$ in. thick. Find the lowest voltage that will break down the insulation if the dielectric material is (a) air, (b) rubber, and (c) mica.   *Ans.* (a) 19.05 kV, (b) 171.45 kV, (c) 1.27 MV

10-4.2 Repeat Exercise 10-4.1 if the insulation is $\frac{1}{16}$ in. thick.
     Ans. (a) 4.76 kV, (b) 42.86 kV, (c) 317.5 kV

10.5 SWITCHES AND FUSES

Two common devices found in electric circuits that function as both ideal conductors (short circuits) and ideal insulators (open circuits), depending on their position or state, are *switches* and *fuses,* which we consider in this section. Another such device is a *circuit breaker,* which performs the function of a fuse but can be reset and used repeatedly.

Make and Break: We have already used switches in a few of the circuits considered in earlier chapters, and have used the fact, which everyone knows, that a switch may be closed, or in the *make* position, or it may be open, or in the *break* position. An example is the switch of Fig. 10.5(a) and its circuit symbol of Fig. 10.5(b), both of which are open. Closing the switch, as indicated by the arrow in Fig. 10.5(a), connects the two conductors shown and thus changes the open circuit to a short circuit.

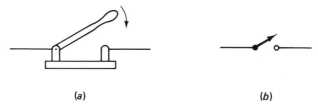

(a)   *(b)*

FIGURE 10.5   *(a) Switch and (b) its circuit symbol.*

Poles and Throw: The part of the switch that is moved to open or close a circuit is called the switch *pole*. A switch may be *single-pole, double-pole,* or have any number of poles. The example of Fig. 10.5 is a single-pole switch.

If each contact of a switch opens or closes only one circuit, as in Fig. 10.5, the switch is a *single-throw* switch. A *double-throw* switch is one that opens one circuit while simultaneously closing another. We may therefore have a single-pole, single-throw (SPST) switch; a single-pole, double-throw (SPDT) switch; a double-pole, single-throw (DPST) switch; a double-pole, double-throw (DPDT) switch; or miltipole, single- and double-throw switches. Examples of the four most common types are shown in Fig. 10.6. Photographs of actual SPST switches with 2 to 10 available positions are shown in Fig. 10.7, and a DPDT switch is shown in Fig. 10.8. In the latter case, red and orange conductors (R and O) are connected to

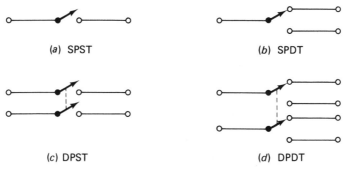

(a) SPST   *(b)* SPDT

(c) DPST   *(d)* DPDT

FIGURE 10.6   *Circuit symbols for switches.*

FIGURE 10.7   *SPST switches with 2 to 10 positions (Courtesy, Grayhill, Inc.).*

yellow or green (Y or G) and brown or black (BR or BL) conductors by the switching action.

Toggle Switch: A switch in which a knob moves through a small arc to open or close the contacts of a circuit is called a *toggle* switch. The contacts are closed or opened suddenly and a tight connection is made in the closed position. Lighting switches in most houses are toggle switches, for this reason. A toggle switch in the off and on positions is shown in Fig. 10.9, and a photograph of four actual SPDT toggle switches is shown in Fig. 10.10.

Other Switches: The switching action may be accomplished in a number of ways other than by the on–off knob of the toggle switch. For example, we may have *pushbutton* switches, as shown in Fig. 10.11, or in Fig. 10.12 for multiple contacts. We may also have *slide* switches, where the circuit is closed by the motion of a

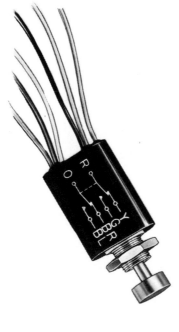

FIGURE 10.8 *DPDT switch (Courtesy, Grayhill, Inc.).*

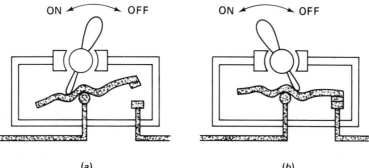

ON ◄────► OFF   ON ◄────► OFF

(a)   (b)

FIGURE 10.9 *(a) Off and (b) on positions of a toggle switch.*

sliding contact, and *mercury* switches, which contain a pool of mercury that connects two conductors when the vial containing the mercury is horizontal, and which breaks the contact when the vial is tilted.

A *rotary* switch is one that makes or breaks circuits as its shaft is rotated. A knob is used to turn the shaft of the switch, which turns an electrical contact. This rotating contact makes connections with conducting terminals which are located on one or more insulated *wafers* or *decks* that surround the shaft. Thus the rotary switch has many poles, which may be set to many different positions. Three rotary switches are shown in Fig. 10.13, and a slightly different type is shown in Fig. 10.14.

FIGURE 10.10 *SPDT toggle switches (Courtesy, Grayhill, Inc.).*

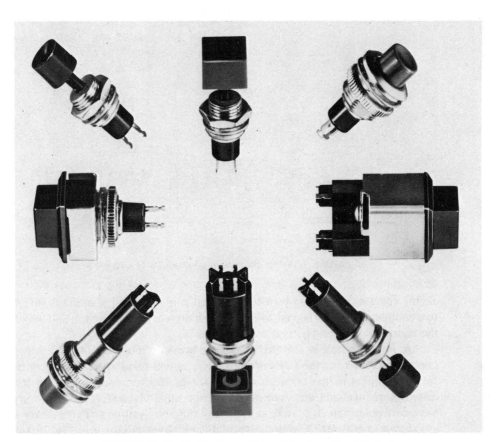

FIGURE 10.11 *Pushbutton switches (Courtesy, Grayhill, Inc.).*

FIGURE 10.12 *Pushbutton switches with multiple contacts (Courtesy, Centralab Electronics Division, Globe-Union, Inc.).*

FIGURE 10.13 *Rotary switches with three, one, and two wafers (Courtesy, Centralab Electronics Division, Globe-Union, Inc.).*

Fuses: Another element that acts first as a conductor (short circuit) and then as an insulator (open circuit) is a fuse, the circuit symbol for which is shown in Fig. 10.15. The purpose of the fuse is to conduct current when the circuit is performing normally and to act as an insulator when there is a large surge of current that might damage the circuit, due to an overload or a short circuit.

A fuse is basically a fine wire that heats up and melts, thereby breaking the circuit, when its maximum current rating is exceeded. A typical fuse is the *glass-cartridge* type of Fig. 10.16, which is used, for example, in circuits of automobiles. External connections are made to the metal contacts which are connected to the wire element inside the glass. The fuse "blows," of course, when the element melts.

Fuses are rated from a small fraction of an ampere to many hundreds of amperes. They are used to protect circuits in such diverse places as automobiles, houses, and

FIGURE 10.14 *Rotary switch (Courtesy, Grayhill, Inc.).*

FIGURE 10.15 *Circuit symbol of a fuse.*

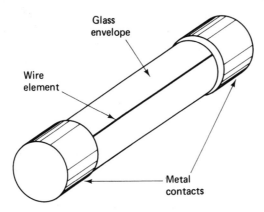

Glass envelope

Wire element

Metal contacts

FIGURE 10.16 *Glass-cartridge fuse.*

TV sets. A typical house fuse may be rated at 15 A, while that protecting the high-voltage circuit in a television set may be of the glass-cartridge type with a $\frac{1}{4}$-A rating.

Circuit Breakers: When a fuse blows, it is destroyed because its wire element is burned out. A device that operates like a fuse but may be reset and used again is a *circuit breaker.* The breaker is a switch which opens by means of either a magnetic field or by a thermal action when there is a large current. In the first case, the field magnetizes an iron bar, which is drawn into a coil carrying the current. This action opens the switch. The thermal-type breaker has a spring that expands with the heat of the current, tripping open the circuit.

PRACTICE EXERCISES

10-5.1 The fuse in the circuit shown is rated at 5 A. Find the minimum voltage V of the source that will cause the fuse to blow.   *Ans.* 150 V

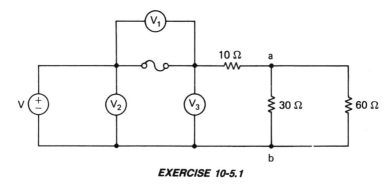

EXERCISE 10-5.1

10-5.2 If $V = 100$ V in the circuit of Exercise 10-5.1, find the readings of voltmeters V_1, V_2, and V_3. (*Suggestion:* The fuse is good in this case and acts like a short circuit.)   *Ans.* 0 V, 100 V, 100 V

10-5.3 Repeat Exercise 10-5.2 if $V = 200$ V. (*Suggestion:* The fuse has blown and acts like an open circuit.)   *Ans.* 200 V, 200 V, 0 V

10.6 SUMMARY

Materials may be classified as conductors, semiconductors, or insulators, in accordance with how easily they conduct electric current. Conductors have small, and often negligible, resistance and readily conduct current. Insulators have extremely high resistance and are used to prevent the flow of current. In between, semiconductors conduct current but not nearly as easily as conductors, although much more easily than insulators.

For a given shape and type of material, its resistance may be calculated. If the cross-sectional area is uniform, the resistance is proportional to the length of the material and inversely proportional to its cross-sectional area. The constant of proportionality is the resistivity, which is fixed for a given material. In the case of copper conductors, a wire-gage table is available for making quick computations of resistance. The resistance of a conductor increases with temperature in a manner dependent on its temperature coefficient, which is also fixed for a given material.

Switches, fuses, and circuit breakers act as insulators at certain times (open circuits) and as conductors at other times (short circuits). Switches are opened or closed manually, or by some deliberate action, whereas fuses and circuit breakers open automatically in reaction to an abnormally high current that could damage the circuit.

PROBLEMS

10.1 Find the cross-sectional area in circular mils of a wire having a diameter of (a) 0.017 in., (b) 0.001 ft, (c) 0.02 cm, and (d) 51 mils.

10.2 Find the diameter in inches of a circle having an area in circular mils of (a) 1225, (b) 174.24, (c) 11.56, and (d) 12,100.

10.3 Find the resistance of a conductor with a cross-sectional area of 10 sq cm and a length of 50,000 cm if the material is (a) silver, (b) copper, and (c) carbon. The temperature is 20°C.

10.4 Find the length in cm of a conductor having a resistance of 0.5 Ω and a cross-sectional area of 0.2 cm² if the material is (a) copper and (b) aluminum. The temperature is 20°C.

10.5 Find the resistance of a copper conductor at 20°C with a 1 in. × 4 in. rectangular cross section and a length of 1000 ft. (*Suggestion:* Note that 1 sq mil = 4/π cmil.)

10.6 Solve Problem 10.5 if the material is aluminum.

10.7 Find the resistance at 20°C of 5000 ft of (a) No. 12 copper wire, (b) No. 24 copper wire, and (c) No. 40 copper wire.

10.8 Find the resistance at 20°C of 472 ft of (a) No. 6 copper wire, (b) No. 19 copper wire, and (c) No. 38 copper wire.

10.9 A 100-V battery is connected to a 250-Ω resistor by two No. 10 copper conductors, each of which is 100 ft long. Find the current that flows at a temperature of 20°C.

10.10 Solve Problem 10.9 if the conductors are both No. 40 copper wire.

10.11 Find the resistance at 50°C of the conductors of Problem 10.7.

10.12 Find the resistance at 100°C of the conductors of Problem 10.7.

10.13 A material has a resistance of 20 Ω at 20°C. Find its resistance at 100°C if the material is (a) copper, (b) aluminum, and (c) nickel.

10.14 Find the resistance at 0°C of the conductors of Problem 10.13.

10.15 If V = 120 V in the circuit of Exercise 10-5.1, find the voltmeter readings V_1, V_2, and V_3. The fuse is rated, as before, at 5 A.

10.16 Solve Problem 10.15 if a short is placed across points a and b in the circuit.

11

CAPACITORS

In previous chapters we have concentrated on resistive circuits: that is, circuits containing only resistors and sources. The terminal characteristics of the circuit elements were simple expressions, such as a number like $v = 6$ V in the case of a source, or $v = Ri$ (Ohm's law) in the case of a resistor. The circuit equations consequently were relatively simple algebraic equations and could be solved by a number of methods for the currents and voltages.

Resistive circuit equations are relatively simple because for a resistor the value of the voltage is determined by the value of the current (and vice versa). There are two other very important circuit elements, *capacitors* and *inductors,* for which one of the two quantities (voltage and current) is determined by the *rate of change* of the other. Thus circuits with capacitors and inductors will not be governed by the simple algebraic equations of resistive circuits. In contrast to resistive circuits, however, these more general circuits may *store* energy at one time to be retrieved and used at a later time. In other words, circuits with capacitors and inductors have *memory* (the stored energy can be recalled), whereas resistive circuits can produce only instantaneous results.

As an example, an electronic flash attachment on a camera can store the energy, by means of a capacitor, to produce at a later time the light needed for the picture. As we shall see, the energy is stored in the capacitor by means of a voltage, which is held at a fixed value until it is needed to provide power for the light.

In this chapter we will consider capacitors, their associated property of *capacitance,* their voltage–current relationship, the energy they store, and their series and parallel connections in a circuit. We will also briefly discuss the concept of an *electric field,* in which the capacitor's energy is stored, and consider some practical aspects of various types of capacitors and their physical construction. Inductors will be considered in Chapter 14 after we have discussed the concept of a *magnetic field.*

11.1 DEFINITIONS

A *capacitor* is a two-terminal device consisting of two conducting bodies (conductors) separated by a dielectric material (an insulator). An example is the *parallel-plate* capacitor of Fig. 11.1, for which the conducting bodies are two flat, rectangular conductors, and the dielectric material is the air between them. The voltage *v* across the terminals at the left is also, by KVL, the voltage across the plates, as indicated.

The applied voltage *v* causes a movement of positive charge *q* from the bottom plate to the top plate of the capacitor, as indicated. If the dielectric is perfect (a perfect insulator), no charges can flow between the two plates. Therefore, the transfer of charge takes place through the external wire. The top plate thus has a charge of +*q* deposited on it and the bottom plate, having lost the +*q* charge, has a charge of −*q*. (In the actual case of electron current, *q* coulombs of electrons have been moved from the top plate, leaving it positively charged, to the bottom plate, giving it a negative charge.) The capacitor thus has been *charged* to a voltage *v,* which will remain if the terminals are opened.

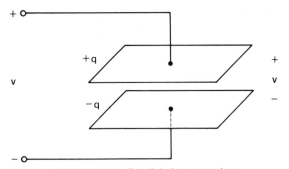

FIGURE 11.1 *Parallel-plate capacitor.*

Capacitance: The amount of charge *q* that is placed on the conducting bodies of a capacitor (+*q* on one body and −*q* on the other) is proportional to the voltage *v* that is applied. The more voltage we have, the more charge is moved. Therefore, we may write

$$q = Cv \qquad (11.1)$$

where *C* is a constant of proportionality, called the *capacitance* of the device. Since by (11.1) we have

$$C = \frac{q}{v} \qquad\qquad (11.2)$$

we see that the units of capacitance are coulombs/volt (C/V). The unit of 1 C/V is known in the SI as the *farad* (abbreviated F), in honor of the famous British physicist Michael Faraday, whom we have referred to on previous occasions.

Capacitance is a measure of the capacitor's ability to store charge on its plates. That is, if the capacitance is 1 F, a 1-V potential difference across the plates results in 1 C of charge being placed on the plates. The farad, however, is much too large in practical applications, because typical capacitors store charges in the range of microcoulombs or less. Therefore, common values of capacitance are in microfarads (μF) and picofarads (pF). Their relations to farads are

$$1 \ \mu F = 10^{-6} \ F$$

$$1 \ pF = 10^{-12} \ F$$

Example 11.1: Find the charge stored in a 2-μF capacitor if the voltage across its plates is 25 V.

Solution: By (11.1) we have

$$q = Cv = (2 \ \mu F)(25 \ V) = 50 \ \mu C$$

Electric Field: As we noted in Section 1.2, two charged bodies exert forces on each other—forces of attraction if the bodies have unlike charges and forces of repulsion if they have like charges. It may help to understand these forces if we consider that an *electric field*, consisting of electric lines of force, or *flux*, exists in the region surrounding the charges. For example, in Fig. 11.2(a) the lines in the field are shown emanating from two positive charges which repel each other. In Fig. 11.2(b) the charges are unlike and attract each other. This is illustrated by the lines of force, which originate on the positive charge and terminate on the negative charge.

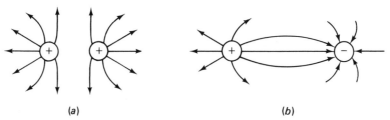

(a)   (b)

FIGURE 11.2   *Electric field of (a) like charges and (b) unlike charges.*

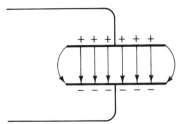

FIGURE 11.3 *Electric field in a parallel-plate capacitor.*

A capacitor is constructed so that the flux lines, rather than extending in all directions, are concentrated between the capacitor plates. For example, a cross section of a parallel-plate capacitor is shown in Fig. 11.3 with its top plate positively charged relative to its bottom plate. Thus, except for the *fringing* at the edges of the capacitor, the flux is kept between the two plates.

Capacity to Store Energy: Capacitance may be thought of as the capacity to store charge, or equivalently, to store energy in the electric field between its plates. In the case of the parallel-plate capacitor, the strength of the electric field, and therefore the energy stored and the capacitance, depends on (1) the material used for the dielectric, (2) the area of the plates, and (3) the distance between the plates. In the first case, it is easier to establish an electric field in some materials than it is in others, as we shall see. Second, the larger the area of the plates, the more lines of flux are possible. Thus *capacitance is directly proportional to the area A of the plates.* Finally, the closer the charges are, the more intense is the field, and the higher the capacitance. Therefore, *capacitance is inversely proportional to the distance d between the plates.*

Dielectric Constant: The foregoing discussion allows us to obtain a formula for the capacitance of a parallel-plate capacitor in terms of its physical characteristics, illustrated in Fig. 11.4. Since the capacitance C is directly proportional to the area A and inversely proportional to the distance d between the plates, we may write

$$C = \frac{\epsilon A}{d} \tag{11.3}$$

where ϵ (the lowercase Greek letter *epsilon*) is a constant of proportionality that

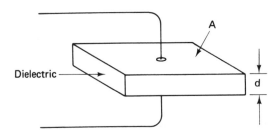

FIGURE 11.4 *Parallel-plate capacitor with area A and width d.*

depends on the dielectric. Since ϵ is a measure of the ease with which the dielectric *permits* the establishment of the electric field (the higher ϵ, the higher the capacitance), it is called the *permittivity* of the material.

Solving for ϵ in (11.3), we have

$$\epsilon = \frac{Cd}{A} \qquad (11.4)$$

so that in the SI, the units of permittivity are farad-meter/(meter)², or farad/meter (F/m). As an example, the permittivity of a vacuum is designated ϵ_0, and is given by

$$\epsilon_0 = 8.85 \times 10^{-12} \text{ F/m} = 8.85 \text{ pF/m} \qquad (11.5)$$

The permittivity ϵ_0 of a vacuum may be used as a yardstick for measuring ϵ for any other material. This is done by defining a quantity K, known as the *relative permittivity* or *dielectric constant,* by the ratio

$$K = \frac{\epsilon}{\epsilon_0} \qquad (11.6)$$

Thus the permittivity ϵ of a material is given by

$$\epsilon = K\epsilon_0 \qquad (11.7)$$

where K is the dielectric constant of the material. Average values of dielectric constants for some common materials are given in Table 11.1.

TABLE 11.1 *Average values of dielectric constants K.*

Dielectric	$K = \epsilon/\epsilon_0$
Vacuum	1.0
Air	1.0006
Teflon	2.0
Polystyrene	2.5
Mica	5.0
Porcelain	6.0
Bakelite	7.0
Barium-strontium titanate (ceramic)	7500.0

Example 11.2: A parallel-plate capacitor has a plate area $A = 0.1$ m², a distance between plates of $d = 3$ mm, and its dielectric is air. Find its capacitance.

Solution: From Table 11.1 we have the permittivity given by

$$\epsilon = K\epsilon_0 = 1.0006 \ (8.85 \ \text{pF/m}) = 8.85531 \ \text{pF/m}$$

In the SI we have $A = 0.1 \ \text{m}^2$ and

$$d = (3 \ \text{mm}) \left(\frac{1}{1000} \ \text{m/mm} \right) = 0.003 \ \text{m}$$

Therefore, by (11.3) the capacitance is

$$C = \frac{(8.85531 \ \text{pF/m})(0.1 \ \text{m}^2)}{0.003 \ \text{m}} = 295.2 \ \text{pF}$$

Example 11.3: Repeat Example 11.2 if the dielectric is barium-strontium titanate.

Solution: In this case by Table 11.1 the permittivity is

$$\epsilon = K\epsilon_0 = (7500)(8.85 \times 10^{-12}) = 6.6375 \times 10^{-8} \ \text{F/m}$$

and therefore the capacitance is

$$C = \frac{(6.6375 \times 10^{-8})(0.1)}{0.003} = 221.25 \times 10^{-8} \ \text{F}$$

or

$$C = 2.2 \ \mu\text{F}$$

We note from the foregoing examples that much higher values of capacitance can be obtained with dielectrics having high permittivities. However, we are still limited to practical capacitances that are small compared to the standard unit of 1 F, as the following example shows.

Example 11.4: Find the plate dimensions in feet of a parallel-plate capacitor having $C = 1 \ \text{F}$, $d = 0.1 \ \text{m}$, and a dielectric of air, if the plates are square.

Solution: The permittivity is $\epsilon = 8.85531 \times 10^{-12} \ \text{F/m}$, so that by (11.3), solved for A, we have

$$A = \frac{Cd}{\epsilon} = \frac{1(0.1)}{8.85531 \times 10^{-12}} = 1.1293 \times 10^{10} \ \text{m}^2$$

Therefore, each plate is a square with each dimension $\sqrt{1.1293 \times 10^{10}} = 1.063 \times 10^5$ m. Since we have

$$1 \text{ m} = (100 \text{ cm})\left(\frac{1}{2.54} \text{ in/cm}\right)\left(\frac{1}{12} \text{ ft/in.}\right)$$

$$= 3.281 \text{ ft}$$

each plate dimension is

$$(3.281)(1.063 \times 10^5) = 348{,}770 \text{ ft}$$

which is over 66 miles!

PRACTICE EXERCISES

11-1.1 A constant current of 3 μA *charges* a capacitor (flows from one plate to the other) for 10 s. Find (a) the charge stored by the current and (b) the voltage across the capacitor after 10 s if the capacitance is $C = 2$ μF. (*Suggestion:* Recall that the charge Q delivered in t seconds by a constant current I is $Q = I \times t$.)

Ans. (a) 30 μC, (b) 15 V

11-1.2 Find the capacitance of a parallel-plate capacitor with $A = 8$ cm², $d = 4$ mm, and a dielectric of mica.   *Ans.* 8.85 pF

11-1.3 Find the charge stored in a capacitor with 40 V across its terminals if the capacitance is (a) 2 μF and (b) 200 μF.   *Ans.* (a) 80 μC, (b) 8 mC

11.2 *CIRCUIT RELATIONSHIPS*

Although capacitors have many sizes and shapes, as we shall see in Section 11.4, the parallel-plate capacitor is used as the model for the general circuit symbol of a capacitor, shown in Fig. 11.5. The symbol resembles the cross section of a parallel-plate capacitor except that one of the lines is slightly curved. We have shown the

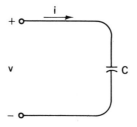

FIGURE 11.5 *Circuit symbol for a capacitor.*

straight line with the positive polarity; however, except in the case of certain special types that we consider later, it does not matter which way the symbol is drawn.

The circuit relationship between the charge q and the voltage v was given previously in (11.1), which for convenience we repeat as

$$q = Cv \qquad (11.8)$$

This relationship, however, is not convenient for use in circuits because it does not consider the current i. What we need is a relationship like Ohm's law which relates i and v.

Rate of Change: As we know, current is the rate at which charge is changing. That is, if the current i is constant for a time t, the charge q is given by

$$q = it$$

or the current in coulombs per second is

$$i = \frac{q}{t} \qquad (11.9)$$

That is, q coulombs of charge move through some point every t seconds. In the example of Fig. 11.6(a) it may be seen that $i = q/t$ is the *slope* of the q curve and is constant since the q curve is a straight line. In fact, in this example the current i (the slope) is the rise $10 - 4 = 6$ C divided by the run $5 - 2 = 3$ s. Therefore, we have

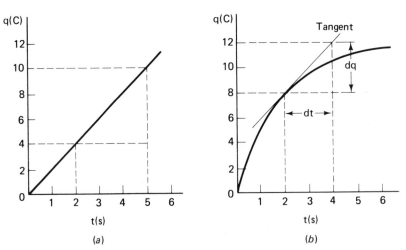

FIGURE 11.6   *Graphs of charge versus time with (a) constant slope and (b) changing slope.*

$$i = \frac{6}{3} = 2 \text{ C/s} = 2 \text{ A}$$

Now let us consider the case, such as that of Fig. 11.6(b), where the q curve is not a straight line. What is the current i in this case? Obviously, i is not constant but is changing because the slope of the q curve is changing with time. For a given time we may approximate i by considering a small change (a rise) in q and measuring the corresponding change (a run) in t. For extremely small values of rise and run, the ratio rise/run is the slope of the *tangent* line at the point at which we are interested in the current.

If we let dq represent the change in q corresponding to the change dt in t, relative to the tangent line at $t = 2$, shown in Fig. 11.6(b), the slope of the tangent line is the value of the current i at $t = 2$. In this case i is the rise dq divided by the run dt, given by

$$i = \frac{dq}{dt} = \frac{12 - 8}{4 - 2} = \frac{4}{2} = 2 \text{ A}$$

If the changes dq and dt are extremely small, the portion of the tangent line forming the hypotenuse of the triangle with legs dq and dt will coincide, for all practical purposes, with the q curve. Thus there is justification for the statement that the current at any time is the rate of change,

$$i = \frac{dq}{dt} \tag{11.10}$$

The quantity dq/dt is called the *derivative* and may be found directly from the equation of the q curve using the methods of calculus. However, we will be content to note that i is the *rate of change* of charge q with respect to time t, and measure it by the slope of the tangent line.

Voltage–Current Relation for a Capacitor: We are now in a position to find the voltage–current *(v–i)* relation for the capacitor. Since by (11.8) q is Cv, and by (11.10) i is the rate of change of q, it follows that i is the rate of change of Cv. However, since C does not change, the rate of change of Cv is C times the rate of change of v. That is,

$$i = C \frac{dv}{dt} \tag{11.11}$$

is the v–i relationship for a capacitor with current and voltage polarities as shown in Fig. 11.5. If either the current or the voltage (but not both) are reversed in polarity, we must change the sign of one of the members of (11.11).

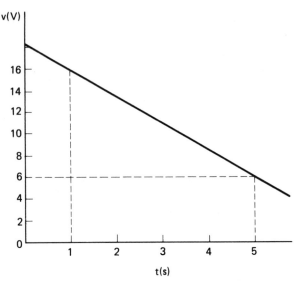

FIGURE 11.7 *Voltage curve for a capacitor.*

Example 11.5: The voltage shown in Fig. 11.7 is that across a 2-μF capacitor. Find the current i through the capacitor.

Solution: Since the voltage curve is a straight line, the rate of change dv/dt is constant. Its value, using the dashed lines, is

$$\frac{dv}{dt} = \frac{\text{rise}}{\text{run}} = \frac{-(16-6)}{5-1} = -2.5 \text{ V/s}$$

(Note that the rise is negative—from 16 to 6—and thus the slope is negative.) Therefore, by (11.11) we have

$$i = (2 \ \mu\text{F})(-2.5 \text{ V/s})$$

$$= -5 \ \mu\text{A}$$

Thus the current into the positively marked voltage terminal is negative. Therefore, a positive current of 5 μA is flowing *out* of the positively marked voltage terminal.

Energy Stored in Capacitors: As we have seen, a capacitor may be *charged* to a voltage v by connecting a voltage source to its terminals. The source may then be disconnected leaving the terminals open, and the charge, and thus the voltage, will remain. If at any subsequent time the capacitor is connected to a load, such as a resistor, the capacitor will *discharge*. That is, its voltage will drive a current through the resistor until the voltage has returned to zero. Thus as long as it had a voltage across its terminals, the capacitor was storing energy. This energy was dissipated in the resistor in the form of heat.

An ideal capacitor does not dissipate energy like a resistor. It merely stores energy during the charging phase and returns the energy to the external circuit during the discharging phase. The energy stored can be found by considering the power, $p = vi$, delivered to the capacitor and the time in which it is delivered. However, because the voltage v and the current i are generally both changing, we must use calculus to obtain the result.

If w_c is the energy stored in a capacitor with capacitance C and a terminal voltage of v, it may be shown that

$$w_c = \frac{1}{2}\, Cv^2 \tag{11.12}$$

If C is in farads and v is in volts, w_c is in joules.

Example 11.6: A 1-μF capacitor is charged to 300 V. Find the energy stored in the capacitor.

Solution: Since 1 μF = 10^{-6} F, we have by (11.12)

$$w_c = \frac{1}{2}\, (10^{-6})(300)^2 = 0.045 \text{ J} = 45 \text{ mJ}$$

PRACTICE EXERCISES

11-2.1 A 100-μF capacitor has a voltage v as shown. Find the current into the positively marked voltage terminal at the times (a) $t = 0.5$ s, (b) $t = 1.5$ s, and (c) $t = 2.5$ s.
Ans. (a) 200 μA, (b) 0, (c) -200 μA

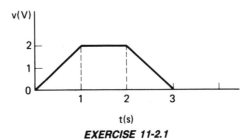

t(s)

EXERCISE 11-2.1

11-2.2 Find the energy stored in a 10-μF capacitor when its terminal voltage is 100 V.
Ans. 0.05 J

11-2.3 Find the energy stored in the capacitor of Exercise 11-2.1 at (a) $t = 1$ s, (b) $t = 2$ s, (c) $t = 2.5$ s, and (d) $t = 3$ s.   *Ans.* (a) 200 μJ, (b) 200 μJ, (c) 50 μJ, (d) 0

11.3 SERIES AND PARALLEL CAPACITORS

Like resistors, capacitors may be connected in series or in parallel, and equivalent circuits may be obtained. For example, two capacitors are connected in series in Fig. 11.8(a) and charged as shown by the voltage source v. If q_1 is the charge on C_1 and q_2 is the charge on C_2, with polarities as shown, q_1 must be the charge q_T seen by the source at its positive terminal. Also q_2 on the upper plate of C_2 must have come from the lower plate of C_1, and thus must be q_1. In other words,

$$q_T = q_1 = q_2 \tag{11.13}$$

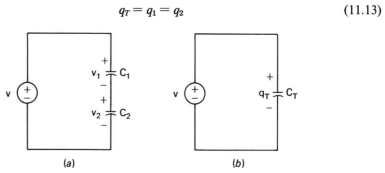

(a) $\hphantom{xxxxxxxxxxxxxxxxx}$ (b)

FIGURE 11.8 *(a) Series capacitors and (b) their equivalent.*

Equivalent Circuit for Series Capacitors: Using (11.13) and (11.8), we may obtain an equivalent circuit for the two series capacitors. Such a circuit is shown in Fig. 11.8(b), where for equivalence we must have, by (11.8),

$$v = \frac{q_T}{C_T} \tag{11.14}$$

Also, by KVL we have in Fig. 11.8(a),

$$v = v_1 + v_2$$

which by (11.8) and (11.13) is

$$v = \frac{q_1}{C_1} + \frac{q_2}{C_2} \tag{11.15}$$

$$= \frac{q_T}{C_1} + \frac{q_T}{C_2}$$

Comparing (11.14) and (11.15), we have

$$\frac{q_T}{C_T} = \frac{q_T}{C_1} + \frac{q_T}{C_2}$$

which, upon dividing out q_T, becomes

$$\frac{1}{C_T} = \frac{1}{C_1} + \frac{1}{C_2} \tag{11.16}$$

This result is similar to (4.20), which held for two *parallel* resistances. As we did in that case, we solve (11.16) for the equivalent capacitance C_T, resulting in the product divided by the sum,

$$C_T = \frac{C_1 C_2}{C_1 + C_2} \tag{11.17}$$

General Case of Series Capacitors:　In the general case of N series capacitors, shown in Fig. 11.9, we may obtain an equivalent circuit of one capacitance C_T, as shown in Fig. 11.8(b). The procedure is the same as in the two-capacitor case, except that there are more terms to consider. By KVL in Fig. 11.9, we have

$$v = v_1 + v_2 + \cdots + v_N$$

which becomes, since the charge on each capacitor is q_T,

$$\frac{q_T}{C_T} = \frac{q_T}{C_1} + \frac{q_T}{C_2} + \cdots + \frac{q_T}{C_N}$$

Dividing through by q_T results in

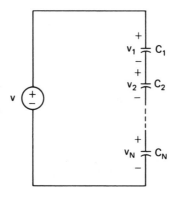

FIGURE 11.9 *Series circuit with N capacitors.*

$$\frac{1}{C_T} = \frac{1}{C_1} + \frac{1}{C_2} + \cdots + \frac{1}{C_N} \qquad (11.18)$$

Example 11.7: Two capacitors, one of 3 μF and one of 6 μF, are connected in series. Find the equivalent capacitance C_T.

Solution: By (11.17) we have

$$C_T = \frac{3 \times 6}{3 + 6} = 2 \ \mu F$$

Example 11.8: Solve Example 11.7 if there are four series capacitances of 4 μF, 8 μF, 12 μF, and 24 μF.

Solution: By (11.18) we have

$$\frac{1}{C_T} = \frac{1}{4} + \frac{1}{8} + \frac{1}{12} + \frac{1}{24} = \frac{12}{24} = \frac{1}{2}$$

Therefore, the equivalent capacitance is $C_T = 2 \ \mu F$.

Parallel Capacitors: In the general case of N capacitors connected in parallel, shown in Fig. 11.10, we may also obtain an equivalent capacitance C_T, as in Fig. 11.8(b). To see this, we note that the voltage v is across each capacitor, so that q/C is the same for each capacitor. That is, if q_1 is placed on C_1, q_2 on C_2, and so on, then

$$v = \frac{q_1}{C_1} = \frac{q_2}{C_2} = \cdots = \frac{q_N}{C_N} \qquad (11.19)$$

In order for Figs. 11.8(b) and 11.10 to be equivalent, the charge q_T seen at the top plate of C_T in Fig. 11.8(b) must be the sum of the charges q_1, q_2, ... , q_N on the top plates in Fig. 11.10. That is,

$$q_T = q_1 + q_2 + \cdots + q_N \qquad (11.20)$$

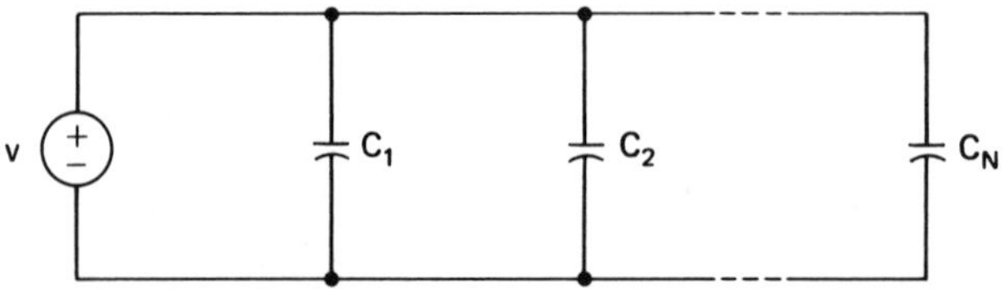

FIGURE 11.10 *Parallel circuit with N capacitors.*

Combining (11.14) and (11.19), we have

$$v = \frac{q_T}{C_T} = \frac{q_1}{C_1} = \frac{q_2}{C_2} = \cdots = \frac{q_N}{C_N}$$

or

$$
\begin{aligned}
q_1 &= C_1 v \\
q_2 &= C_2 v \\
&\vdots \\
q_N &= C_N v \\
q_T &= C_T v
\end{aligned}
\tag{11.21}
$$

Substituting these results in (11.20) yields

$$C_T v = C_1 v + C_2 v + \cdots + C_N v$$

Finally, dividing out v results in

$$C_T = C_1 + C_2 + \cdots + C_N \tag{11.22}$$

Therefore, parallel capacitances are combined into an equivalent, like series resistances, by adding them together.

Example 11.9: Find the equivalent C_T of three parallel capacitances of 2 μF, 3 μF, and 12 μF.

Solution: By (11.22) we have

$$C_T = 2 + 3 + 12 = 17 \ \mu\text{F}$$

Example 11.10: Find the equivalent capacitance C_T seen at the terminals of the circuit of Fig. 11.11, if $C_1 = 6 \ \mu$F, $C_2 = 4 \ \mu$F, $C_3 = 9 \ \mu$F, and $C_4 = 12 \ \mu$F.

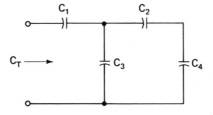

FIGURE 11.11 *Circuit of series-parallel capacitors.*

Solution: . The series capacitances C_2 and C_4 may be combined into an equivalent C_5, given by

$$C_5 = \frac{4 \times 12}{4 + 12} = 3 \ \mu F$$

In turn, C_5 and C_3 are in parallel and equivalent to C_6, given by

$$C_6 = 9 + 3 = 12 \ \mu F$$

Finally, C_6 and C_1 are in series, and their combination is C_T, given by

$$C_T = \frac{12 \times 6}{12 + 6} = 4 \ \mu F$$

PRACTICAL EXERCISES

11-3.1 Find the equivalent capacitance of two 100-pF capacitors connected (a) in series and (b) in parallel.   *Ans.* (a) 50 pF, (b) 200 pF

11-3.2 Find the equivalent capacitance of five 10-μF capacitors connected (a) in series and (b) in parallel.   *Ans.* (a) 2 μF, (b) 50 μF

11-3.3 Find C_T in Fig. 11.11 if $C_1 = 12 \ \mu F$, $C_2 = 10 \ \mu F$, $C_3 = 3 \ \mu F$, and $C_4 = 90 \ \mu F$.
Ans. 6 μF

11.4 TYPES OF CAPACITORS

Capacitors come in many sizes and shapes, as illustrated in Fig. 11.12. They may be in the form of ceramic disks as in Fig. 11.13, rectangular cans as in Fig. 11.14, or *dual in-line packages* (DIP) for use in printed circuit boards as in Fig. 11.15. They may range in size from integrated-circuit capacitors on a small IC chip, to the microminiature tantalum capacitors of Fig. 11.16, to the power capacitors shown earlier in Fig. 2.3.

Capacitors may be constructed for many purposes, such as the *feedthrough* capacitors that screw into a chassis, shown in Fig. 11.17, or the *auto-generator* type of Fig. 11.18. They may be found in circuits ranging from tiny IC chips to large power systems.

Capacitors are generally classified by the type of dielectric used, such as mica, paper, polystyrene, Mylar, and so on, and the capacitance is determined by the dielectric and the physical geometry of the device. Simple capacitors are often constructed by using two sheets of metal foil which are separated by a dielectric material. The

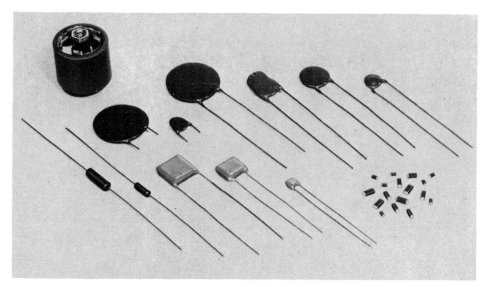

FIGURE 11.12 *Various sizes and shapes of capacitors (Courtesy, Centralab Electronics Division, Globe-Union, Inc.).*

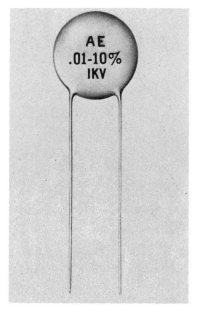

FIGURE 11.13 *Ceramic-disk capacitor (Courtesy, Arco Electronics, Inc.).*

FIGURE *11.14 Motor-run, rectangular size capacitors (Courtesy, Cornell-Dubilier Electronics/Subsidiary of Federal Pacific Electric Company).*

FIGURE *11.15 Dual-in-line package capacitors (Courtesy, Sprague Electric Company).*

FIGURE 11.16 *Microminiature tantalum capacitors (Courtesy, Corning Glass Works).*

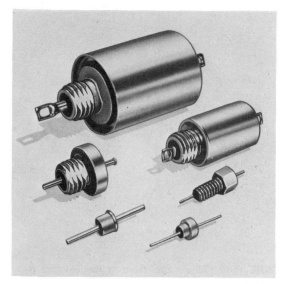

FIGURE 11.17 *Feedthrough capacitors (Courtesy, Centralab Electronics Division, Globe-Union, Inc.).*

foil and dielectric are pressed together in a laminar form and rolled into a compact package, with conductors attached to each metal-foil sheet to constitute the terminals. Examples of *tubular* capacitors of this type are shown in Fig. 11.19. Examples of *dipped* capacitors are the tantalum capacitors of Fig. 11.20 and the dipped-mica capacitors of Fig. 11.21.

FIGURE 11.18 Auto generator type capacitor (Courtesy of Sprague Electric Company).

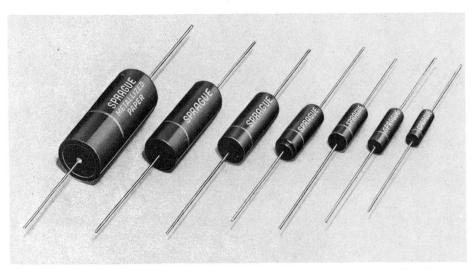

FIGURE 11.19 Tubular capacitors of various sizes (Courtesy, Sprague Electric Company).

Electrolytic Capacitors: The largest capacitance values are generally those of *electrolytic* capacitors, which are usually made of an aluminum or tantalum sheet as one conductor and an electrolytic paste as the other. A dc voltage is applied which polarizes the metal sheet, forming a thin layer of aluminum oxide or tantalum oxide, which serves as the dielectric. The sandwich of conductors and dielectric is rolled into a cylinder for compactness. A disadvantage of this method of construction is that the voltage polarity of the capacitor must be observed, for if the incorrect

FIGURE 11.20 *Minidip tantalum capacitors (Courtesy, Corning Glass Works).*

FIGURE 11.21 *Dipped-mica capacitors (Courtesy, Cornell-Dubilier Electronics/Subsidiary of Federal Pacific Electric Company).*

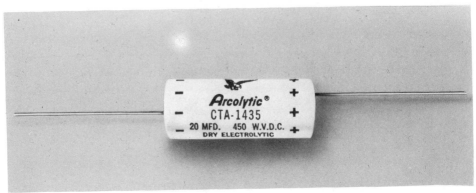

FIGURE 11.22 Electrolytic capacitor (Courtesy, Arco Electronics, Inc.).

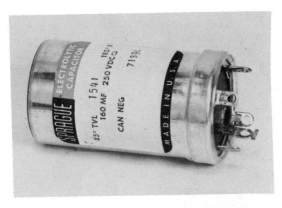

FIGURE 11.23 Can-type aluminum electrolytic capacitor (Courtesy, Sprague Electric Company).

polarity is used, the oxide will be reduced and conduction will take place between the plates.

Examples of cylindrically shaped electrolytic capacitors are shown in Figs. 11.22 and 11.23. As may be seen, their capacitances are the relatively high values of 20 μF and 160 μF, respectively. Other examples, previously shown, are the minidip tantalum capacitors of Fig. 11.20, whose available range of values is 0.68 to 22 μF in one case and 2.2 to 68 μF in the other.

Variable Capacitors: All the capacitors we have considered up to now have fixed values of capacitance. However, capacitors, like resistors, are also available with variable values. The most common of the variable-type capacitors is the variable air capacitor, an example of which is shown in Fig. 11.24. The capacitor consists of a number of metal plates which rotate with respect to each other. The rotation changes the area by which the plates overlap each other, thereby changing the capacitance. The plates are "ganged" to form a number of parallel capacitances so that the overall capacitance is larger (the sum of the individual capacitances). The dielectric is, of course, the air between the plates.

To illustrate the different sizes available in variable air capacitors, a "micro" capacitor is shown in Fig. 11.25.

FIGURE 11.24   *Variable air capacitor (Courtesy, E. F. Johnson Company).*

FIGURE 11.25   *"Micro" variable air capacitor (Courtesy, E. F. Johnson Company).*

FIGURE 11.26 *Capacitor decade boxes (Courtesy, Cornell-Dubilier Electronics/ Subsidiary of Federal Pacific Electric Company).*

Variable capacitors are also available in decade boxes, like resistors, in which the capacitance is selected by turning a knob. Examples are the five decade boxes shown in Fig. 11.26, for which the ranges are 0.0001 to 0.11 μF in steps of 0.0001, 0.01 to 1.1 μF in steps of 0.01, 1 to 10 μF in steps of 1, and 10 to 150 μF in steps of 10.

The circuit symbol for a variable capacitor is shown in Fig. 11.27.

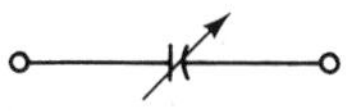

FIGURE 11.27   Circuit symbol for a variable capacitor.

11.5 PROPERTIES OF CAPACITORS

The most important property of a capacitor is, of course, its capacitance. However, there are many other factors that determine which capacitor should be selected to perform a certain function. We shall consider briefly some of these other properties of capacitors in this section.

Working Voltage: The voltage rating, or *working voltage,* of a capacitor is the maximum voltage it can withstand without destroying or breaking down the dielectric. This value, also known as the *breakdown voltage,* discussed in Chapter 10, is the product of the dielectric strength of the insulating material and the thickness of its dielectric layer. As examples, the mica capacitors of Fig. 11.21 vary in size from 1 to 390 μF, with voltage ratings of 500, 300, 100, and 50 V dc. The tantalum capacitors of Fig. 11.16 range from 0.1 μF at 50 V to 47 μF at 10 V, and the ceramic disk capacitor of Fig. 11.13 has a capacitance of 0.01 μF at 1 kV.

Leakage Current: Practical capacitors, unlike ideal capacitors, generally dissipate a small amount of energy. This is because the dielectric is not perfect (its resistance is not infinite) and consequently a small so-called *leakage current* flows between its conducting plates. The capacitor may be thought of as having a *leakage resistance* R_c, to account for the leakage current, as well as having a capacitance C. The product R_cC, a quantity specified by the manufacturer, may be used as a measure of the quality of the capacitor. The higher R_cC, the better the capacitor.

As examples, ceramic disk capacitors may have resistance–capacitance products of 10^3 Ω-F, while for high-quality Teflon capacitors the product may be 2×10^6 Ω-F.

A list of some common capacitor types, with capacitance ranges, working voltages, and leakage resistances is given in Table 11.2.

TABLE 11.2   Capacitors and their properties

Dielectric	Capacitance	Leakage Resistance ($M\Omega$)	Working Voltage (V)
Mica	10–5000 pF	1000	10,000
Ceramic	1000 pF–1 μF	30–1000	100–2000
Polystyrene	500 pF–10 μF	10,000	1000
Mylar	5000 pF–10 μF	10,000	100–600
Air-variable	5–500 pF		500
Electrolytic			
Tantalum	0.01–3000 μF	1	6–50
Aluminum	0.1–100,000 μF	1	10–500

Color Codes: Capacitance values are stamped on the bodies of many types of capacitors, but in many cases capacitors, like resistors, have a color code for identifying capacitance values, tolerances (possible deviations from nominal values), and working voltages. In the case of tubular capacitors, the color code is very similar to that of resistors, with colored stripes or bands used to indicate the capacitance values. The colors used are the same as for resistors, from black for 0 to white for 9.

Rectangular-shaped mica capacitors and ceramic disk capacitors use a code of colored dots strategically placed on the body of the capacitor. Three of the dots give the capacitance in picofarads, with the same color values as resistors. Other dots indicate tolerance and temperature coefficient.

A color-coded tubular capacitor is shown in Fig. 11.28 with six color bands. The color code for the bands is given in Table 11.3. The first three bands, marked *a*, *b*, and *d*, are used for the nominal capacitance value *C*, given by

$$C = (10a + b) \times 10^d \text{ pF} \tag{11.23}$$

The fourth band is the tolerance band, which indicates a maximum percentage that the actual capacitance may deviate from the nominal value. Bands *e* and *f* are used for the working voltage *V*, which is given by

$$V = 100(10e + f) \text{ V} \tag{11.24}$$

if *V* is more than 900 V, and by

$$V = 100e \text{ V} \tag{11.25}$$

if *V* is 900 V or less. In this case there is no *f* band.

In other words, the capacitance is the two-digit number *ab* multiplied by 10 to the power *d*, and the working voltage is 100 times the two-digit number *ef* or, if *f* is missing, it is 100 times the one-digit number *e*.

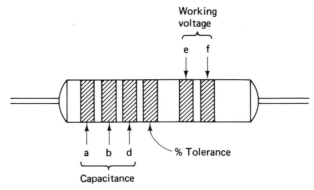

FIGURE 11.28 *Color-coded tubular capacitor.*

TABLE 11.3 *Color code for tubular capacitors*

Color	Bands a, b, d, e, and f	Tolerance (±%)
Black	0	20
Brown	1	—
Red	2	—
Orange	3	30
Yellow	4	40
Green	5	5
Blue	6	—
Violet	7	—
Gray	8	—
White	9	10

Example 11.11: A capacitor with color code given by Table 11.3 has bands *a, b, d, e,* and *f,* which are yellow, violet, red, brown, and black, respectively. The tolerance band is white. Find the nominal capacitance, the working voltage, and the range in which the actual capacitance lies.

Solution: By Table 11.3 we have $a = 4$, $b = 7$, $d = 2$, $e = 1$, and $f = 0$. Also, the tolerance is ±10%. Therefore, by (11.23) the nominal capacitance is

$$C = [10(4) + 7] \times 10^2 = 4700 \text{ pF}$$

By (11.24) we have the working voltage,

$$V = 100[10(1) + 0] = 1000 \text{ V}$$

Finally, since the tolerance is ±10%, the actual capacitance may deviate from the nominal value by $\pm 0.1(4700) = \pm 470$ pF. The range in which the actual capacitance lies is therefore $4700 - 470 = 4230$ pF to $4700 + 470 = 5170$ pF.

PRACTICE EXERCISES

11-5.1 Find the nominal value of capacitance, the working voltage, and the range in which the actual capacitance lies if the capacitor is color-coded as in Table 11.3, with bands from *a* to *f* colored blue, gray, red, white, brown, and green.

Ans. 6800 pF, 1500 V, 6120 to 7480 pF

11-5.2 Repeat Exercise 11-5.1 if the bands are orange, orange, brown, black, and green. (The *f* band is missing.)

Ans. 330 pF, 500 V, 264 to 396 pF

11.6 SUMMARY

Capacitors are two-terminal elements consisting of two conductors separated by a dielectric. They are capable of storing energy in the electric field produced between the conductors when a voltage is applied to the terminals. A measure of the ability of a capacitor to store energy is its capacitance, measured in the standard units of farads, or in the more practical units of microfarads or picofarads.

The current through a capacitor is proportional to the rate of change of its voltage. The energy stored in a capacitor is proportional to the square of the voltage, and in both cases the capacitance is a constant of proportionality.

Capacitors in series may be combined like parallel resistors into a single equivalent capacitor. Likewise, parallel capacitors are combined into an equivalent like series resistors.

The most common capacitors are those with dielectrics of air, mica, ceramic, Mylar, and polystyrene. Electrolytic capacitors have the highest capacitances and they are the only capacitors with polarity that must be observed.

A capacitor is characterized by a working voltage, which is the maximum voltage it can withstand without a breakdown of the dielectric. Another property is a leakage resistance which accounts for a leakage current due to the nonideal nature of the dielectric.

Finally, some capacitors have color codes similar to those of resistors. The code enables us to determine the nominal capacitance, the working voltage, and the possible deviation of the actual capacitance from the nominal value.

PROBLEMS

11.1 A constant current of 5 μA is charging a 3-μF capacitor. If the capacitor was previously uncharged, find the charge and the voltage on it after 30 s.

11.2 Find the capacitance of a capacitor if a charge of 100 μC on its plates results in a terminal voltage of (a) 10 V, (b) 20 V, and (c) 50 V.

11.3 A parallel-plate capacitor has square plates 10 cm on a side and a separation between plates of 2 mm. If the dielectric is air, find the capacitance.

11.4 A parallel-plate capacitor, with plates that are 4 cm $\times$ 6 cm, has a capacitance of 68 pF. If the dielectric is Bakelite, find the separation between the plates.

11.5 Repeat Problem 11.4 if the dielectric is barium-strontium titanate and the capacitance is 0.1 μF.

11.6 Find the charge stored in a 2-μF capacitor if the voltage across its terminals is (a) 10 V and (b) 100 V.

11.7 The voltage v across a 10-μF capacitor is as shown. Find the current into the positively marked voltage terminal at the times (a) $t = 2$ s and (b) $t = 5$ s.

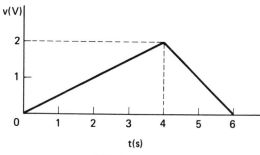

PROBLEM 11.7

11.8 Repeat Problem 11.7 if the t axis is measured in milliseconds.

11.9 Find the energy stored in a 0.1-μF capacitor with a terminal voltage of 200 V.

11.10 Find the energy stored in the capacitor of Problem 11.7 at (a) $t = 4$ s and (b) $t = 5$ s.

11.11 The energy stored in a 2-μF capacitor is 400 μJ. Find the terminal voltage.

11.12 Find the equivalent capacitance of four 2-μF capacitors connected in series.

11.13 Find the equivalent capacitance of a 12-μF capacitor and five 20-μF capacitors all connected in series.

11.14 Repeat Problem 11.13 if the capacitors are all connected in parallel.

11.15 Find C_T in Fig. 11.11 if $C_1 = 40$ μF, $C_2 = 6$ μF, $C_3 = 6$ μF, and $C_4 = 12$ μF.

11.16 Find C_T in Fig. 11.11 if all the capacitances are 100 pF.

11.17 Find C in the circuit shown so that (a) $C_T = 6$ μF and (b) $C_T = 8$ μF.

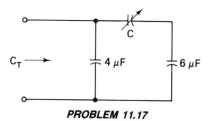

PROBLEM 11.17

11.18 Find the nominal capacitance, the working voltage, and the range in which the actual capacitance lies if the capacitor is color-coded as in Fig. 11.28 with bands (a) $a =$ gray, $b =$ red, $d =$ orange, tolerance $=$ white, $e =$ brown, and $f =$ red, and (b) $a =$ orange, $b =$ orange, $d =$ orange, tolerance $=$ black, and $e =$ white (no f band).

$$C = \frac{q}{V} \quad C_T = \frac{4\times6}{4+6} = \frac{24}{10} = 2.4$$

$$C_T = \frac{(C+6)\times4}{C+6+4}$$

$$6 = \frac{4C+24}{C+6+4}$$

$$6C + 36 + 24 = 4C + 24 \qquad 2C = -36$$
$$6C - 4C = -36 \qquad C = \frac{36}{2} = 18$$

12

RC *CIRCUITS*

As we have seen in Chapter 11, a capacitor may be *charged* to a certain voltage by connecting a voltage source to its terminals. This establishes a charge on the capacitor plates and a voltage across the plates. The capacitor may then be *discharged* by disconnecting the source and connecting the terminals of the charged capacitor across a load, such as a resistor. A typical example, also discussed earlier, is that of an electronic flash attached to a camera. The capacitor is charged by a battery and then discharged into a load, which in this case is the light-producing element, by a switch that simultaneously opens the shutter of the camera.

The capacitor is charged, of course, by the action of the battery, which causes the current to flow through the external circuit from one plate of the capacitor to the other plate. The discharging process is the current flowing back (in the opposite direction) through the load, which has replaced the battery in the external circuit.

The circuits produced by the switching action in both the charging phase and the discharging phase are simple *RC* circuits. In the charging phase the circuit is a *driven RC* circuit consisting of the capacitor, a resistor, and the source which *drives* the circuit (causes the current). In the discharging phase the battery is switched out of the circuit, so that it is a *source-free RC* circuit (an *RC* circuit without a source).

In this chapter we will consider both source-free *RC* circuits and driven *RC* circuits. We will analyze these circuits by finding the currents and the voltages that result. In the process we will define a parameter called an *RC time constant,* which, as we will see, determines the shape of the currents and voltages, as well as the speed with which the charging and discharging processes take place. Finally, we will consider more complicated RC circuits that may be reduced to simple RC circuits by combining several elements into a single equivalent element.

12.1 CHARGING AND DISCHARGING A CAPACITOR

The charging and discharging of a capacitor may be illustrated by means of Fig. 12.1. During the charging phase, the switch is in position 1 and the voltage V_b causes a current i in the direction shown through the resistor R. This causes an accumulation of charge to build up on the capacitor C, with a corresponding capacitor voltage v_c, as shown.

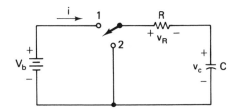

FIGURE 12.1 *Circuit for charging and discharging a capacitor.*

The voltage v_R on the resistor is, by KVL,

$$v_R = V_b - v_c \qquad (12.1)$$

so that if the switch is kept in position 1, the buildup of charge will continue until v_c reaches V_b, the battery voltage. At this point, by (12.1), we have $v_R = 0$, so that the current i, given by

$$i = \frac{v_R}{R}$$

becomes zero, and the charging phase is complete.

DC Steady State: The time after the capacitor voltage v_c of Fig. 12.1 reaches the battery voltage V_b and the current ceases to flow, is illustrated in Fig. 12.2. The current and all the voltages in the circuit are now constants ($i = 0$, $v_c = V_b$,

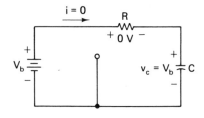

FIGURE 12.2 *RC circuit in dc steady state.*

$v_R = 0$, and the battery voltage is V_b, which is constant throughout). In this case we say that the circuit is in *dc steady state,* or simply *steady state.* All the currents and voltages are constant and *none of them is in the process of changing.* As long as the circuit of Fig. 12.2 is connected as shown, no current will flow and no voltages will change.

Capacitor like an Open Circuit to DC: In the dc steady state, as we have seen, no current flows through a capacitor. Thus a capacitor acts like an *open circuit* to dc (in the steady state). This may be seen also from the voltage–current relation,

$$i = C\frac{dv}{dt} \tag{12.2}$$

which we discussed in Chapter 11. The quantity dv/dt is the rate of change of the capacitor voltage v, and at dc steady state the voltage is constant and thus not changing. Therefore, the rate of change of v is zero, so that $i = 0$.

Example 12.1: Find the capacitor voltage v in Fig. 12.3(a) when the circuit is in dc steady state.

Solution: The capacitor acts like an open circuit in dc steady state, so that an equivalent circuit is that of Fig. 12.3(b), with the capacitor replaced by an open circuit. Since there is no current and thus no *IR* drop, we have $v = 6$ V.

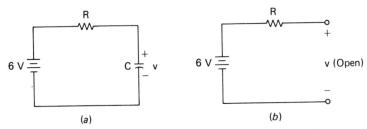

(a)   (b)

FIGURE 12.3 (a) RC circuit and (b) its equivalent in dc steady state.

Discharging the Capacitor: If the capacitor of Fig. 12.1 is charged to a voltage $v_c = V_0$ and then the switch is moved to position 2, we will have the circuit shown in Fig. 12.4. (The battery is switched out of the circuit and thus need not be shown.) The capacitor voltage v_c will now appear across the resistor, since by KVL in Fig. 12.4 we have

$$-v_R + v_c = 0$$

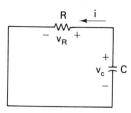

FIGURE 12.4 *Capacitor discharging through a resistor.*

or $v_R = v_c$. (We have changed the polarity of v_R from what it was in Fig. 12.1, because the current will now be in the opposite direction.)

By Ohm's law there will be a current

$$i = \frac{v_R}{R} = \frac{v_c}{R} \tag{12.3}$$

in the direction shown. Thus charge will move from the top plate to the bottom plate of the capacitor through the resistor, which discharges the capacitor. The initial current is

$$i = \frac{V_0}{R}$$

since initially $v_c = V_0$, but as time passes v_c continually drops, because of the transfer of charge, until eventually v_c (and thus i) reaches zero. There is now no charge and no voltage on the capacitor, which is at this point completely discharged.

Driven and Source-free Circuits: The circuit of Fig. 12.2 is an RC circuit with a *driver* or *source* V_b, and is called a *driven RC* circuit. During the charging stage, the voltage v_c and the current i are changing in a certain manner and will eventually reach the *final value* shown of $v_c = V_b$ and $i = 0$.

On the other hand, the circuit of Fig. 12.4 is a *source-free RC* circuit, consisting of the resistor and capacitor but no source. In this case both i and v_c will drop from some initial value to zero when the discharging phase is completed.

The remainder of the chapter will be devoted to finding the currents and voltages in source-free and driven RC circuits. We will consider first the source-free circuit because of its relative simplicity, and then discuss the driven circuit.

PRACTICE EXERCISES

12-1.1 The switch in the circuit shown is in position 1. Find the current i_c and the voltage v_c when (a) $v_1 = 9$ V, (b) $v_1 = 5$ V, and (c) dc steady state is reached.

Ans. (a) 9 mA, 1 V, (b) 5 mA, 5 V, (c) 0, 10 V

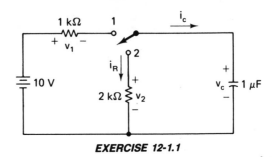

EXERCISE 12-1.1

12-1.2 The switch in the circuit of Exercise 12-1.1 is moved to position 2 at a time when v_c = 10 V. Find i_R and v_2 (a) initially and (b) in dc steady state.

Ans. (a) 5 mA, 10 V, (b) 0, 0

12.2 *SOURCE-FREE* RC *CIRCUITS*

We will begin our study of the source-free *RC* circuit by considering the series connection of the resistor *R* and the capacitor *C*, shown in Fig. 12.5. This circuit is a *simple* source-free *RC* circuit because it may be described by a single equation, as we shall see.

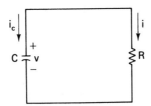

FIGURE 12.5 *Simple source-free RC circuit.*

Circuit Equation: We see by KVL that the voltage across the resistor is the voltage *v* across the capacitor. Therefore, the current *i* through the resistor is

$$i = \frac{v}{R} \tag{12.4}$$

Also, the current i_c through the capacitor is, by (12.2),

$$i_c = C\frac{dv}{dt} \tag{12.5}$$

Finally, applying KCL at the top node, we have

$$i_c + i = 0$$

which by (12.4) and (12.5) is

$$C\frac{dv}{dt} + \frac{v}{R} = 0 \tag{12.6}$$

This is the circuit equation which describes the source-free RC circuit.

In order to find v in Fig. 12.5, we must have more information than (12.6). We must also know how much voltage is on the capacitor in the beginning. If we take the beginning or initial time to be $t = 0$, we must know the *initial* voltage $v = v(t)$ at $t = 0$, which we denote by $v(0)$. If this initial value is given by $v(0) = V_0$, some specified constant, $v = v(t)$ must satisfy the two equations

$$\frac{dv}{dt} + \frac{1}{RC}v = 0 \tag{12.7}$$

$$v(0) = V_0$$

[The first of these is (12.6) divided through by C.]

Capacitor Voltage: As we saw in Chapter 11, the energy stored in a capacitor with capacitance C and voltage v is

$$w_c = \frac{1}{2}Cv^2 \tag{12.8}$$

It is not possible to change the energy abruptly (in zero time) because it takes time to move charge. Therefore, since w_c cannot change instantaneously, we see by (12.8) that the *capacitor voltage v cannot change instantaneously.* Thus v satisfying (12.7) will start at V_0 and change continuously with time. In other words, there will not be any jumps, or discontinuities, in v.

The first of (12.7) corresponds to the algebraic equations describing the resistive circuits in the earlier chapters. However, because of the presence of the derivative, we must use calculus to solve for v. It may be shown that the solution of (12.7) is

$$v = V_0 e^{-t/RC} \tag{12.9}$$

which is an *exponential* function of t.

The number e is an irrational number given to five decimal places by

$$e = 2.71828 \tag{12.10}$$

It may be found with most hand calculators, such as the Hewlett-Packard 31E shown in Fig. 12.6, by calculating e^x when $x = 1$. Of course, the e^x key can also be used in finding v in (12.9) for a given time t.

FIGURE 12.6 Hewlett-Packard 31E hand calculator (Courtesy, Hewlett-Packard).

RC Time Constant: The voltage v of (12.9) may be put in the compact form

$$v = V_0 e^{-t/\tau} \tag{12.11}$$

where τ (the lowercase Greek letter *tau*) is given by

$$\tau = RC \tag{12.12}$$

The number τ is thus the resistance–capacitance product, and is very simple to find. Its units of measurement may be found by noting those of R and C. By Ohm's law, the units of $R = v/i$ are V/A, and by (12.2) we have

$$C = \frac{i}{dv/dt}$$

Therefore, the units of C are A/(V/s) or A-s/V. Thus τ is measured in

$$(V/A)(A\text{-s}/V) = \text{seconds}$$

Since τ has the unit for time (second) and is a constant depending on R and C, it is called the *RC time constant,* or the *time constant* of the *RC* circuit. Its value, as we will see, determines how fast the capacitor discharges.

The values of $e^{-t/\tau}$ are listed in Table 12.1 for t given in multiples of the time constant τ. From the table we may find the voltage v of (12.9) at various times. For example, the initial voltage (at $t = 0$) is

$$v(0) = V_0$$

and the voltage after one time constant $(t = \tau)$ is

$$v(\tau) = 0.368 \, V_0$$

Thus after one time constant the capacitor has discharged to the point where its voltage is 0.368 times its initial value, or 36.8% of its initial value. Therefore, the smaller the time constant, the faster the capacitor discharges. If $\tau = 1$ s, for example, it takes 1 s to reach 36.8% of the initial value, but if $\tau = 0.001$ s, it takes only 1 ms to reach this value.

In the ideal case $e^{-t/\tau}$ never mathematically reaches zero in a finite time, but as a practical matter, we may see from Table 12.1 that it is extremely small after a few time constants. Usually, the capacitor is considered to be completely discharged (with zero voltage) after five time constants $(t = 5\tau)$.

Example 12.2: Find the voltage v across the capacitor of Fig. 12.5 for (a) all time t, (b) $t = 1\tau$, (c) $t = 3\tau$, and (d) $t = 5\tau$, if the voltage at $t = 0$ is $V_0 = 10$ V, $R = 100$ kΩ, and $C = 30$ μF.

Solution: The time constant is

$$\tau = RC = (100 \times 10^3)(30 \times 10^{-6}) = 3 \text{ s}$$

and thus by (12.9) the voltage for all time t is

$$v = v(t) = 10e^{-t/3} \text{ V} \tag{12.13}$$

TABLE 12.1   *Values of* $e^{-t/\tau}$

t	$e^{-t/\tau}$
0	$e^0 = 1.0$
τ	$e^{-1} = 0.368$
2τ	$e^{-2} = 0.135$
3τ	$e^{-3} = 0.0498$
4τ	$e^{-4} = 0.0183$
5τ	$e^{-5} = 0.00674$
6τ	$e^{-6} = 0.00248$

At $t = 1\tau = 3$ s we have by Table 12.1

$$v(\tau) = v(3) = 10e^{-1} = 3.68 \text{ V}$$

Similarly, at $t = 3\tau$ and $t = 5\tau$ we have

$$v(3\tau) = v(9) = 10e^{-3} = 0.5 \text{ V}$$

and

$$v(5\tau) = v(15) = 10e^{-5} = 0.07 \text{ V}$$

The graph of the voltage for all positive time t may be found by plotting (12.13). This may be done by calculating v for a number of values of t using the hand calculator and connecting the points. (We recall that there are no jumps in the curve.) The result for $0 < t < 18$ s is the solid curve of Fig. 12.7, which decreases rapidly and essentially reaches zero after $t = 15$ s (5 time constants).

The dashed curve of Fig. 12.7 is the capacitor voltage for $\tau = 6$ s. As expected, it decays (decreases with time) more slowly, reaching 3.68 V (36.8% of its initial value) at $\tau = 6$ s, as opposed to $\tau = 3$ s for the solid curve. These values are shown by the dashed construction.

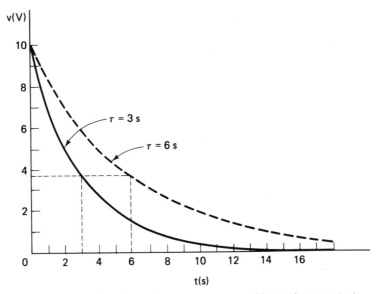

FIGURE 12.7 *Capacitor voltage curves for two different time constants.*

Capacitor Current: From (12.4) and (12.11) we may find the current i flowing out of, or the current i_c flowing into, the positive terminal of the capacitor in Fig. 12.5. The results are

$$i = \frac{V_0}{R} e^{-t/\tau} \qquad\qquad (12.14)$$

and

$$i_c = -i = -\frac{V_0}{R} e^{-t/\tau} \qquad\qquad (12.15)$$

Thus the capacitor current resembles the capacitor voltage in a source-free RC circuit. Both quantities are decreasing exponentials.

Example 12.3: The circuit of Fig. 12.8 is in dc steady state with the switch in position 1. If the switch is moved to position 2 at $t = 0$, find the voltage v and the current i for all $t > 0$.

Solution: Before the switch is moved from position 1 to position 2 the circuit is as shown in Fig. 12.9(a). Since the circuit is in dc steady state, the current is zero (the capacitor is acting like an open circuit) and thus the initial capacitor voltage equals the source voltage. That is,

$$v(0) = V_0 = 20 \text{ V}$$

For $t > 0$ the switch is in position 2, resulting in the source-free RC circuit of Fig. 12.9(b). The time constant is

$$\tau = RC = (10 \times 10^3)(20 \times 10^{-6}) = 0.2 \text{ s}$$

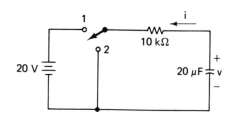

FIGURE 12.8 *RC circuit.*

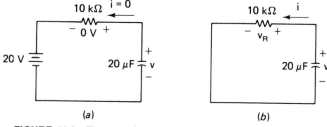

(a)   (b)

FIGURE 12.9 *The circuit of Fig. 12.8 for (a) $t < 0$ and (b) $t > 0$.*

Therefore, for $t > 0$ we have by (12.9)

$$v = 20e^{-t/0.2} \text{ V}$$

or

$$v = 20e^{-5t} \text{ V}$$

The current i may be found from Fig. 12.9(b) or from (12.14). Since by KVL the resistor voltage is $v_R = v$, we have

$$i = \frac{v_R}{R} = \frac{20e^{-5t} \text{ V}}{10 \text{ k}\Omega} = 2e^{-5t} \text{ mA} \tag{12.16}$$

The current, therefore, is an exponential function like the voltage, as expected.

General Case: As we have seen, both the current and voltage in a source-free *RC* circuit are exponential functions of the form

$$y = Ke^{-t/\tau} \tag{12.17}$$

where $\tau = RC$ is the time constant, y may be either i or v, and K is the initial value of i or v. As was pointed out earlier, we may use the hand calculator to find i or v for any t, but in the general case (12.17) may be sketched and the values read off the graph. To be completely general, let us take $K = 1$ and sketch

$$y = e^{-t/\tau} \tag{12.18}$$

with t given in multiples of τ. The result is the exponential curve, which decays from 1 to 0 as t increases, as shown in Fig. 12.10. (The other curve shown is one that increases with time, which we will consider in the case of the driven *RC* circuit in the next section.) To obtain (12.17) we merely find y (a voltage or a current) from the graph of (12.18) and multiply the result by K.

Example 12.4: A capacitor voltage v is given by

$$v = 5e^{-2t} \text{ V}$$

Find v at (a) $t = 1$ s, (b) $t = 1.5$ s, and (c) $t = 4\tau$.

Solution: Comparing e^{-2t} with $e^{-t/\tau}$, we see that $\tau = 0.5$ s. Therefore, each unit on the time t scale of Fig. 12.10 is 0.5 s. Thus in part (a) we want $v(1)$, which is $e^{-t/\tau}$ at $t = 2$ units ($t = 2 \times 0.5 = 1$ s $= 2\tau$), multiplied by 5. The result from the graph is approximately

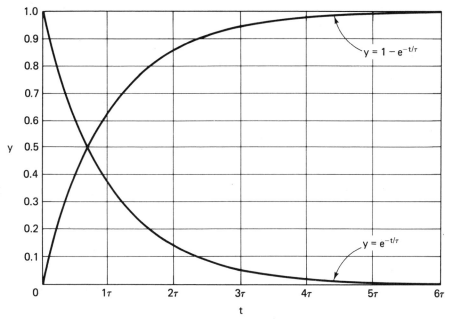

FIGURE 12.10 *General exponential curves for a time constant τ.*

$$v(1) = 5 \times 0.14 = 0.7 \text{ V}$$

This compares with the exact value (to two decimal places) using the hand calculator, of

$$v(1) = 5e^{-2} = 0.68 \text{ V}$$

For part (b), 1.5 s $= 3\tau$, and from Fig. 12.10 we have approximately

$$v(1.5) = 5 \times 0.05 = 0.25 \text{ V}$$

(The exact value is 0.249 V.) Finally, for part (c), we have approximately

$$v(4\tau) = 5 \times 0.02 = 0.1 \text{ V}$$

which compares favorably with the exact value of 0.092 V.

PRACTICE EXERCISES

12-2.1   The circuit of Fig. 12.5 has $R = 25$ kΩ, $C = 4$ μF, and an initial voltage of $v(0) = V_0 = 5$ V. Find (a) the time constant, (b) the voltage v for all positive time, and (c) v at $t = 3\tau$.   *Ans.* (a) 0.1 s, (b) $5e^{-10t}$ V, (c) 0.249 V

12-2.2 Find the current i in the circuit of Fig. 12.5 for the circuit elements and initial voltage
given in Exercise 12-2.1 for (a) all positive time, (b) $t = 0.3$ s, and (c) $t = 4\tau$.

Ans. (a) $0.2e^{-10t}$ mA, (b) 0.01 mA, (c) 0.004 mA

12-2.3 Repeat Exercise 12-2.1 if R and C are changed to $R = 2$ kΩ and $C = 0.5$ μF.

Ans. (a) 1 ms, (b) $5e^{-1000t}$ V, (c) 0.249 V

12-2.4 Find the approximate value (using Fig. 12.10) and the exact value of

$$v = 12e^{-4t} \text{ V}$$

at (a) $t = 0.5$ s, (b) $t = 1$ s, and (c) $t = 3\tau$.

Ans. (a) 1.68, 1.62 V, (b) 0.24, 0.22 V, (c) 0.6, 0.597 V

12.3 DRIVEN RC CIRCUITS

If the switch in Fig. 12.1 is in position 1, the resulting circuit is a *driven RC* circuit,
as shown in Fig. 12.11. That is, the circuit is *driven* by the driver, or source, V_b, as
discussed earlier.

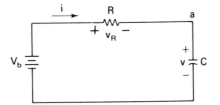

FIGURE 12.11 Driven RC circuit.

Circuit Equation: We may obtain the equation for the capacitor voltage v in Fig.
12.11 by writing a nodal equation at the node labeled *a*. If the reference node is
the bottom node, the node voltage at *a* is v and the current i entering *a* is given by

$$i = \frac{v_R}{R} = \frac{V_b - v}{R} \tag{12.19}$$

The current leaving *a* through the capacitor is also i and is given by

$$i = C\frac{dv}{dt} \tag{12.20}$$

Thus from (12.19) and (12.20) we have

$$C\frac{dv}{dt} = \frac{V_b - v}{R}$$

which may be rearranged in the form

$$\frac{dv}{dt} + \frac{1}{RC} v = \frac{1}{RC} V_b \tag{12.21}$$

which is the circuit equation for Fig. 12.11.

Capacitor Voltage: If the initial voltage on the capacitor is

$$v(0) = 0 \tag{12.22}$$

then it may be shown by calculus that (12.21) has the solution

$$v = V_b(1 - e^{-t/RC}) \text{ V} \tag{12.23}$$

Again, we recognize the time constant

$$\tau = RC \tag{12.24}$$

considered previously for the source-free circuit. Substituting (12.24) into (12.23), we have

$$v = V_b(1 - e^{-t/\tau}) \text{ V} \tag{12.25}$$

From the expression for v, we see that the capacitor voltage in the driven case is a *growing*, or *increasing*, exponential function. At $t = 0$, we have

$$v = v(0) = V_b(1 - e^0) = V_b(1 - 1) = 0$$

as it should be. However, as t increases, the term $e^{-t/\tau}$ decreases exponentially, as in Fig. 12.10. Thus $1 - e^{-t/\tau}$ increases with time and approaches 1 as t gets larger.

General Case: This last result may be considered generally by plotting the function

$$y = 1 - e^{-t/\tau} \tag{12.26}$$

versus t in multiples of τ. This is the rising exponential curve of Fig. 12.10, which was mentioned in Section 12.2. Therefore, to find a voltage or current of the form

$$y = K(1 - e^{-t/\tau}) \tag{12.27}$$

we may read the points off the curve of Fig. 12.10 and multiply the results by K.

Of course, we may also find the exact value of the voltage or current, by using the hand calculator, if we are only interested in a few points on the curve.

Example 12.5: Find the capacitor voltage v in Fig. 12.11 for (a) all positive time t, (b) $t = 0.1$ s, and (c) $t = 5\tau$, if $R = 10$ kΩ, $C = 25$ μF, $V_b = 6$ V, and the initial capacitor voltage is $v(0) = 0$.

Solution:: The time constant is

$$\tau = RC = (10 \times 10^3)(25 \times 10^{-6}) = 0.25 \text{ s}$$

so that by (12.25) we have, for all positive time t,

$$v = 6(1 - e^{-t/0.25}) \text{ V} \tag{12.28}$$

or

$$v = 6(1 - e^{-4t}) \text{ V} \tag{12.29}$$

Thus for part (a) we see that v has the rising exponential shape of Fig. 12.10.
    For part (b) we have from (12.29)

$$v(0.1) = 6(1 - e^{-0.4}) = 1.98 \text{ V}$$

and for part (c), by (12.28), we have

$$v = 6(1 - e^{-5}) = 5.96 \text{ V} \tag{12.30}$$

The capacitor will have the full 6 V of the battery across its terminals when it is fully charged. Therefore, by (12.30), we see that it is virtually fully charged and the circuit is in dc steady state for $t \geq 5$ time constants.

Example 12.6: Find the current i in the circuit of Example 12.5 for (a) all positive time t, (b) $t = 0.1$s, and (c) $t = 5\tau$.

Solution: By (12.19) we have

$$i = \frac{V_b - v}{R} = \frac{6 - v}{10} \text{ mA}$$

Substituting for v from (12.29), we have

$$i = \frac{6 - 6(1 - e^{-4t})}{10}$$

or

$$i = 0.6e^{-4t} \text{ mA} \tag{12.31}$$

which is the answer to part (a). For parts (b) and (c) we have

$$i(0.1) = 0.6e^{-0.4} = 0.402 \text{ mA}$$

and

$$i(5\tau) = 0.6e^{-4(5 \times 0.25)} = 0.004 \text{ mA}$$

General Current Case: We may find the current i in the general case of Fig. 12.11 by substituting (12.25) into (12.19). The result is

$$i = \frac{V_b - V_b(1 - e^{-t/\tau})}{R}$$

or

$$i = \frac{V_b}{R} e^{-t/\tau} \tag{12.32}$$

Thus we see that the current in the driven RC circuit with no initial stored energy $[v(0) = 0]$ is a decaying exponential like the voltage in the source-free circuit. As dc steady state is approached, the current tends to zero.

Transient and Steady-State Values: The current in the driven RC circuit, given by (12.32), and both the voltage v and the current in the source-free circuit, given by (12.11) and by v/R, respectively, are decaying exponentials. That is, they are *transitory* functions that are present for a short time and then are gone. Such functions are sometimes called *transient* functions, and have the property that they are only present in a circuit for a short time.

On the other hand, a function that remains after the transient is gone is called a *steady-state* function. It may be a varying function such as the sine wave of an ac current, or it may be a constant, or dc, steady-state function. We may note from (12.23) that the capacitor voltage v of the driven RC circuit contains both a transient and a steady-state function. It is the sum

$$v = v_{tr} + v_{ss} \tag{12.33}$$

of the transient function

$$v_{tr} = -V_b e^{-t/RC} \tag{12.34}$$

and the steady-state function

$$v_{ss} = V_b \tag{12.35}$$

PRACTICE EXERCISES

12-3.1 Find the capacitor voltage v and the current i for all positive time t in Fig. 12.11 if $R = 200$ kΩ, $C = 1$ μF, $V_b = 20$ V, and $v(0) = 0$.

Ans. $20(1 - e^{-5t})$ V, $0.1e^{-5t}$ mA

12-3.2 Calculate v and i in Exercise 12-3.1 at (a) $t = 0.1$ s, (b) $t = 0.5$ s, and (c) $t = 5\tau$. Check by using Fig. 12.10.

Ans. (a) 7.87 V, 0.06 mA, (b) 18.36 V, 0.008 mA, (c) 19.87 V, 0.0007 mA

12-3.3 Find the charge on the capacitor in Exercise 12-3.1 at (a) $t = 0.1$ s and (b) $t = 0.5$ s.

Ans. (a) 7.87 μC, (b) 18.36 μC

12.4 MORE GENERAL CIRCUITS

In many cases it is possible to analyze RC circuits that are more general than simple circuits by using the concepts of equivalent resistances and capacitances. For example, let us consider the more general circuit of Fig. 12.12, which contains three resistors and a capacitor. Suppose that we want to find the capacitor voltage v for all positive time given that $v(0) = 30$ V.

The three resistances of Fig. 12.12 may be combined into a single equivalent resistance R_T seen at the terminals of the capacitor. Since R_T is the equivalent of an 8-kΩ resistor in series with a parallel combination of a 3-kΩ and a 6-kΩ resistor, we have

$$R_T = 8 + \frac{3 \times 6}{3 + 6} = 10 \text{ k}\Omega$$

Thus the circuit of Fig. 12.12 is equivalent at the capacitor terminals to that of Fig. 12.13, which is a simple source-free RC circuit obtained by replacing the resistor combination in Fig. 12.12 by R_T.

The time constant for the circuit of Fig. 12.13 is

$$\tau = R_T C$$
$$= (10 \times 10^3)(0.25 \times 10^{-6})$$
$$= 0.25 \times 10^{-2} \text{ s}$$

Therefore, since $V_0 = 30$ V, we have by (12.11)

$$v = 30e^{-t/(0.25 \times 10^{-2})}$$

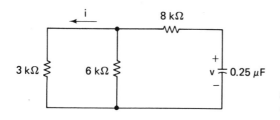

FIGURE 12.12 *RC circuit with four elements.*

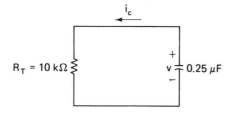

FIGURE 12.13 *Equivalent circuit of Fig. 12.12.*

or

$$v = 30e^{-400t} \, \text{V} \tag{12.36}$$

Now that we know the capacitor voltage we may find all the other voltages and currents. We may find all the resistor voltages in Fig. 12.12 by applying voltage division to the capacitor voltage, and from the resistor voltages we may use Ohm's law to get the resistor currents.

Example 12.7: Find the current i through the 3–kΩ resistor of Fig. 12.12 for all positive time.

Solution: From Fig. 12.13 we may find the capacitor current i_c by Ohm's law and (12.36). The result is

$$i_c = \frac{v}{R_T} = \frac{30e^{-400t} \, \text{V}}{10 \, \text{k}\Omega} = 3e^{-400t} \, \text{mA} \tag{12.37}$$

Since the circuit of Fig. 12.13 is equivalent, as far as the capacitor is concerned, to that of Fig. 12.12, we see that the current leaving the positive terminal of the capacitor in Fig. 12.12 is i_c. Thus by current division the current i is

$$i = \left(\frac{6}{3+6} \right) i_c = \frac{2}{3}(3e^{-400t}) = 2e^{-400t} \, \text{mA}$$

Example 12.8: Find v for all positive time in Fig. 12.14 if $v(0) = 0$.

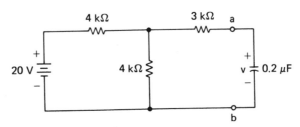

FIGURE 12.14 *Driven RC circuit with five elements.*

Solution: The circuit cannot be simplified by combining resistances into an equivalent, but we may replace the resistive network to the left of terminals *a–b* by its Thévenin equivalent circuit and solve for *v*. This is done by means of Fig. 12.15(a) and (b). The dead circuit is shown in Fig. 12.15(a), for which we see that the Thévenin resistance is

$$R_{\text{th}} = 3 + \frac{4 \times 4}{4 + 4} = 5 \text{ k}\Omega$$

Fig. 12.15(b) may be used to obtain the open-circuit voltage v_{oc}, which by voltage division is

$$v_{\text{oc}} = \frac{4}{4 + 4} \cdot 20 = 10 \text{ V}$$

The Thévenin equivalent circuit with the 0.2-μF capacitor is shown in Fig. 12.16. As far as terminals *a–b* are concerned, this circuit is the equivalent of that of Fig. 12.14.

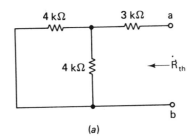

(a)

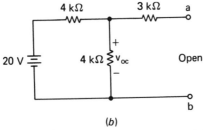

(b)

FIGURE 12.15 *Circuits for obtaining (a) R_{th} and (b) v_{oc} for the circuit of Fig. 12.14.*

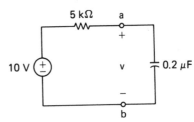

FIGURE 12.16 *Equivalent circuit of Fig. 12.14.*

From Fig. 12.16 we have the time constant,

$$\tau = (5 \times 10^3)(0.2 \times 10^{-6}) = 10^{-3} \text{ s}$$

and by (12.25) the voltage is

$$v = 10(1 - e^{-t/10^{-3}})$$
$$= 10(1 - e^{-1000\,t}) \text{ V}$$

PRACTICE EXERCISES

12-4.1 Find the voltage v for all positive time in the circuit shown if $v(0) = 30$ V and the element E is a 6–kΩ resistor.   *Ans.* $30e^{-2000\,t}$ V

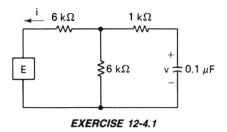

EXERCISE 12-4.1

12-4.2 Find the current i for all positive time in the circuit of Exercise 12-4.1.
                                                                       Ans. $2e^{-2000\,t}$ mA

12-4.3 Find the voltage v for all positive time in the circuit of Exercise 12-4.1 if the element E is a 20–V battery with positive terminal at the top and $v(0) = 0$.
                                                                       Ans. $10(1 - e^{-2500\,t})$ V

12.5 SHORTCUT PROCEDURE

As we saw in Section 12.3, the capacitor voltage v in the driven circuit of Fig. 12.11 is the sum of a transient voltage v_{tr} and a steady-state voltage v_{ss}. That is, we may write

$$v = v_{tr} + v_{ss} \qquad (12.38)$$

This fact may be used to develop a shortcut procedure for finding v, which may be used when $v(0) = 0$, to which we have been restricted in the previous cases, or for any given $v(0)$.

Since we know that the transient term is a decaying exponential with a time constant $\tau = RC$, we may write in the general case

$$v_{\text{tr}} = Ke^{-t/\tau} \qquad (12.39)$$

where K is an *arbitrary* constant. (That is, K may have any value, depending on the initial capacitor voltage.) Moreover, since τ is the same for a source-free circuit as it is for a driven circuit, we may kill the source (make it zero) and find τ directly from the source-free circuit which results.

We also know that v_{ss} is the dc steady-state voltage that is left after the transient is gone. Therefore, we may find it directly from the circuit when it reaches steady-state, at which time the capacitor is an open circuit.

All the currents and voltages in an *RC* circuit have the general form of (12.38) and (12.39), and may be found by the shortcut procedure. However, capacitor voltages are the easiest to obtain because they are continuous and thus their initial values are known.

In summary, we may kill the source in a driven circuit and find the time constant from the source-free circuit which results. The time constant is then used in (12.39) to obtain the transient term in the capacitor voltage. We may replace the capacitor in the driven circuit by an open circuit and find its voltage, which is v_{ss}. Then the capacitor voltage is the sum of the transient term v_{tr} and the steady-state term v_{ss}. That is, we have

$$v = Ke^{-t/\tau} + v_{\text{ss}} \qquad (12.40)$$

or

$$v = Ke^{-t/RC} + v_{\text{ss}} \qquad (12.41)$$

The constant K is then found from the initial voltage $v(0)$.

Example 12.9: Find the capacitor voltage v for all positive time in the circuit of Fig. 12.17 if the initial voltage is $v(0) = 20$ V.

Solution: Killing the source (replacing it by a short circuit) results in the source-free circuit of Fig. 12.18, from which we have

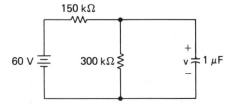

FIGURE 12.17 *Driven circuit with a parallel resistor-capacitor combination.*

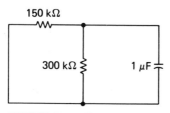

FIGURE 12.18 The circuit of Fig. 12.17 with the source killed.

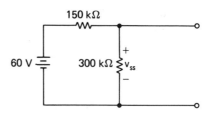

FIGURE 12.19 The circuit of Fig. 12.17 at dc steady state.

$$R_T = \frac{150 \times 300}{150 + 300} = 100 \text{ k}\Omega$$

and the time constant

$$\tau = R_T C = (100 \times 10^3)(10^{-6}) = 0.1 \text{ s}$$

Therefore, the transient term in the voltage is

$$v_{\text{tr}} = Ke^{-t/0.1} = Ke^{-10t} \text{ V} \tag{12.42}$$

To find the steady-state term in v we replace the capacitor by an open circuit, as shown in Fig. 12.19. In this case we have by voltage division

$$v_{\text{ss}} = \frac{300}{150 + 300} \cdot 60 = 40 \text{ V} \tag{12.43}$$

From (12.42) and (12.43) we have

$$v = v(t) = v_{\text{tr}} + v_{\text{ss}} = Ke^{-10t} + 40 \tag{12.44}$$

We are given that

$$v(0) = 20 \text{ V}$$

and from (12.44) with $t = 0$ we have

$$v(0) = K + 40$$

Equating these two results, we have

$$K + 40 = 20$$

from which $K = -20$. Therefore, the capacitor voltage for all positive time is by (12.44)

$$v = 40 - 20e^{-10t} \text{ V}$$

Example 12.10: To illustrate the charging and discharging of a single capacitor, let us consider the circuit of Practice Exercise 12-1.1. If the switch is moved to position 1 at $t = 0$ when $v_c = 0$ and kept in that position for 10 ms before being moved to position 2, find the voltage v_c for all positive time.

Solution: From $t = 0$ to $t = 10$ ms the circuit is a driven RC circuit as shown in Fig. 12.20(a). The time constant is

$$\tau_1 = 10^3 \times 10^{-6} = 10^{-3} \text{ s} = 1 \text{ ms}$$

obtained by killing the source. The transient term is therefore

$$v_{tr} = Ke^{-1000t} \text{ V}$$

The steady-state term is

$$v_{ss} = 10 \text{ V}$$

obtained by open-circuiting the capacitor. Thus we have

$$v_c = Ke^{-1000t} + 10 \text{ V} \tag{12.45}$$

At $t = 0$ we have

$$v_c(0) = 0 = K + 10$$

Thus $K = -10$, so that (12.45) becomes

$$v_c = 10(1 - e^{-1000t}) \text{ V} \tag{12.46}$$

At $t = 10$ ms we have

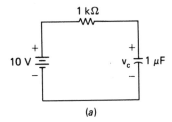

(a)

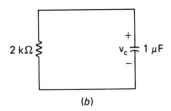

(b)

FIGURE 12.20 Equivalent circuits of the circuit of Exercise 12.1.1 for (a) $0 < t < 10$ ms and (b) $t > 10$ ms.

$$v_c(0.01) = 10(1 - e^{-10}) = 10 \text{ V}$$

so that the capacitor is fully charged.

For $t > 10$ ms the switch is in position 2, so that the circuit is that of Fig. 12.20(b). The time constant is

$$\tau_2 = (2 \times 10^3)(10^{-6}) \text{ s} = 2 \text{ ms}$$

and the voltage at the beginning ($t = 10$ ms) is 10 V. Thus we see that v_c rises, like the increasing function of Fig. 12.10, to its fully charged value of 10 V in the first 10 ms. It then decays like the decreasing function of Fig. 12.10, but with a different time constant. In the decaying stage ($t > 10$ ms), the voltage essentially reaches zero after five time constants, or $5 \times 2 = 10$ ms. The sketch of v_c for $0 \leq t \leq 20$ ms is shown in Fig. 12.21.

As we see, the voltage first rises from 0 to 10 V in the charging phase with a time constant of 1 ms. It is then discharged (from 10 to 0 V) with a time constant of 2 ms. In the charging phase it takes about five time constants, or 5 ms, to reach its final, fully charged value, and after five time constants (10 ms) in the discharging phase it is completely discharged.

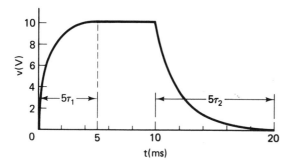

FIGURE 12.21 Sketch of the capacitor voltage for the circuit of Exercise 12.1.1.

PRACTICE EXERCISES

12-5.1 Find the capacitor voltage v for all positive time in the circuit shown if the initial voltage is 15 V.   *Ans.* $30 - 15e^{-500t}$ V

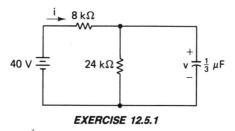

EXERCISE 12.5.1

12-5.2 Find i for all positive time in the circuit of Exercise 12-5.1. (*Suggestion:* Use the result of Exercise 12-5.1.)   *Ans.* $\frac{1}{8}$ $(10 + 15e^{-500\,t})$ mA

12-5.3 Find the steady-state values of v_1 and v_2 in the circuit shown.   *Ans.* 10 V, 40 V

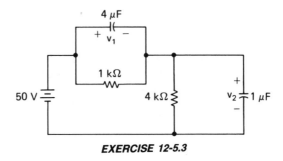

EXERCISE 12-5.3

12.6 SUMMARY

A simple source-free RC circuit is one containing a resistor R in series with a capacitor C. If the capacitor is initially charged (at $t = 0$) to some voltage V_0, a current will flow through the resistor, discharging the capacitor. The capacitor voltage v for all positive time is given by

$$v = V_0 e^{-t/\tau} \text{ V}$$

where τ is the RC time constant.

A driven RC circuit is one containing a source, such as a battery. The capacitor voltage v in this case is given by

$$v = v_{tr} + v_{ss}$$

where v_{tr} is a transient voltage given by

$$v_{tr} = K e^{-t/\tau}$$

and v_{ss} is the dc steady-state capacitor voltage. Again τ is the RC time constant and K is an arbitrary constant, which depends on the initial voltage $v(0)$. In the case where $v(0) = 0$ (an initially uncharged capacitor) and $v_{ss} = V_b$, the voltage v is given by

$$v = V_b(1 - e^{-t/\tau}) \text{ V}$$

A shortcut procedure is to kill the source and find τ from the resulting source-free circuit. Then v_{ss} is found as the dc steady-state voltage across the capacitor replaced by an open circuit.

More general circuits may be solved by reducing them to simple circuits by means of combining resistors and/or capacitors into single equivalent elements, or by means of Thévenin's theorem.

PROBLEMS

12.1 A simple source-free RC circuit has $R = 200$ kΩ, $C = 10$ μF, and an initial capacitor voltage of $v(0) = 10$ V. Find (a) the time constant, (b) the capacitor voltage v for all positive time, and (c) the current for all positive time leaving the positive terminal of the capacitor.

12.2 Repeat Problem 12.1 if $R = 40$ kΩ and $C = 0.5$ μF.

12.3 A simple source-free RC circuit has $R = 100$ kΩ, $C = 10$ μF, and $v(0) = 20$ V, where v is the capacitor voltage. Find v for (a) $t = 1$ s, (b) $t = 2$ s, and (c) $t = 5$ τ, where τ is the RC time constant.

12.4 A 2-μF capacitor is in series with a 10-kΩ resistor. If the energy stored in the capacitor is 100 μJ at $t = 0$, find the capacitor voltage (a) for all positive time, (b) at $t = 2$ ms, and (c) at $t = 100$ ms.

12.5 A simple source-free RC circuit has a capacitor voltage of

$$v = 20e^{-4t} \text{ V}$$

Find (a) the time constant, (b) the energy stored at $t = 0$, and (c) the charge on the plates at $t = 0$, if the capacitance is 2 μF.

12.6 The circuit shown is in dc steady state when the switch is moved from position 1 to position 2 at $t = 0$. Find v for all positive time.

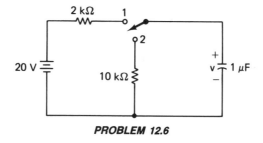

PROBLEM 12.6

12.7 Repeat Problem 12.6 if the switch is moved from position 2 to position 1 at $t = 0$.

12.8 Find v for $t > 0$ if the circuit shown is in dc steady state when the switch is opened at $t = 0$.

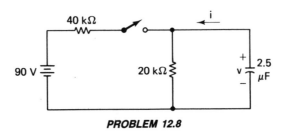

PROBLEM 12.8

12.9 Find i for $t > 0$ in Problem 12.8.

12.10 Find v for $t > 0$ in the circuit shown if $v(0) = 12$ V.

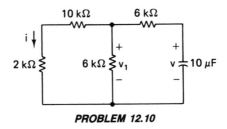

PROBLEM 12.10

12.11 Find v_1 for $t > 0$ in the circuit of Problem 12.10.

12.12 Find i for $t > 0$ in the circuit of Problem 12.10.

12.13 Repeat Problem 12.8 if the switch is *closed* at $t = 0$.

12.14 The circuit shown is in dc steady state when the switch is opened at $t = 0$. Find v for all positive time.

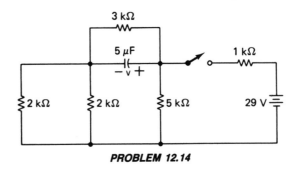

PROBLEM 12.14

12.15 Find i for $t > 0$ in the circuit shown if $v(0) = 12$ V. (*Suggestion:* Find v first.)

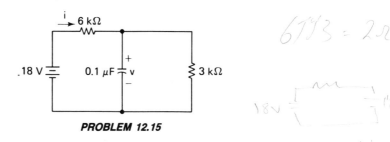

PROBLEM 12.15

12.16 Find *v* and *i* for *t* > 0 in the circuit shown if *v*(0) = 0. (*Suggestion:* Replace the source and resistor by the equivalent Thévenin circuit, or use the shortcut procedure.)

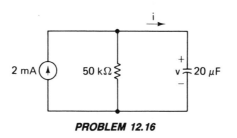

PROBLEM 12.16

13

MAGNETISM

In previous chapters we have discussed the force of gravity, or the *gravitational field,* which holds us on the earth, and the force exhibited by an *electric field,* which exists between the conductors of a charged capacitor. These are unseen forces that act without physical contact but are clearly present, as their effects show. Such forces, as we have noted, are called *field forces,* or *fields.* In this chapter we will consider another such field effect, known as *magnetism,* or a *magnetic field,* which is closely related to electricity.

Magnetism is a force that acts between certain objects called *magnets,* some of which we have mentioned earlier in connection with dc meters. The compass, used as early as A.D. 1000 by Chinese and Mediterranean navigators, is perhaps the most popular use of a magnet. It consists of a small magnet in the shape of a pointer, or needle, mounted on a pivot so it can move freely. Unless there are other metal objects near, one end of the needle will point toward the north and the other end toward the south, because the earth itself is surrounded by a magnetic field.

A *permanent* magnet is one made of a material such as iron, steel, or magnetite (loadstone) that can retain its magnetic properties without the aid of external means. *Electromagnets,* on the other hand, are those whose magnetic properties are induced by electric currents in a manner that we will consider in this chapter. Electromagnets retain their magnetic properties only as long as the current is present. This relation between electricity and magnetism, or *electromagnetism,* as it is sometimes called, will be our main concern in the chapter.

The laws of magnetism were developed before the adoption of the SI units, and consequently the old units honoring the pioneers in the field are being abandoned. Some of these pioneers were the English physician William Gilbert (1540–1603), the great German mathematician Karl Friedrich Gauss (1777–1855), James Clerk Maxwell (1831–1879), the great British scientist, and the Danish physicist and chemist Hans Christian Oersted (1777–1851). Gilbert, official physician to Queen Elizabeth I of England, was one of the earliest researchers in magnetism, showing in 1600 how the electric attraction of amber differed from the magnetic attraction of loadstone. He was one of the first to use the Latin word *electrum* for amber, and he coined the word *electrica* for other substances that behave like amber. Gauss, the "Prince of Mathematicians," contributed greatly to the mathematical theory of electromagnetics. Maxwell worked out the exact mathematical laws of electricity and magnetism, and Oersted was the discoverer, in 1819, of electromagnetism.

There are countless uses of magnetism in our everyday lives. Magnets are found in television sets, telephones, radios, motors, generators, transformers, and circuit breakers. Other applications are magnetic lifts, magnetic door latches, loudspeakers, magnetic tapes and disks for handling data in a computer, and magnetic vibrators, to name a few. Without electromagnetism our standard of living would be quite different, to say the least.

13.1 THE MAGNETIC FIELD

If a permanent magnet in the vicinity of a compass is moved, the needle of the compass will move with it. Also, as Oersted first showed in 1819, the needle of a compass will deflect if brought near a current-carrying wire. These effects, as we have noted, may be explained by postulating that a *magnetic field* exists in the region surrounding a magnet (the permanent magnet or the current-carrying wire, which is an electromagnet).

North and South Poles: The magnet may be thought of as having *magnetic poles* called *north* and *south,* which correspond to the opposite polarities of electric charges. The magnetic field is then postulated to consist of invisible *lines of force,* or *magnetic flux,* which radiate in continuous loops in the direction north to south. The magnetic field is symbolized by the lines of force surrounding the bar magnet (a bar of iron with magnetic properties), as shown in Fig. 13.1. The north and south poles are labeled N and S. The symbolism is very similar to that of the electric field, as shown earlier in Fig. 11.3.

The magnetic field may be strengthened by concentrating the flux lines in a smaller area. This may be done, as shown by the *horseshoe* magnet of Fig. 13.2, which is a bar magnet bent in the shape of a horseshoe. Other examples of horseshoe magnets are shown in Figs. 13.3 and 13.4. In both cases a *keeper* is shown, which

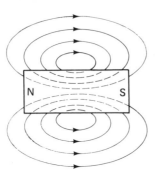

FIGURE 13.1 Symbol of a magnetic field around a bar magnet.

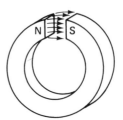

FIGURE 13.2 Horse-shoe magnet.

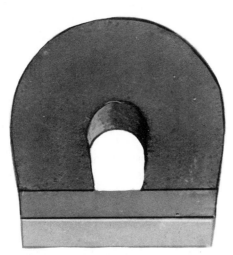

FIGURE 13.3 Horseshoe magnet (Courtesy, Sargent-Welch Scientific Company).

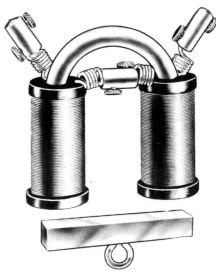

FIGURE 13.4 Horseshoe magnet using electric current to produce the magnetizing force (Courtesy, Sargent-Welch Scientific Company).

is placed across the poles of the magnet to maintain its strength when it is not in use.

Attraction and Repulsion: Magnets, like electric charges, may attract or repel each other. If unlike poles of two magnets are brought together, the magnets will attract, with fields as shown in Fig. 13.5. If like poles are brought together, the magnets will repel, as shown in Fig. 13.6.

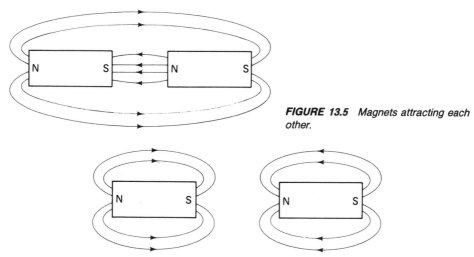

FIGURE 13.5 *Magnets attracting each other.*

FIGURE 13.6 *Magnets repelling each other.*

Electromagnets: Oersted discovered in 1819 that every current in a wire is surrounded by a magnetic field, as symbolized in Fig. 13.7 by the current *i* and the flux lines as shown. Thus electric current can be used to obtain a magnet (produce a magnetic flux), and such a magnet, which we called earlier an *electromagnet,* can be turned "on" and "off" with the current. This is an advantage it has over the permanent magnet, which is always "on." An example of an electromagnet is the horseshoe magnet shown earlier in Fig. 13.4.

Flux lines

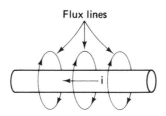

FIGURE   13.7 *Current-carrying wire with accompanying magnetic field.*

Right-Hand Rule: We may determine the direction of the flux around a current-carrying wire by means of a so-called *right-hand rule.* If we imagine grasping the wire with the right hand so that the thumb points in the direction of the current (we are considering conventional current only), the fingers circling the wire point in the direction of the lines of flux around the wire. The right-hand rule may be illustrated by means of Fig. 13.7.

If the conductor is formed in the shape of a loop, as shown in Fig. 13.8, by the right-hand rule we see that all the flux lines pass through the loop in the same direction. Thus for a given current, the total number of flux lines has not changed, but we have strengthened the field by concentrating it in a smaller area.

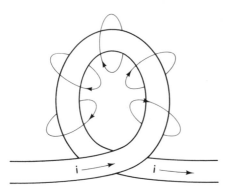

FIGURE 13.8 *Loop of wire with a flux-producing current.*

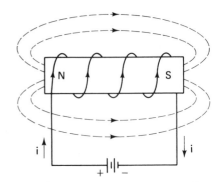

FIGURE 13.9 *Coil of wire with a flux-producing current.*

We may concentrate the flux even more by winding the wire into a *coil* of many loops around a *core* of metal or even a column of air. Such an arrangement is shown in Fig. 13.9, which is evidently an electromagnet with north and south poles as shown. To distinguish the current from the magnetic flux we have used dashed lines for the flux. The right-hand rule is easier to apply in Fig. 13.9 if the fingers grasp the core in the direction of the current. The thumb will then point in the direction of the flux. (This statement of the right-hand rule is equivalent to the earlier statement, as may be seen in Fig. 13.9.) The coil of wire is the general shape of the *inductor,* which we will consider in Chapter 14.

13.2 MAGNETIC FLUX

The symbol we will use for the magnetic flux is Φ (the capital Greek letter *phi*), and its unit of measurement in the SI is the *weber* (Wb), named for the German physicist Wilhelm Eduard Weber (1804–1891). One weber is 10^8 lines of flux, or 10^8 *maxwells,* as it was called in the earlier cgs system (centimeter-gram-second). Thus the weber is a large unit and for typical fields the flux may be measured in such smaller multiples as microwebers (μWb).

Example 13.1: Find the number of lines of flux in 1 μWb.

Solution: Since 1 Wb = 10^8 lines, we have

$$1 \ \mu\text{Wb} = (1 \times 10^{-6} \ \text{Wb})(10^8 \ \text{lines/Wb})$$

$$= 100 \ \text{lines}$$

We note that the unit Wb cancels, leaving the answer in lines.

Magnetic Circuits: As we have noted, magnetic flux lines always form closed loops. Thus we may think of the path these loops trace as a *magnetic circuit,* analogous to the electric circuit with charges flowing around closed paths. Although there is no *flow* in the case of flux (it simply exists in the form of a loop), there are many analogies, as we will see, between magnetic flux in a magnetic circuit and current, or moving charges, in an electric circuit. An example is the doughnut-shaped magnetic circuit of Fig. 13.10, having a flux Φ produced by the current I.

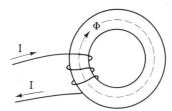

FIGURE 13.10 *Magnetic circuit.*

Permeability: It is easier to set up magnetic flux in, or *magnetize,* some materials such as iron than in other materials such as air. If, for example, the magnetic circuit of Fig. 13.10 is an easily magnetized, or *magnetic* material, the flux Φ produced by the current I will be largely confined to the circuit and there will be relatively little flux *leaking* through the surrounding air. This is because the flow will naturally follow the course of least resistance and set up in the magnetic material rather than through the nonmagnetic air.

A measure of how easily magnetic flux can be established in a material is its *permeability,* denoted by μ (the lowercase Greek letter *mu*). The higher the permeability, the more easily the material can be magnetized. Thus permeability is a measure of the material's ability to permit the setting up of magnetic flux. As examples, the permeability of iron may be 200 times that of air, and that of permalloy may be 100,000 times that of air. The counterpart of permeability in the electric field is permittivity, which, as we may recall from Chapter 11, is a measure of how easily electric flux lines can be established in the material.

Relative Permeability: The SI units of permeability, as we will see, are webers per ampere-meter (Wb/Am). As an example, the permeability of a vacuum (which is very nearly that of air) is denoted by $\mu = \mu_0$ and given by

$$\mu_0 = 4\pi \times 10^{-7} \text{ Wb/Am} \tag{13.1}$$

The ratio of the permeability μ of a material to that of a vacuum is called its *relative permeability,* and denoted by μ_r. Thus we have

$$\mu_r = \frac{\mu}{\mu_0} \tag{13.2}$$

The relative permeability is a dimensionless quantity since it is the ratio of two permeabilities.

Nonmagnetic materials have relative permeabilities very nearly equal to 1. These include materials called *paramagnetic* materials, such as aluminum, platinum, and manganese, with μ_r slightly greater than 1, and *diamagnetic materials,* such as copper, zinc, gold, and silver, with μ_r slightly less than 1. Also, paramagnetic and diamagnetic materials are magnetized in opposite directions by the field of the magnetizing current. On the other hand, magnetic materials have values of μ_r much greater than 1. *Ferromagnetic materials* are those, such as iron, steel, nickel, and permalloy, that have very high relative permeabilities, ranging from 100 to 100,000.

Example 13.2: Find the permeability μ of iron if its relative permeability is $\mu_r = 50$.

Solution: From (13.2) we have

$$\mu = \mu_r \mu_0 \qquad (13.3)$$

which in this case is

$$\mu = (4\pi \times 10^{-7})(50)$$
$$= 2\pi \times 10^{-5} \text{ Wb/Am}$$

Flux Density: In a magnetic field with magnetic flux Φ, the number of flux lines per unit area is called the *flux density,* denoted by B. If the cross-sectional area perpendicular to the flux is A, as in Fig. 13.11, the flux density is therefore

$$B = \frac{\Phi}{A} \qquad (13.4)$$

If Φ is in webers and A in m², then B is in Wb/m². In the SI, by definition, a Wb/m² is a *tesla* (symbolized by T), named for the Yugoslav-born American inventor Nikola Tesla (1857–1943).

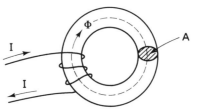

FIGURE 13.11 Magnetic circuit with cross-sectional area A.

Example 13.3: A magnetic field with a flux of 10 mWb passes through an area of 4 cm². Find the flux density B in teslas.

Solution: The cross-sectional area A is given in m² by

$$A = (4 \text{ cm}^2)(0.01 \text{ m/cm})^2$$
$$= 4 \times 10^{-4} \text{ m}^2$$

Therefore, by (13.4) we have

$$B = \frac{\Phi}{A}$$

$$= \frac{10 \times 10^{-3} \text{ Wb}}{4 \times 10^{-4} \text{ m}^2} = 25 \text{ T}$$

PRACTICE EXERCISES

13-2.1 A magnetic field has a flux $\Phi = 500{,}000$ lines. Find Φ in SI units.   *Ans.* 5 mWb

13-2.2 Find the relative permeability of a material whose permeability is $16\pi \times 10^{-6}$ Wb/Am.   *Ans.* 40

13-2.3 In Exercise 13.2.1, find the flux density in teslas if the flux cross-sectional area is (a) 0.002 m² and (b) 0.25 in².   *Ans.* (a) 2.5 T, (b) 31 T

13-2.4 If the flux density of a magnetic field is 1.5 T and the cross-sectional area is 6 cm², find the flux.   *Ans.* 0.9 mWb

13.3 OHM'S LAW FOR MAGNETIC CIRCUITS

As we have seen, magnetic flux in a magnetic circuit is analogous to current in an electric circuit. There is, in fact, a close analogy between magnetic circuits and electric circuits that we will discuss in this section. In addition to current and flux, there are analogies in magnetic circuits to voltage, resistance, and even Ohm's law in electric circuits.

Magnetomotive Force: To complete the analogy between magnetic and electric circuits, let us consider the magnetic circuit of Fig. 13.12. The external force, analogous to voltage in the electric circuit, is produced by the current I, which establishes the magnetic field with flux Φ. However, as was pointed out in Section 13.1, the wire carrying the current may be shaped in the form of a coil with N turns, as in Fig.

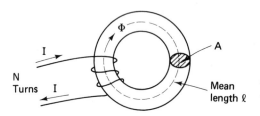

FIGURE 13.12 *Magnetic circuit with flux produced by a current in N turns of wire.*

13.12, and the current I can be made to do the work of N currents NI. The quantity NI, called the *magnetomotive force* (mmf), is thus the driving agent analogous to voltage or emf in the electric circuit. The SI unit for mmf is *ampere-turns*, abbreviated At, and the symbol for magnetomotive force is $\mathscr{F}$.

Example 13.4: A coil with 20 turns is to produce a magnetomotive force of $\mathscr{F} = 500$ At. Find the current I required.

Solution: Since we have

$$\mathscr{F} = NI \tag{13.5}$$

we may write

$$I = \frac{\mathscr{F}}{N} = \frac{500}{20} = 25 \text{ A}$$

We note in this example that the units of ampere-turns divided by turns gives amperes, as we should expect. Actually, *turns* is a dimensionless number that divides into amperes to yield amperes.

Reluctance: Just as it is more difficult for electric current to flow in some materials than it is in others, it is also more difficult for a magnetic field to be set up in some materials than in others. The measure of the difficulty of establishing the magnetic flux is called the *reluctance* of the material. For a given mmf, the higher the reluctance, the lower the flux, and vice versa. Thus reluctance in a magnetic circuit is the opposition to the establishment of the magnetic flux just as resistance in an electric circuit is the opposition to the flow of current.

Reluctance is symbolized by the script letter $\mathscr{R}$ to distinguish it from resistance. As in the case of resistance, the reluctance is directly proportional to the length l of the magnetic circuit and inversely proportional to the cross-sectional area A, both of which are illustrated in Fig. 13.12. The expression for reluctance is

$$\mathscr{R} = \frac{l}{\mu A} \tag{13.6}$$

where μ is the permeability considered earlier. Since μ is measured in Wb/Am, the units of reluctance are, by (13.6), m/(Wb–m²/A–m) = amperes/weber (A/Wb), or ampere-turns/weber (At/Wb). Some authors use the unit *rel* for reluctance, where 1 rel is defined as 1 At/Wb.

Equation (13.6) is consistent with the definition of permeability. The lower the permeability, the more difficult the establishment of the flux, as we recall, and, by (13.6), the higher the reluctance. The converse is also true. High permeability corresponds to low reluctance and thus less opposition to the flux.

Ohm's Law: By comparison with electric circuits, as we have seen, the magnetomotive force $\mathscr{F} = NI$ corresponds to voltage, the flux Φ corresponds to current, and the reluctance $\mathscr{R}$ corresponds to resistance. To complete the analogy, there is an *Ohm's law* for magnetic circuits, given by

$$\Phi = \frac{\mathscr{F}}{\mathscr{R}} \tag{13.7}$$

or

$$\Phi = \frac{NI}{\mathscr{R}} \tag{13.8}$$

which is identical in form to that of electric circuits.

Example 13.5: For the circuit of Fig. 13.12 the coil has 200 turns, the current is $I = 3$ A, the permeability of the magnetic material is $\mu = 5 \times 10^{-5}$ Wb/Am, the *average* or *mean* length of the circuit is $l = 2$ m, and the cross-sectional area is $A = 0.008$ m². Find the magnetic flux Φ.

Solution: By (13.6) the reluctance is given by

$$\mathscr{R} = \frac{l}{\mu A} = \frac{2}{(5 \times 10^{-5})(8 \times 10^{-3})}$$

$$= 5 \times 10^6 \text{ A/Wb}$$

The mmf is given by

$$\mathscr{F} = NI = 200 \times 3 = 600 \text{ At}$$

so that by Ohm's law we have

$$\Phi = \frac{\mathscr{F}}{\mathscr{R}} = \frac{600}{5 \times 10^6} \text{ Wb}$$

$$= 120 \ \mu\text{Wb}$$

Example 13.6: A magnetic circuit has $NI = 300$ At, $l = 0.5$ m, and $\mu = 6 \times 10^{-5}$ Wb/Am. Find the flux density B.

Solution: By means of (13.6) and (13.8) we may write

$$\Phi = \frac{NI}{\mathcal{R}}$$

$$= \frac{NI}{l/\mu A}$$

or

$$\Phi = \frac{\mu A NI}{l}$$

Since $\Phi = BA$ this may be written

$$BA = \frac{\mu A NI}{l}$$

which becomes, upon canceling A,

$$B = \frac{\mu NI}{l} \tag{13.9}$$

Therefore, in the case under consideration we have

$$B = \frac{(6 \times 10^{-5})(300)}{0.5} = 0.036 \text{ T}$$

PRACTICE EXERCISES

13-3.1 Find the reluctance of the magnetic circuit of Fig. 13.12 if the relative permeability of the material is 20, the length l is 0.5 m, and the area A is 10^{-3} m².
Ans. 2×10^7 A/Wb

13-3.2 If the reluctance of the circuit of Fig. 13.12 is as given in Exercise 13.3.1, find (a) the flux Φ if $N = 100$ and $I = 4$ A and (b) the current I if $\Phi = 20$ μWb and $N = 50$.
Ans. (a) 2×10^{-5} Wb, (b) 8 A

13-3.3 In Fig. 13.12 the flux density is 200 μT, the relative permeability is 50, and the length is $l = 0.6$ m. Find the ampere-turns.
Ans. 1.91 At

13.4 MAGNETIC FIELD INTENSITY

The magnetomotive force of a magnetic circuit is specified by the ampere-turns, $\mathscr{F} = NI$, but the *intensity* of the magnetic field depends on the length of the circuit. Evidently, the field will be less intense for a long length of material than for a short length with the same mmf. Specifically, by intensity, or *magnetic field intensity,* we mean the *magnetomotive force per unit length.* This quantity is symbolized by H and thus is given by

$$H = \frac{\mathscr{F}}{l} = \frac{NI}{l} \qquad (13.10)$$

The units of H are evidently At/m or A/m.

Other terms used for H are *magnetic field strength* or *magnetizing force.* It is of interest to note from (13.10) that H depends only on the ampere-turns and the length of the circuit. It is independent, for example, of the magnetic material.

B–H Curve: The flux density B and the field intensity H may be shown to be related by using the results we have obtained thus far. By (13.4) and (13.8) we have

$$B = \frac{\Phi}{A} = \frac{NI}{\mathscr{R}A} \qquad (13.11)$$

Substituting for $\mathscr{R}$ from (13.6), we may write

$$B = \frac{NI}{(l/\mu A)\,A} = \mu \frac{NI}{l}$$

Finally, by (13.10), we have

$$B = \mu H \qquad (13.12)$$

If the permeability μ were constant, the plot of B versus H, by (13.12), would be a straight line. However, permeability varies not only with the material, but also with the field intensity H in a typical magnetic material. Thus the *B–H curve* (the plot of B versus H) is not a straight line, but typically resembles that of Fig. 13.13.

As H initially increases, B increases more or less linearly, but at a certain point the *B–H* curve begins to level off. *Saturation* is eventually reached where further increases in H have very little effect on B. Thus the permeability $\mu = B/H$ decreases as H increases.

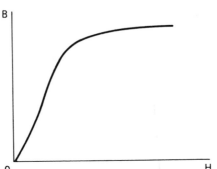

B

0   H   **FIGURE 13.13** *Typical B-H curve.*

Example 13.7: A magnetizing force is obtained with a coil of 200 turns and a current *I*. The length of the magnetic circuit is 0.25 m and the relative permeability μ_r of the magnetic material is 100 when $I = 1$ A, 80 when $I = 4$ A, and 65 when $I = 5$ A. Find *B* for these three values of current.

Solution: When $I = 1$ A, the magnetizing force is $NI = 200$ At and

$$H = \frac{NI}{l} = \frac{200}{0.25} = 800 \text{ At/m}$$

In this case the flux density is

$$B = \mu H = (100)(4\pi \times 10^{-7})(800) = 0.1 \text{ T}$$

For $I = 4$ A, we have $NI = 800$ At, $H = 800/0.25 = 3200$ At/m, and

$$B = (80)(4\pi \times 10^{-7})(3200) = 0.322 \text{ T}$$

Finally, for $I = 5$ A, we have $NI = 1000$ At, $H = 1000/0.25 = 4000$ At/m, and

$$B = (65)(4\pi \times 10^{-7})(4000) = 0.327 \text{ T}$$

Hysteresis: If the magnetic material is initially unmagnetized and the magnetizing current *I* is zero, then $H = NI/l = 0$ and $B = 0$. This situation is represented by the point 0 in Fig. 13.14. If the material is ferromagnetic and an increasing current is then applied, *H* will, of course, increase with *I*, as will *B*, and the *B–H* curve will be traced along *0–a* in the direction shown. At point *a* the material is saturated and any further increase in *H* (or *I*) will not cause an appreciable increase in *B*. If it is attempted to *demagnetize* the material by reversing the current, the *B–H* curve will not follow the line *a–0* but will decrease along the line *a–b* in the direction shown. This *lagging* of *B* behind *H* (*H* reaches zero at point *b* but *B* is still not zero) is called *hysteresis* and is a consequence of the magnetic properties of the iron.

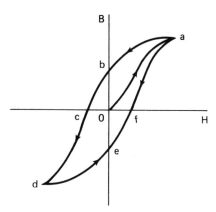

FIGURE 13.14 *Hysteresis loop.*

The flux density *B* can be made zero only by a negative current (or *H*) occurring at point *c.* Any further increase in *H* in the negative direction will lead to saturation at point *d,* and a reversal of the magnetizing current, or *H,* will lead to the curve *def,* at the end of which *B* is again zero. Increasing *H* still further leads to saturation again at *a* and the process will be repeated by decreasing *H* at this point. The value of *B* at point *b* is called the *residual flux density* and, of course, the negative of this value occurs at *e.* This is the flux density *residing* in the material when the magnetizing current is zero.

The closed path traced in Fig. 13.14 is called a *hysteresis loop.* The energy lost (supplied by the source) in tracing the loop is called the *hysteresis loss* and may be shown to be related to the area inside the hysteresis loop. For some materials, such as a ferrite switching core, used in memory banks in computers, the hysteresis loop is very nearly rectangular, as shown in Fig. 13.15. There are two stable states, as required in computer memories, the saturated positive *B* state and the saturated negative *B* state. When *H* changes sufficiently in either of the states, the situation changes almost immediately to the other state.

The device shown in Fig. 13.16 is a *hysteresis loop tracer,* which may be used to project a hysteresis loop of a given specimen of magnetic material on the screen of an oscilloscope, such as the one shown in Fig. 13.17.

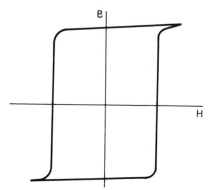

FIGURE 13.15 *Rectangular hysteresis loop.*

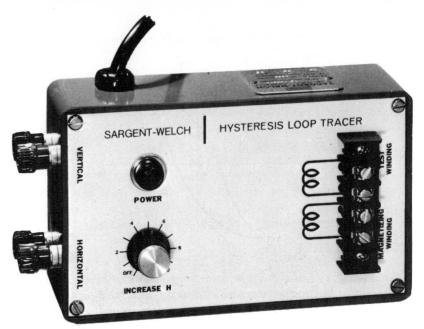

FIGURE 13.16 *Hysteresis loop tracer (Courtesy, Sargent-Welch Scientific Company).*

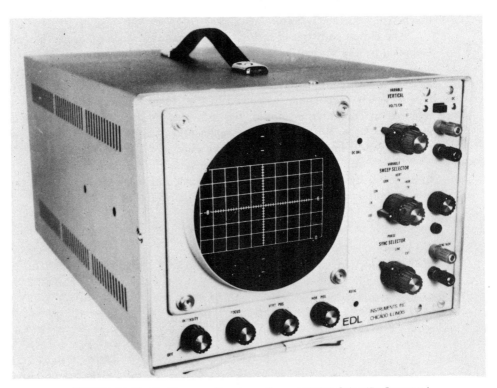

FIGURE 13.17 *Oscilloscope (Courtesy, Sargent-Welch Scientific Company).*

PRACTICE EXERCISES

13-4.1 Find the flux density B in a magnetic circuit if $N = 100$ turns, $I = 4$ A, the length of the circuit is $l = 0.2$ m, and the relative permeability is $\mu_r = 100$.

Ans. 0.251 T

13-4.2 In the circuit of Exercise 13.4.1 the current is increased to 8 A with the result that μ_r drops to 85. Find B in this case. *Ans.* 0.427 T

13.5 SIMPLE MAGNETIC CIRCUITS

As we have seen, Ohm's law for electric circuits has its counterpart in magnetic circuits. The same is true for Kirchhoff's laws. The flux lines exist in closed loops and do not flow in the same sense as charges in an electric circuit. However, to complete the analogies between electric and magnetic circuits, we may think of the flux as the counterpart of current and assign polarity arrows as shown in the example of Fig. 13.18. The analogy with Kirchhoff's current law in this case is

$$\Phi_a = \Phi_b + \Phi_c \qquad (13.13)$$

which states that the sum of the *fluxes* entering a junction is equal to the sum of those leaving the junction.

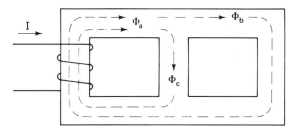

FIGURE 13.18 *Magnetic circuit with alternate flux paths.*

Ampère's Circuital Law: If we think of an applied mmf *NI* as an mmf *rise* and the force required to establish the flux in the rest of the circuit as an mmf *drop,* we have the analogy of voltage rises and drops in an electric circuit. Using this analogy, we may state the counterpart of Kirchhoff's voltage law, which was given by Ampère in 1820. This law, known as *Ampère's circuital law,* states that the sum of the mmf rises equals the sum of the mmf drops around any closed path of a magnetic circuit. (Ampère, one of the early researchers in electric circuits, was also one of the pioneers in magnetism. It is said that he learned on September 11, 1820, of Oersted's discovery that a magnetic needle is influenced by an electric current, and seven days later he presented a paper with an explanation.)

If we take the applied mmf as NI and let $\mathscr{F}_1$, $\mathscr{F}_2$, . . . , $\mathscr{F}_n$ be the mmf drops around the closed path, Ampère's circuital law takes the form

$$NI = \mathscr{F}_1 + \mathscr{F}_2 + \cdots + \mathscr{F}_n \tag{13.14}$$

It is often most convenient to represent $\mathscr{F}$ by Hl, in which case (13.14) becomes

$$NI = H_1 l_1 + H_2 l_2 + \cdots + H_n l_n \tag{13.15}$$

where H_1, H_2, . . . , H_n are the magnetizing forces required for sections 1, 2, . . . , n in the circuit, and l_1, l_2, . . . , l_n are the section lengths.

As an example, the circuit of Fig. 13.12, considered in Section 13.3, consists of a single closed path of one type of material. Thus (13.15) in this case is

$$NI = Hl$$

which, since $H = B/\mu$, may be written

$$NI = \frac{Bl}{\mu}$$

$$= BA \frac{l}{\mu A}$$

$$= \Phi \mathscr{R}$$

Therefore we have, as before,

$$\Phi = \frac{NI}{\mathscr{R}}$$

By analogy with electric circuits, the circuit of Fig. 13.12 is a *series* circuit, because the flux Φ is the same throughout. A more complex series circuit is that of Fig. 13.19, in which two different materials are used. As our next example we will analyze this circuit.

Example 13.8: Find the flux Φ in the series circuit of Fig. 13.19 if the current I and the turns N are known. The mean length of the part made of material 1 is l_1, the mean length of the part made of material 2 is l_2, the cross-sectional area is A, and μ_1 and μ_2 are the permeabilities of materials 1 and 2.

Solution: By Ampére's law we have

$$NI = H_1 l_1 + H_2 l_2$$

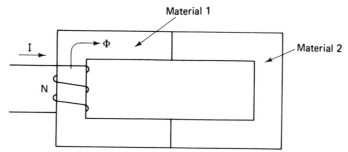

FIGURE 13.19 *Series magnetic circuit.*

which, since $H_1 = B/\mu_1$ and $H_2 = B/\mu_2$, becomes

$$NI = B\left(\frac{l_1}{\mu_1} + \frac{l_2}{\mu_2}\right)$$

(Since the area A is the same everywhere in the circuit, the flux density B is also the same.) Substituting $B = \Phi/A$ and solving for Φ results in

$$\Phi = \frac{ANI}{l_1/\mu_1 + l_2/\mu_2} \tag{13.16}$$

which is our answer.

Series Reluctances: We may rewrite (13.16) in the form

$$\Phi = \frac{NI}{l_1/\mu_1 A + l_2/\mu_2 A}$$

Using the expression for reluctance, this is equivalent to

$$\Phi = \frac{NI}{\mathscr{R}_1 + \mathscr{R}_2} \tag{13.17}$$

where

$$\mathscr{R}_1 = \frac{l_1}{\mu_1 A}$$

and

$$\mathscr{R}_2 = \frac{l_2}{\mu_2 A}$$

are the reluctances of the paths in materials 1 and 2.

Therefore, we may write (13.17) in the form

$$\Phi = \frac{NI}{\mathscr{R}_T} \qquad (13.18)$$

where

$$\mathscr{R}_T = \mathscr{R}_1 + \mathscr{R}_2 \qquad (13.19)$$

This illustrates another analogy with electric circuits. In a series circuit the *equivalent* reluctance is the sum of the series reluctances.

Example 13.9: Find the ampere turns needed to produce a flux of Φ in Fig. 13.20 if the core of the coil is air with a cross-sectional area of A.

Solution: By Ampère's law we have

$$NI = Hl + H_a l_a \qquad (13.20)$$

where l is the length of the core, as shown, and l_a is the length (on the average) of the remainder of the closed flux loops. The flux inside the coil is concentrated in a relatively small area A, so that the intensity H is relatively high. On the other hand, outside the coil the flux is dispersed over a very large area so that H_a is very small. In most cases, therefore, the term $H_a l_a$ may be neglected by comparison with Hl, so that we have approximately

$$NI = Hl$$

$$= \frac{Bl}{\mu_0}$$

or

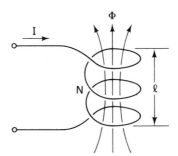

FIGURE 13.20 *Coil with an air core.*

$$NI = \frac{\Phi l}{\mu_0 A} \tag{13.21}$$

Since for the magnetic core the reluctance is

$$\mathcal{R} = \frac{l}{\mu_0 A} \tag{13.22}$$

we have

$$NI = \Phi \mathcal{R} \tag{13.23}$$

Relay: As a final example of a magnetic circuit, let us consider the *relay* of Fig. 13.21. A relay is a device which uses magnetism produced by an electric current to move a piece of metal that opens and closes electric contacts. In the case of Fig. 13.21, when a current I flows in the coil the iron core is magnetized, which attracts the moving bar, closing the air gap. When the current returns to zero, the spring reopens the gap, moving the bar back to its original position.

We note in Fig. 13.21 the fringing of the flux in the air gap, as was also true of the electric field lines. If the gap is sufficiently small, we may approximate its cross-sectional area as equal to that of the iron core. If fringing is appreciable, we must take it into account by using a larger air gap area.

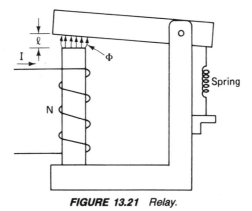

FIGURE 13.21 *Relay.*

Example 13.10: Find the current I needed to establish a flux Φ of 10^{-4} Wb in the magnetic circuit of Fig. 13.21 if the mean length of the iron is $l_i = 500$ mm, the air gap length is $l = 5$ mm, the cross-sectional area is $A = 2 \times 10^{-4}$ m^2, N = 200, and the relative permeability of the iron is $\mu_r = 1000$.

Solution: By Ampère's law we have

$$NI = Hl + H_l l_l \tag{13.24}$$

where Hl is the force required for the air gap and $H_l l_l$ is that required for the iron. Since we have

$$B = \frac{\Phi}{A} = \frac{10^{-4}}{2 \times 10^{-4}} = 0.5 \text{ T}$$

$$H = \frac{B}{\mu_0} = \frac{0.5}{4\pi \times 10^{-7}} = 3.98 \times 10^5 \text{ A/m}$$

and

$$H_l = \frac{B}{1000\mu_0} = 3.98 \times 10^2 \text{ A/m}$$

we may write (13.24) in the form

$$200I = (3.98 \times 10^5)(5 \times 10^{-3}) + (3.98 \times 10^2)(500 \times 10^{-3})$$

$$= 2189$$

Therefore, the current is

$$I = \frac{2189}{200} = 10.945 \text{ A}$$

We note in this example that H required for the air gap is much greater than that required for the magnetic material. This is a typical case in that respect.

As a final note in this section, we observe that only series circuits have been considered. The circuit of Fig. 13.18 is a parallel circuit which could be analyzed by means of the counterpart of Kirchhoff's current law (13.13), Ampère's circuital law, and Ohm's law. However, for our purposes in Chapter 14, we shall have need for only series circuit theory.

PRACTICE EXERCISES

13-5.1 Find the equivalent reluctance of the series circuit shown if the relative permeabilities of materials 1 and 2 are 50 and 100, respectively; their lengths are 200 mm and 300 mm; the length of the air gap is 2 mm; and the cross-sectional area is 2×10^{-4} m². (*Suggestion:* There are three reluctances in series.)   *Ans.* 3.581×10^7 At/Wb

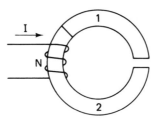

EXERCISE 13-5.1

13-5.2   If $N = 200$ and $I = 3$ A in Exercise 13.5.1, find the flux Φ.   *Ans.* 16.76 μWb

13-5.3   Find the current I required to produce a flux of 20 μWb in the coil of Fig. 13.20 if $N = 400$, $l = 10$ cm, and $A = 2 \times 10^{-4}$ m². [*Suggestion:* Use the shortcut approximation of Eq. (13.21).]   *Ans.* 19.9 A

13.6 SUMMARY

Current flowing in a wire produces a magnetic field consisting of closed loops of magnetic flux around the wire. To produce a more concentrated field the wire may be wound in a coil around a core of air or of some magnetic material such as iron or steel. The core is then an electromagnet which will attract other magnets. Permanent magnets are those which retain their magnetic properties without the aid of external forces produced by electric currents.

The permeability μ of a material is a measure of how easily a magnetic field can be established in the material. For example, the reluctance of the material is

$$\mathscr{R} = \frac{l}{\mu A}$$

where l is the length and A is the cross-sectional area of the material. Thus the higher the permeability, the lower the reluctance. Ohm's law for the magnetic circuit of the core is

$$\Phi = \frac{NI}{\mathscr{R}}$$

where Φ is the magnetic flux, N is the number of turns of the wire, and I is the current in the wire. The quantity $NI = \mathscr{F}$ is the magnetomotive force corresponding to emf in the electric circuit.

The magnetic field intensity is

$$H = \frac{NI}{l}$$

and is related to the flux density $B = \Phi/A$ by

$$B = \mu H$$

This is the equation of the $B–H$ curve, which traces a closed loop, called the hysteresis loop, when the material is a ferromagnetic material and the current is varied repeatedly from plus to minus.

In addition to Ohm's law for magnetic circuits, there are also magnetic circuit analogies to Kirchhoff's current law and Kirchhoff's voltage law. In the latter case, the magnetic circuit law is called Ampère's circuital law, and states that the sum of the magnetomotive force rises NI equals the sum of the Hl drops in any closed path. These laws enable us to analyze magnetic circuits in the same way we analyze resistive electric circuits.

PROBLEMS

13.1 For the magnetic circuit shown, the cross-sectional area A is 2×10^{-4} m². Find the flux density if the flux Φ is (a) 5 mWb, (b) 200 μWb, and (c) 0.006 Wb.

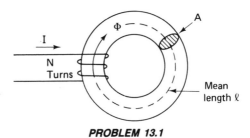

PROBLEM 13.1

13.2 Find Φ in the circuit of Problem 13.1 if $B = 2$ T, and the area A is (a) 2 cm², (b) 0.5 in², and (c) that of a circle of radius 5 mm.

13.3 Find the reluctance of the magnetic circuit and the flux in Problem 13.1 if $A = 3 \times 10^{-4}$ m², $l = 20$ cm, $N = 500$, $I = 4$ A, and the relative permeability of the material is 100.

13.4 Find Φ in the circuit of Problem 13.1 if the inner circle of the circuit is of radius 14 cm and the outer circle is of radius 18 cm. The other quantities are $A = 10^{-4}$ m², $N = 200$, $I = 4$ A, and $\mu_r = 1000$. (*Suggestion:* The mean length is that of a circle whose diameter is the average of 14 and 18 cm.)

13.5 A magnetic circuit has $NI = 500$ At, $l = 0.25$ m, $\mu = 0.02$ Wb/Am, and $A = 0.001$ m². Find Φ.

13.6 Find the mean length l of the magnetic circuit shown. (*Suggestion:* The mean path is that shown dashed through the center.)

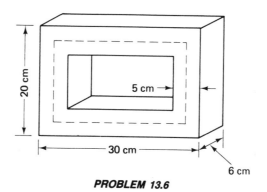

PROBLEM 13.6

13.7 If an mmf of 1000 At is applied to the circuit of Problem 13.6 and the permeability of the material is 0.008 Wb/Am, find the flux. (*Suggestion:* Take A as the area of a rectangle 5 cm $\times$ 6 cm.)

13.8 Repeat Problem 13.7 if the outside dimensions of the circuit are changed from 20 cm $\times$ 30 cm to 25 cm $\times$ 50 cm.

13.9 Find the flux density in a magnetic circuit of length 0.5 m and $N = 400$ turns if (a) the current is $I = 1$ A when the relative permeability is $\mu_r = 100$, (b) $I = 5$ A when $\mu_r = 75$, and (c) $I = 6$ A when $\mu_r = 63$.

13.10 Find the flux Φ in a magnetic circuit with $l = 0.25$ m, $A = 10^{-3}$ m², $N = 200$ turns, and (a) $I = 2$ A, (b) $I = 4$ A, and (c) $I = 6$ A if for these values of current μ_r is 100, 80, and 60, respectively.

13.11 Find the flux Φ in the series circuit of Fig. 13.19 if $N = 200$, $I = 3$ A, the mean length of the part made of material 1 is 0.4 m with a permeability of 2×10^{-4} Wb/Am, and the mean length of the part made of material 2 is 0.6 m with a permeability of 3×10^{-4} Wb/Am. The cross-sectional area is 2×10^{-4} m² for both materials.

13.12 Find the flux Φ in the circuit of Fig. 13.20 if the core is air, the cross-sectional area of the core is 10^{-4} m², $l = 20$ cm, $N = 100$, and $I = 2$ A.

13.13 Find the current needed to produce a flux of 1 μWb in the coil of Problem 13.12.

13.14 Find the flux Φ in the series magnetic circuit of Fig. 13.21 if the mean length of the air gap is $l = 2$ cm, that of the rest of the circuit is 50 cm with relative permeability $\mu_r = 10^4$, the cross-sectional area of the air gap and the rest of the circuit is 10^{-4} m², and $NI = 800$ At.

13.15 Find the reluctance of the series magnetic circuit of Practice Exercise 13.5.1 if material 1 has $\mu_r = 100$ and $l = 10$ cm, material 2 has $\mu_r = 500$ and $l = 20$ cm, the air gap has $l = 1$ cm, and the cross-sectional area for all three parts of the circuit is $A = 0.2$ cm².

13.16 Find Φ in the circuit of Problem 13.15 if $N = 200$ and $I = 4$ A.

13.17 Find the flux Φ in the given circuit if $N_1 = 200$, $I_1 = 3$ A, $N_2 = 400$, $I_2 = 2$ A, $A = 10^{-4}$ m^2, $l = 20$ cm, and $\mu = 2 \times 10^{-3}$ Wb/Am. (*Suggestion:* The magnetomotive force is $\mathscr{F} = N_1I_1 + N_2I_2$ since by the right-hand rule, the mmfs are producing flux in the same direction.)

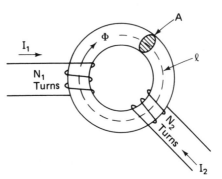

PROBLEM 13.17

13.18 If the current I_1 is reversed in Problem 13.17, find Φ.

14

INDUCTORS

Thus far we have considered circuits with three types of elements—sources, resistors, and capacitors. With the completion of the chapter on magnetism we are now ready to consider the fourth circuit element, the *inductor*. As we shall see in this chapter, an inductor has many properties that are very similar to those of a capacitor. Like the capacitor, the inductor has the capability of storing energy, with the difference that the field associated with the inductor is a magnetic field, whereas that associated with the capacitor is an electric field. The voltage–current terminal relation of an inductor is based on a rate of change of one of the variables, as was the case for the capacitor. However, the inductor case is the dual of the capacitor, its voltage being proportional to the rate of change of its current.

Also, like the capacitor, the inductor, when connected with a resistor to form a simple *RL* circuit is associated with a time constant. The voltages and currents in *RL* circuits are identical in form to those of *RC* circuits, since both may be expressed in terms of exponential functions. The only difference, as we will see, is the value of the time constant.

In this chapter we will discuss inductors, their associated property of *inductance*, their voltage–current relationship, their stored energy, and their equivalent series and parallel connections. We will see that these topics arise from the concept of the magnetic field, just as those of the capacitor were based on the electric field.

14.1 DEFINITIONS

An *inductor* is a two-terminal element consisting of a conductor wound in the shape of a coil, as shown in Fig. 14.1. The number of turns N of the coil may vary from a fraction of a single turn to many hundreds of turns, depending on the application for which the inductor is designed. There is a terminal current i, as shown, which establishes the magnetic flux Φ in the manner discussed in Chapter 13. The terminal voltage v is an *induced* voltage, caused by the action of the magnetic field, as we will see.

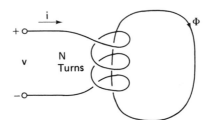

FIGURE 14.1 *Inductor.*

Faraday's Law: Two scientists, the English physicist Michael Faraday (1791–1867) and the American physicist Joseph Henry (1797–1878), discovered independently and at almost the same time in 1831 the principle of the voltage induced by a magnetic field. Their discovery was that a changing flux Φ, as in Fig. 14.1, produces a voltage such as v across the terminals of the coil. This principle, known as *Faraday's law of electromagnetic induction,* may be stated mathematically in the form

$$v = N\frac{d\Phi}{dt} \qquad (14.1)$$

where N is the number of turns of the inductor and $d\Phi/dt$ is the *rate of change* of the flux Φ. The quantity $d\Phi/dt$ is also called the *derivative* of Φ with respect to t, and may be found, as discussed in Chapter 11, by the methods of calculus. However, we will find the rate of change as the slope of the flux curve, as we did in Chapter 11 for the quantity dv/dt in the case of the voltage curve.

Electric Generator Principle: Faraday's law is also the principle of the electric generator, since the flux linking the coil may be changed by moving the coil as well as by changing the current. Thus if the coil, shown in Fig. 14.2 as a single loop, is moved, or rotated, as indicated, in the magnetic field produced by the magnet, the flux Φ linking the coil will be changing and therefore a voltage v will appear across its terminals. This is precisely the action of a generator where the coils are wound on an armature or a rotor, which is turned by an external prime mover. The *sliprings* and *brushes* shown provide a path from the generated voltage to any external load. The sliprings rotate with the coil and rub against the brushes.

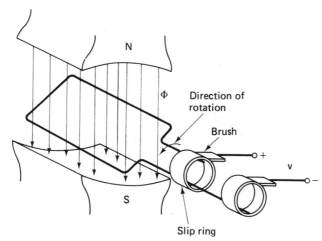

FIGURE 14.2 *Simplified action of a generator.*

The situation, therefore, is exactly as Faraday thought in 1820 when he first heard of Oersted's discovery of electromagnetism: "If a current can produce a magnetic field, a field should be able to produce a current."

The action illustrated in Fig. 14.2 is a highly simplified version of the actual case. At the other extreme is the armature of a 2500-hp dc motor produced by Westinghouse Electric Corporation, shown in Fig. 14.3. In the case of ac generators, the coils that produce the voltage are stationary, and those that produce the magnetic field rotate, but the principle is the same as that of Fig. 14.2.

Lenz's Law: Our discussion thus far has concerned the *magnitude* of the induced voltage rather than its *polarity*. The polarity may be determined by a principle known as *Lenz's law,* which is attributed to the German scientist Heinrich F. E. Lenz. Lenz's law states that *an induced effect, such as a voltage, is always such as to oppose the cause that induces it.* Thus in the case of the induced voltage of a coil, the polarity is such as to produce a current, and therefore a flux, which will oppose any change in the original flux. If this were not the case, the current would produce a flux, which would aid the current, producing more flux, which would further aid the current, producing still more flux, and so on.

Referring to Fig. 14.1, we may see how the coil is wound and, by the right-hand rule, which way the flux produced by the current is oriented. The voltage polarity is as shown since the flux tends to produce a current that limits, or "chokes," the change in the current through the coil. Thus the coil appears like a voltage source to the external circuit with a polarity that opposes the entering current. Because of this "choking" of the change in current, the inductor is sometimes called a *choke* or *choke coil.* Therefore, as a consequence of Lenz's law, the inductor current cannot change instantaneously and is thus a continuous function like the capacitor voltage discussed in Chapter 11.

FIGURE 14.3 *Armature of a 2500-horsepower dc motor (Courtesy, Westinghouse Electric Corporation).*

Examples of inductors with capabilities of carrying 5 to 20 A of current are shown in Fig. 14.4. These are so-called *power line chokes* that are used to prevent high-frequency currents from going out over power lines and interfering with nearby radio receiving sets.

PRACTICE EXERCISES

14-1.1 Find the induced voltage of an inductor with 200 turns if the flux Φ increases in a straight line from 0 to 50 μWb in 2 ms. (*Suggestion:* The rate of change of Φ is $d\Phi/dt = 50$ μWb/2 ms $= 25$ mWb/s.)   *Ans.* 5 V

14-1.2 Repeat Exercise 14.1.1 if Φ is a constant value of 40 μWb.   *Ans.* 0 V

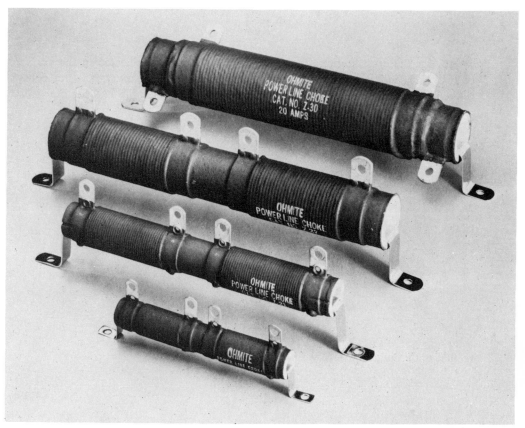

FIGURE 14.4 *Power-line choke coils (Courtesy, Ohmite Manufacturing Company).*

14.2 INDUCTANCE AND CIRCUIT RELATIONSHIPS

We define a *linear* inductor as one whose flux $N\Phi$ is directly proportional to the current i which produces it. That is, we have

$$N\Phi = Li \tag{14.2}$$

where L is a constant of proportionality. We will consider only linear inductors and refer to them simply as inductors.

Inductance: The quantity L in (14.2) is called the *inductance* of the inductor, and its SI unit is the henry (H), named, of course, for Joseph Henry. We may relate the henry to other SI units by means of (14.2), and we may show by eliminating Φ and i that L depends only on the physical characteristics of the inductor. To see this, we note that

$$\Phi = \frac{Ni}{\mathscr{R}} = \frac{Ni\mu A}{l} \tag{14.3}$$

where A is the cross-sectional area and l is the length, as indicated in the example of Fig. 14.5, and μ is the permeability of the core. The quantity $\mathscr{R}$ is, of course, the reluctance of the core path, as illustrated earlier in Example 13.9 for Fig. 13.20.

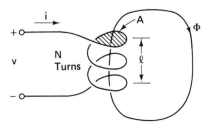

FIGURE 14.5 *Inductor with indicated dimensions.*

Substituting (14.3) into (14.2) results in

$$\frac{N^2 i\mu A}{l} = Li$$

so that upon dividing out i, we have the inductance

$$L = \frac{N^2 \mu A}{l} \tag{14.4}$$

Example 14.1: Find the inductance of the coil of Fig. 14.5 if $N = 50$, $l = 0.05$ m, $A = 0.003$ m². and the core is air.

Solution: By (14.4) we have

$$L = \frac{(50)^2 (4\pi \times 10^{-7})(0.003)}{0.05}$$

$$= 1.885 \times 10^{-4} \text{ H}$$

or $L = 0.1885$ mH.

Range of Inductance Values: Inductors are available with values of inductance as low as a few nanohenrys (nH). These are made of very fine wire in a coil of less than a single turn and are used for very high frequency work. Inductors used for radio frequencies have typical values in the microhenry (μH) and millihenry (mH) range, and those with iron cores used in the power industry may be as large as 1 to 100 H. As examples, the inductors of Fig. 14.4 have inductances of 14, 15, 18, and 110 μH.

Voltage–Current Relationship: The circuit symbol for an inductor is shown in Fig. 14.6. It is, of course, impossible to tell from the symbol how the coil is wound, but this information is not needed if we specify the voltage–current relationship.

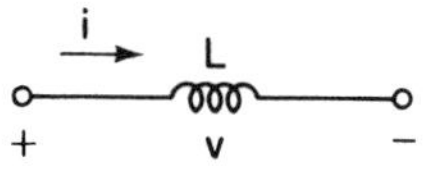

FIGURE 14.6 *Circuit symbol for an inductor.*

This may be done by means of (14.1) and (14.2). Since N and L are not functions of time, the rate of change of $N\Phi$ and Li are, respectively, $Nd\Phi/dt$ and Ldi/dt, so that by (14.2) we may write

$$N\frac{d\Phi}{dt} = L\frac{di}{dt} \qquad (14.5)$$

Substituting this result into (14.1), we have the voltage–current relationship for the inductor,

$$v = L\frac{di}{dt} \qquad (14.6)$$

The polarities of v and i are as shown in Fig. 14.6. If either polarity (but not both) is reversed, we must change the sign of one of the members of (14.6).

We note that L and C are duals, since replacing C by L, v by i, and i by v in the terminal relationship (11.11) for the capacitor yields (14.6).

Example 14.2: A 0.5-H inductor carries a current that is changing at a rate of 20 A/s. Find its terminal voltage.

Solution: We have in this case

$$\frac{di}{dt} = 20 \text{ A/s}$$

so that by (14.6) the voltage is

$$v = (0.5)(20) = 10 \text{ V}$$

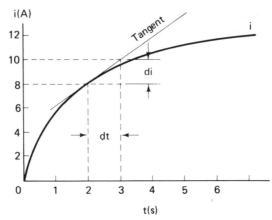

FIGURE 14.7 *Graph of current versus time.*

Example 14.3: A 20-H inductor carries a current i as shown in Fig. 14.7. Find the terminal voltage at $t = 2$ s.

Solution: As in the case of the capacitor in Chapter 11, the rate of change at a point on a curve, such as i or v, is the slope of the tangent to the curve at the point. If the changes di and dt were extremely small, it is seen from Fig. 14.7 that the rate of change of i with respect to t at $t = 2$ s is the slope of the tangent, given by

$$\frac{di}{dt} = \frac{\text{rise}}{\text{run}} = \frac{(10-8)\,\text{A}}{(3-2)\,\text{s}} = 2\text{ A/s}$$

Therefore, the inductor voltage at $t = 2$ s is

$$v = L\frac{di}{dt} = 20 \times 2 = 40\text{ V}$$

Energy Stored in Inductors: An ideal inductor, like an ideal capacitor, does not dissipate energy like a resistor. Energy supplied to the inductor by a current is stored in the magnetic field. The energy is said to be stored *in* the field because when the current changes, the magnetic field flux changes, which produces a voltage.

If w_L is the energy stored in an inductor with inductance L and a current i, it may be shown by calculus that

$$w_L = \frac{1}{2}Li^2 \qquad\qquad (14.7)$$

If L is in henrys and i is in amperes, w_L is in joules. (Note that we also may obtain w_L as the dual of w_C, the energy stored in a capacitor.)

Example 14.4: A 10-H inductor carries a current of 2 A. Find the energy stored in its magnetic field.

Solution: By (14.7) we have

$$w_L = \frac{1}{2} Li^2 = \frac{1}{2}(10)(2)^2 = 20 \text{ J}$$

PRACTICE EXERCISES

14-2.1 Find the inductance of the coil shown if the mean length of the magnetic flux path is 0.1 m, its cross-sectional area is 10^{-4} m², the number of turns is 500, and the permeability of the core is 1.2 mWb/Am.   *Ans.* 0.3 H

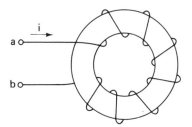

EXERCISE 14-2.1

14-2.2 If the current in the coil of Exercise 14.2.1 is $i = 2$ A, find the flux in the core.
         Ans. 1.2 mWb

14-2.3 If the current in the coil of Exercise 14.2.1 is changing at the rate

$$\frac{di}{dt} = 200 \text{ A/s}$$

find (a) the voltage across terminals *a-b* and (b) its positive terminal.
         Ans. (a) 60 V, (b) terminal *a*

14-2.4 A 5-mH inductor is storing an energy of 40 mJ. Find the current it carries.
         Ans. 4 A

14-2.5 A 0.1-H inductor with terminals *a-b* has a current i entering terminal a given by the graph shown. Find the terminal voltage v_{ab} at the times (a) $t = 1$ s, (b) $t = 3$ s, and (c) $t = 5$ s.   *Ans.* (a) 0.5 V, (b) 0, (c) −0.5 V

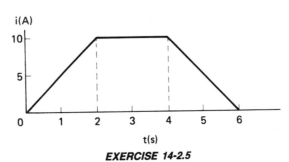

EXERCISE 14-2.5

14-2.6 Show from (14.4) that the unit Wb/Am of permeability is also H/m, and is thus the dual of F/m, the unit of permittivity.

14.3 SERIES AND PARALLEL INDUCTORS

Like resistors and capacitors, inductors may be connected in series or in parallel, and equivalent circuits may be obtained. As examples, the circuits of Figs. 14.8 and 14.9 are series and parallel connections, respectively, of N inductors with inductances $L_1, L_2, \ldots, L_N$.

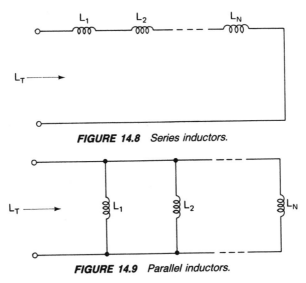

FIGURE 14.8 *Series inductors.*

FIGURE 14.9 *Parallel inductors.*

Series Connection: As we saw in the case of Ohm's law, the voltage $v = Ri$ across a resistor is directly proportional to the resistance. In the case of an inductor, we see by (14.6) that the voltage is proportional to the inductance. Thus in this respect, inductance behaves like resistance and therefore series inductors will combine like series resistors and parallel inductors will combine like parallel resistors. In the case of the series connection of N inductors shown in Fig. 14.8, the total, or *equivalent*, inductance L_T seen at the terminals is therefore the sum of the individual inductances. That is, we have

$$L_T = L_1 + L_2 + \cdots + L_N \qquad (14.8)$$

Parallel Connection: Parallel inductors, as shown in Fig. 14.9, are combined like parallel resistors. That is, the equivalent inductance L_T may be found from

$$\frac{1}{L_T} = \frac{1}{L_1} + \frac{1}{L_2} + \cdots + \frac{1}{L_N} \qquad (14.9)$$

In the special case of two inductances L_1 and L_2, (14.9) becomes

$$\frac{1}{L_T} = \frac{1}{L_1} + \frac{1}{L_2} = \frac{L_1 + L_2}{L_1 L_2}$$

Therefore, we have for two parallel inductors,

$$L_T = \frac{L_1 L_2}{L_1 + L_2} \qquad (14.10)$$

In other words, the equivalent inductance is the product divided by the sum of the two individual inductances. This is, of course, a direct analogy with parallel resistors.

Equations (14.8) and (14.9) may be obtained also as the duals of the corresponding capacitor relations, since L and C are duals, as are series and parallel.

Example 14.5: Find the equivalent inductance L_T in Fig. 14.10.

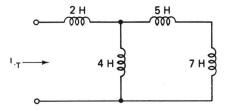

FIGURE 14.10 *Network of inductors.*

Solution: The 5–H and 7–H inductors are in series and thus their equivalent is $5 + 7 = 12$ H, which is in parallel with the 4–H inductor. The equivalent of the parallel 12–H and 4–H inductances is

$$\frac{12 \times 4}{12 + 4} = 3 \text{ H}$$

Finally, the 3–H equivalent inductor is in series with the 2–H inductor, so that we have

$$L_T = 2 + 3 = 5\,H$$

Shielding: When two or more inductors are connected in the vicinity of each other, as in Figs. 14.8 and 14.9, their magnetic fields may interact with one another to produce unwanted effects. (The effects are not always undesirable. In the case of a *transformer,* to be considered in Chapter 21, we deliberately place two coils near each other to use mutually each other's magnetic field.) We may prevent one component from affecting another by *shielding,* or placing metal *shields* over the individual components. A good magnetic metal provides an excellent shield for a steady magnetic field (one produced by a permanent magnet or a dc current), by effectively acting as a short circuit for the magnetic flux lines. A good conducting metal is best for shielding a varying magnetic field. It has induced currents that oppose the inducing field, so that there is little net field strength outside the shield.

In all the circuits involving two or more inductors, except those forming the transformers of Chapter 21, we will assume that the magnetic fields are shielded and do not influence one another.

PRACTICE EXERCISES

14-3.1 Find the equivalent inductance of five 10-mH inductors connected in series.

Ans. 50 mH

14-3.2 Find the equivalent inductance of five 10-mH inductors connected in parallel.

Ans. 2 mH

14.4 SOURCE-FREE RL CIRCUITS

Because of their similarity to *RC* circuits, *RL* circuits may be easily analyzed by using the results of Chapter 12. Thus we will not need to devote an entire chapter to *RL* circuits as we did to *RC* circuits, but instead we will consider source-free *RL* circuits in this section and driven *RL* circuits in Section 14.5.

Inductor a Short Circuit to DC: Just as the capacitor is an open circuit to dc, so the inductor is a short circuit to dc in the steady-state case. To illustrate this, consider the driven *RL* circuit of Fig. 14.11(a). The inductor voltage is

$$v_L = L\frac{di}{dt} \qquad (14.11)$$

and in dc steady state the current i is not changing. Therefore, $di/dt = 0$

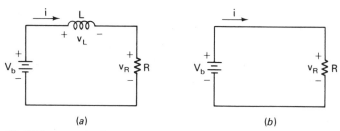

(a)   (b)

FIGURE 14.11 *(a) RL circuit and (b) its equivalent in dc steady state.*

and we have $v_L = 0$. Thus the inductor is a short circuit to dc, so that in the steady state the circuit of Fig. 14.11(b) is equivalent to that of Fig. 14.11(a). The steady-state values of the circuit variables are, from Fig. 14.11(b),

$$v_R = V_b$$

$$i = \frac{v_R}{R} = \frac{V_b}{R}$$

Simple *RL* Circuit: If the battery in Fig. 14.11(a) is replaced by a short circuit, by means of a switching action, the result is the source-free simple *RL* circuit of Fig. 14.12. To obtain its circuit equation we note that by KVL we have

$$v_L + v_R = 0$$

which by (14.11) and Ohm's law may be written

$$L\frac{di}{dt} + Ri = 0 \tag{14.12}$$

If the initial current is $i = i(0) = I_0$, some specified value, the analysis of the source-free circuit requires that we solve the equations

$$\frac{di}{dt} + \frac{R}{L}i = 0 \tag{14.13}$$

$$i(0) = I_0$$

[The first of these is (14.12) divided through by L.]

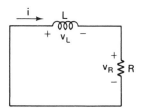

FIGURE 14.12 *Source-free RL circuit.*

Equations (14.13) are almost identical to (12.7), which described the RC source-free circuit. Indeed, if in (12.7) we replace v by i, RC by L/R, and V_0 by I_0, it becomes identical to (14.13). Since RC is the time constant τ in the RC case, it follows that the solution i of (14.13) is identical to that of (12.7), given in (12.11) with V_0 replaced by I_0. The result is

$$i = I_0 e^{-t/\tau} \tag{14.14}$$

where τ is the *RL time constant*,

$$\tau = \frac{L}{R} \quad \text{seconds} \tag{14.15}$$

That is, the solution is the exponential function

$$i = I_0 e^{-Rt/L} \text{ A} \tag{14.16}$$

This result may also be obtained as the dual of (12.11) for the RC circuit.

The values of $e^{-t/\tau}$ were given for $t = 0, \tau, 2\tau, \ldots, 6\tau$ in Table 12.1 and a graph of the function was given in Fig. 12.10. These values and the graph apply to the RL circuit, with τ changed to L/R, of course. Also, as in the RC case, we may find any point on the current curve by means of the hand calculator.

Example 14.6: Find the current i in the source-free circuit of Fig. 14.12 if $L = 2$ H, $R = 10 \ \Omega$, and $I_0 = 3$ A. Compute its value at (a) $t = 0$ s, (b) $t = 0.2$ s, and (c) $t = 1$ s.

Solution: By (14.16) the current for all $t \geq 0$ is given by

$$i = 3e^{-5t} \text{ A}$$

At $t = 0$ we have

$$i(0) = 3e^0 = 3 \text{ A}$$

and at $t = 0.2$ s we have

$$i(0.2) = 3e^{-5(0.2)} = 3e^{-1} = 1.104 \text{ A}$$

Finally, at $t = 1$ s, the current is

$$i(1) = 3e^{-5(1)} = 0.020 \text{ A}$$

In the latter case we note that the time constant is

$$\tau = \frac{L}{R} = \frac{2}{10} = 0.2 \text{ s}$$

Therefore, the time $t = 1$ s is five time constants, so that, as in the RC case, the response is nearly zero.

The graph of i is sketched in Fig. 14.13.

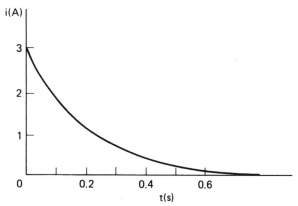

FIGURE 14.13 Graph of the inductor current of the circuit of Example 14.6.

Example 14.7: The circuit of Fig. 14.14 is in dc steady state with the switch in position 1. If the switch is moved to position 2 at $t = 0$, find v and i for $t > 0$.

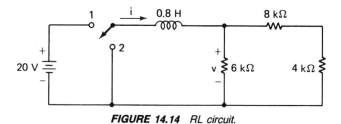

FIGURE 14.14 RL circuit.

Solution: We may replace the three resistors by their equivalent resistance,

$$R_T = \frac{6(8 + 4)}{6 + (8 + 4)} = 4 \text{ k}\Omega$$

and draw the equivalent circuit for $t < 0$ and for $t > 0$. For $t < 0$, the circuit is in dc steady state with the switch at position 1, so that the inductor is like a short circuit. This case is shown in Fig. 14.15(a), from which we see that

$$i = I_0 = \frac{20 \text{ V}}{4 \text{ k}\Omega} = 5 \text{ mA} \tag{14.17}$$

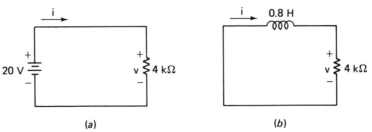

FIGURE 14.15   *The circuit of Fig. 14.14 for (a)* $t < 0$ *and (b)* $t > 0$.

For $t > 0$ the switch is in position 2, disengaging the battery, so that the result is the source-free circuit of Fig. 14.15(b). By (14.16) and (14.17) the current for $t > 0$ is

$$i = 5e^{-4000\,t/0.8}$$

$$= 5e^{-5000\,t}\,\text{mA}$$

The voltage for $t > 0$ is, by Ohm's law,

$$v = 4000i$$

$$= 20e^{-5000\,t}\,\text{V}$$

We note that the currents and voltages have the same general exponential form with the same time constant. This was also true, as we may recall, of the RC circuit.

PRACTICE EXERCISES

14-4.1 The circuit of Fig. 14.12 has $L = 2$ H, $R = 200\ \Omega$, and an initial current of $i(0) = 4$ A. Find (a) the time constant τ, (b) the current i for all positive time, and (c) i at $t = 3\tau$.   *Ans.* (a) 0.01 s, (b) $4e^{-100\,t}$ A, (c) 0.199 A

14-4.2 Find the voltages v_L and v_R in the circuit of Fig. 14.12 if $L = 0.1$ H, $R = 2$ kΩ, and $i(0) = 10$ mA.   *Ans.* $-20e^{-20,000\,t}$ V, $20e^{-20,000\,t}$ V

14-4.3 The circuit shown is in steady state when the switch is moved at $t = 0$ from position 1 to position 2. Find v for $t > 0$. [*Suggestion:* Note that for $t > 0$, we have $\tau = 0.2/(40 + 20)$.]   *Ans.* $20e^{-300\,t}$ V

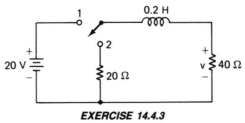

EXERCISE 14.4.3

14.5 DRIVEN RL CIRCUITS

A driven *RL* circuit, like the driven *RC* circuit, is one containing a source, or a driver. An example is the circuit of Fig. 14.16, where the driver is the battery with terminal voltage V_b.

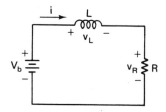

FIGURE 14.16 *Driven RL circuit.*

Circuit Equation: By KVL the circuit equation for Fig. 14.16 is

$$v_L + v_R = V_b$$

which by (14.11) and Ohm's law may be written

$$L\frac{di}{dt} + Ri = V_b$$

or equivalently

$$\frac{di}{dt} + \frac{R}{L} i = \frac{1}{L} V_b \tag{14.18}$$

Inductor Current: Equation (14.18) has almost the identical form of its counterpart (12.21) for the driven *RC* circuit. If the initial current is

$$i(0) = 0 \tag{14.19}$$

(14.18) may be shown by calculus to have the solution

$$i = \frac{V_b}{R}(1 - e^{-Rt/L}) \tag{14.20}$$

which is very similar to the form of the capacitor voltage in (12.23).
If we recognize the time constant

$$\tau = \frac{L}{R} \tag{14.21}$$

then (14.20) may be put in the form

$$i = \frac{V_b}{R} (1 - e^{-t/\tau}) \text{ A} \tag{14.22}$$

Except for the multiplication factor V_b/R, this is the identical function obtained for the voltage of the RC circuit and plotted in Fig. 12.10. From the latter figure we may see that the inductor current in (14.22) rises from zero in an exponential manner to a steady-state value (for large t) of V_b/R. These results are the duals of the corresponding RC cases.

Example 14.8: Find the inductor current i in Fig. 14.16 for (a) all positive time t, (b) $t = 2$ ms, and (c) $t = 5\tau$, if $R = 200$ Ω, $L = 0.2$ H, $V_b = 6$ V, and $i(0) = 0$.

Solution: The time constant is

$$\tau = \frac{L}{R} = \frac{0.2}{200} = 0.001 \text{ s}$$

or $\tau = 1$ ms. Therefore, by (14.22) we have

$$i = \frac{6}{200} (1 - e^{-t/0.001}) \text{ A}$$

or

$$i = 30(1 - e^{-1000 t}) \text{ mA} \tag{14.23}$$

which is valid for all positive t.

For $t = 2$ ms $= 0.002$ s, (14.23) becomes

$$i = 30(1 - e^{-2}) = 25.94 \text{ mA}$$

and for $t = 5\tau = 0.005$ s we have

$$i = 30(1 - e^{-5}) = 29.8 \text{ mA}$$

This checks the earlier statement that at a time of five time constants the response has virtually reached its steady-state value, which in this case is 30 mA.

Shortcut Procedure: As was the case for the capacitor voltage in Chapter 12, the inductor current (14.22) is a sum of a transient component

$$i_{\text{tr}} = -\frac{V_b}{R} e^{-t/\tau}$$

and a steady-state component

$$i_{ss} = \frac{V_b}{R}$$

In the general case, for a constant source and any initial current $i(0)$, we may note from these results that

$$i = i_{tr} + i_{ss} \tag{14.24}$$

where

$$i_{tr} = Ke^{-t/\tau} \tag{14.25}$$

with K an arbitrary constant and i_{ss} is the dc steady-state current left after the transient is gone. Therefore, we may use the shortcut procedure discussed for RC circuits in Section 12.5 for computing τ, obtaining i_{tr} from (14.25) and obtaining i_{ss} directly from the RL circuit. Also, since τ is the same for source-free or driven circuits, we may find it from the circuit with the source killed, as was done in the RC case. We will illustrate the procedure with an example.

Example 14.9: Find the inductor current i in Fig. 14.17 for all positive time if the initial current is $i(0) = 6$ A.

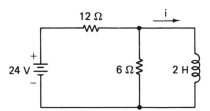

FIGURE 14.17 *Driven circuit with two resistors.*

Solution: To find the time constant we first kill the source (replacing it by a short circuit), as shown in Fig. 14.18(a). We see in this case that the equivalent resistance seen from the inductor is that of the two parallel resistors, given by

$$R_T = \frac{12 \times 6}{12 + 6} = 4 \ \Omega$$

Therefore, we have the time constant

$$\tau = \frac{L}{R_T} = \frac{2}{4} = \frac{1}{2} \ \text{s}$$

and the transient component

$$i_{tr} = Ke^{-2t} \tag{14.26}$$

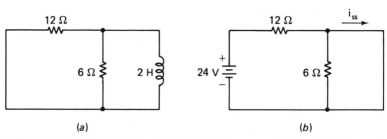

FIGURE 14.18   *Circuit of Fig. 14.17 with (a) the source killed and (b) in steady state.*

The steady-state component i_{ss} is obtained from Fig. 14.18(b), which is the original circuit in steady state. The inductor is a short circuit, as shown, and since the 6-Ω resistor is shorted out, we have

$$i_{ss} = \frac{24}{12} = 2 \text{ A} \tag{14.27}$$

The inductor current is therefore

$$i = i_{tr} + i_{ss}$$

or

$$i = Ke^{-2t} + 2$$

Since the initial current is 6 A, we have

$$i(0) = 6 = K + 2$$

so that $K = 4$. The current for all positive time is thus

$$i = 4e^{-2t} + 2 \text{ A}$$

PRACTICE EXERCISES

14-5.1   Find the steady-state values of i_1 and i_2 in the circuit shown. (*Suggestion:* The inductors become short circuits.)   *Ans.* 3 mA, 2 mA

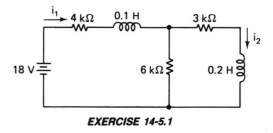

EXERCISE 14-5.1

14-5.2 Find i, v_R, and v_L in Fig. 14.16 for all positive time if $L = 0.5$ H, $R = 1$ kΩ, $V_b = 6$ V, and $i(0) = 0$.   *Ans.* $6 - 6e^{-2000t}$ mA, $6 - 6e^{-2000t}$ V, $6e^{-2000t}$ V

14-5.3 Find i and v for all positive time in the circuit shown if $i(0) = 0$. [*Suggestion:* Note that KCL written at the top node of the inductor yields $v = 20(2 - i)$.]
                                        Ans. $2(1 - e^{-200t})$ A, $40e^{-200t}$ V

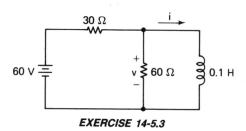

EXERCISE 14-5.3

14.6 SUMMARY

An inductor is a two-terminal element consisting of a coil of wire around a core of air or magnetic material. Its inductance depends upon the number of turns of the coil, its length, its cross-sectional area, and the permeability of its core. The inductance is measured in henrys, and is a measure of the inductor's ability to store energy in its magnetic field.

The voltage across an inductor is proportional to the rate of change of its current. The energy stored in the inductor is proportional to the square of the current, and in the case of voltage and energy, the inductance is a constant of proportionality.

Inductors may be combined in series and parallel exactly as resistors. The equivalent of series inductances is their sum and the reciprocal of the equivalent is the sum of the reciprocals of parallel inductances.

A simple source-free RL circuit is one containing a resistor R in series with an inductor L. If the inductor carries an initial current I_0, the inductor current i for all positive time is

$$i = I_0 e^{-t/\tau} \text{ A}$$

where $\tau = L/R$ is the time constant.

A driven RL circuit is one containing a source, such as a battery. The inductor current in this case is given by

$$i = i_{tr} + i_{ss}$$

where i_{tr} is a transient current given by

$$i_{tr} = K e^{-t/\tau}$$

and i_{ss} is the dc steady-state current. The arbitrary constant K depends on the initial current.

PROBLEMS

14.1   An inductor with 400 turns of wire has a terminal voltage of 10 V. How fast is its flux changing?

14.2   Find the induced voltage of an inductor with 100 turns if the flux varies linearly from 0 to 100 μWb in 4 ms.

14.3   Find the induced voltage of an inductor with 500 turns if the flux is changing at a rate

$$\frac{d\Phi}{dt} = 6e^{-1000t}\,\text{mWb/s}$$

14.4   Find the voltage in Problem 14.3 at (a) $t = 2$ ms, (b) $t = 5$ ms, and (c) $t = 3\tau$. (*Note:* The time constant $\tau = 1/1000$ s in this case.)

14.5   Find the inductance of the coil of Fig. 14.5 if $N = 500$, $l = 0.25$ m, $A = 10^{-3}$ m², and the core is air.

14.6   Find the minimum number of turns necessary for the coil of Fig. 14.5 to have an inductance of at least 1 mH if $l = 0.1$ m, $A = 4 \times 10^{-3}$ m², and the core is air.

14.7   Find the inductance of the coil shown if the radius of the inner circle is 8 cm, the radius of the outer circle is 12 cm, the cross-sectional area of the flux path is 10^{-3} m², the number of turns is 100 and the relative permeability of the core is 1000.

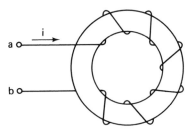

PROBLEM 14.7

14.8   If the current in the coil of Problem 14.7 is $i = 4$ A and is changing at the rate of 100 A/s, find the magnitude of the flux in the core and the voltage v_{ab}.

14.9   Find the energy stored in a 20-mH inductor whose current is (a) 2 A and (b) 3 mA.

14.10   A coil carries a current of 4 A and is storing 24 mJ of energy. Find its inductance.

14.11   A 10-mH inductor with terminals a-b has a current i entering terminal a as shown. Find the voltage v_{ab} and the energy stored at the times (a) $t = 2$ ms and (b) $t = 5$ ms.

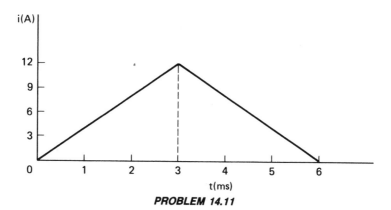

PROBLEM 14.11

14.12 Find the equivalent inductance of four 12-mH inductors connected (a) in series and (b) in parallel.

14.13 Find L_T in the circuit shown.

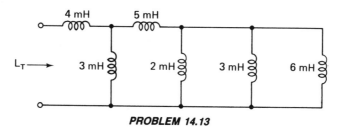

PROBLEM 14.13

14.14 A simple source-free RL circuit has $R = 50\ \Omega$, $L = 0.1$ H, and an initial inductor current of $i(0) = 2$ A. Find i for (a) all positive time, (b) $t = 4$ ms, and (c) $t = 5\tau$, where τ is the time constant.

14.15 A 10-mH inductor is in series with a 1-kΩ resistor. If the energy stored in the inductor at $t = 0$ is 2 μJ, find the inductor current for all positive time.

14.16 Find i for all positive time in the circuit shown if $i(0) = 3$ A.

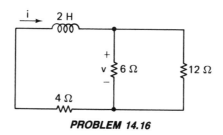

PROBLEM 14.16

14.17 Solve Problem 14.16 if the 4-Ω resistor is replaced by a 2-H inductor.

14.18 Find v for all positive time in Problem 14.16.

14.19 Solve Example 14.7 in Section 14.4 if the 6-kΩ resistor is replaced by a 2400-Ω resistor.

14.20 Find i in Fig. 14.17 for all positive time if $i(0) = 0$ and the 12-Ω resistor is changed to a 3-Ω resistor.

14.21 Solve Problem 14.20 if $i(0) = 4$ A.

14.22 Find i and v for all positive time in Practice Exercise 14.5.3 if the 60-Ω resistor is changed to a 30-Ω resistor.

15

ALTERNATING CURRENT

The only practical source we have discussed thus far is the battery, which provides a constant, or dc, voltage. Batteries are very convenient sources of emf for a variety of uses, such as flashlights, portable radios, and hand calculators, but, of course, they are not adequate for developing the large amounts of energy needed in homes or industries.

Faraday's law of electromagnetic induction, as we noted in Chapter 14, is the principle of the electric generator, which is a second major source of emf. Also, generators may be made large enough to supply the electrical needs of an entire city.

In this chapter we will see how Faraday's law is used with an electric generator to provide power or energy with *alternating current* (ac), by which we will mean *sinusoidal* current. That is, the currents and voltages developed by the generator, or more accurately, the ac generator, will have the shapes of sine waves. We will consider the sine wave in detail, defining what we mean by its *amplitude, frequency,* and *phase.* We will discuss its *average value* and its *effective,* or *root-mean-square, value,* and see how an electric circuit responds to a sinusoidal ac source.

In succeeding chapters we will develop a very elegant method, known as the *phasor* method, for analyzing ac circuits quickly and easily. The equations governing ac circuits contain derivatives, as we will see in this chapter. However, the phasor method allows us to bypass these equations and solve ac circuits in a manner that is very similar to dc resistive circuit analysis.

15.1 AC GENERATOR PRINCIPLE

An ac sinusoidal voltage is generated by the principle of Faraday's law with the rotation of coils of wire in a magnetic field. This was referred to in Section 14.1, where a simplified version using a single coil of one turn was shown in Fig. 14.2. In the actual case, many turns of wire are wrapped on a *rotor,* or rotating cylinder, which is turned at a constant speed by an external prime mover, such as an engine or a turbine using steam, gas, or falling water for its source of mechanical energy. As the rotor turns the flux linking the coils changes, producing a voltage in accordance with Faraday's law.

Generation of a Sine Wave: To illustrate how an ac generator provides a sine wave of voltage, let us reconsider Fig. 14.2. Suppose that the loop of wire is in a horizontal starting position. That is, at time $t = 0$, the plane of the loop is perpendicular to the flux lines. At this point the amount of flux linking the coil is a maximum amount, since the area presented by the coil is a maximum. However, the flux linking the coil is not *changing* at this instant, and therefore the voltage generated is $v = 0$. This situation is illustrated in Fig. 15.1, where at position 1 the coil is horizontal and the voltage is 0.

The speed at which the coil is rotating may be expressed in revolutions per second. Since each revolution is a 360° rotation, the speed could also be expressed in degrees per second. That is, 100 revolutions per second would be $100 \times 360 = 36{,}000$ degrees per second. More often, in circular motion, the angle of rotation is expressed in *radians,* abbreviated *rad,* and defined by

$$2\pi \text{ rad} = 360°$$

or

$$\pi \text{ rad} = 180° \tag{15.1}$$

Thus if ω (the *lowercase* Greek letter *omega*) represents the *angular speed* of rotation in rad/s, then ωt, where t is the time, will be the angle of rotation in radians.

Now let us return to Fig. 15.1. When the coil has rotated $\pi/2$ rad or 90° ($\omega t = \pi/2$), it will be in the vertical position 2. It is not linking any flux at this point, but its *rate* of cutting flux is at its peak. Thus the voltage will be its peak value, shown in this example as 10 V.

As the rotation continues we see that at $\omega t = \pi$ rad (180°), the coil is horizontal again at position 3 and again $v = 0$. In position 4, $\omega t = 3\pi/2$ (270°), and the coil is vertical but in the opposite direction from that of position 2. Therefore, $v = -10$ V. Finally, in position 5 a complete revolution has occurred, $\omega t = 2\pi$ (360°), and the voltage is back to $v = 0$. Since positions 5 and 1 are identical, the coil is ready to repeat the pattern, or *cycle,* executed in one complete revolution.

340

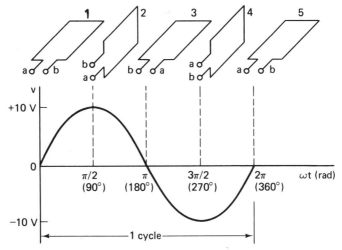

FIGURE 15.1   *One cycle of sinusoidal ac voltage.*

The shape of the voltage in Fig. 15.1 is called a *sine wave* or a *sinusoidal wave.* Its equation is

$$v = 10 \sin \omega t \qquad (15.2)$$

which is read "$v = 10$ times the sine of ωt." We will discuss *sinusoids* or *sinusoidal waves* in some detail in Section 15.2. Because the voltage *alternates* in sign (first positive, then negative, etc.) in a sinusoidal fashion as t increases, it is called an *alternating sinusoidal* voltage. A current it would produce in a resistor has the same general shape and is called an *alternating current,* or ac. A common example is the household voltage in an average home in the United States, where the peak voltage is approximately 170 V and there are 60 cycles per second.

The circuit symbol for an ac generator is shown in Fig. 15.2(a) and (b). Because the polarity of the generator voltage, such as

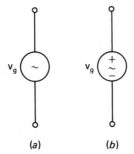

(a)   (b)

FIGURE 15.2   *Circuit symbols for an ac generator.*

$$v_g = 10 \sin \omega t \qquad (15.3)$$

continually changes back and forth as time passes, many authors prefer the symbol of Fig. 15.2(a), which has no polarity marks. However, we will generally use Fig. 15.2(b) to distinguish v_g of (15.3) from

$$v_g = -10 \sin \omega t$$

which would have the opposite polarity.

Example 15.1: Convert (a) 45° to radians and (b) 4 rad to degrees.

Solution: By (15.1) we have

$$1° = \frac{\pi}{180} = 0.0175 \text{ rad} \qquad (15.4)$$

so that

$$45° = 45 \times \frac{\pi}{180} = \frac{\pi}{4} \text{ rad}$$

Also by (15.1) we have

$$1 \text{ rad} = \frac{180}{\pi} = 57.3° \qquad (15.5)$$

and therefore

$$4 \text{ rad} = 4 \times \frac{180}{\pi} = 229.18°$$

These examples may be solved by noting that $\pi = 3.1416$ approximately. However, many hand calculators have a π button that yields its value to 10 significant figures. Thus we may multiply or divide by π as easily as we can by any digit, such as 2, by simply pressing a button. These results, given in (15.4) and (15.5), are important relations to note, but, of course, they too may be found readily with a hand calculator.

Example 15.2: Find the current i in the circuit of Fig. 15.3 for (a) all time t, (b) $t = \pi/400$ s, (c) $t = \pi/200$ s, and (d) $t = 30$ ms. The voltage source is

$$v_g = 100 \sin 200t \text{ V} \qquad (15.6)$$

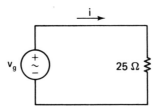

FIGURE 15.3 *Circuit with an ac source.*

Solution: By Ohm's law the current for all time is

$$i = \frac{v_g}{25} = \frac{100 \sin 200t}{25}$$

or

$$i = 4 \sin 200t \text{ A} \qquad (15.7)$$

The current at $t = \pi/400$ s is given by (15.7) to be

$$i = 4 \sin (200)\left(\frac{\pi}{400}\right)$$

$$= 4 \sin \frac{\pi}{2} \qquad (15.8)$$

$$= 4(1)$$

$$= 4 \text{ A}$$

At $t = \pi/200$ s we have

$$i = 4 \sin (200)\left(\frac{\pi}{200}\right)$$

$$= 4 \sin \pi$$

$$= 0$$

and at $t = 30$ ms $= 0.03$ s the current is

$$i = 4 \sin (200)(0.03)$$

$$= 4 \sin 6 \qquad (15.9)$$

$$= 4(-0.2794)$$

$$= -1.118 \text{ A}$$

We note in the latter case that the angle is

$$(200 \text{ rad/s})(0.03 \text{ s}) = 6 \text{ rad}$$

and its sine is −0.2794. This value may be found directly with a hand calculator having the capability of using either degrees or radians, or the 6 rad may be changed first to degrees, given by

$$6 \text{ rad} = 6\left(\frac{180}{\pi}\right) = 343.77°$$

We should also point out that at $t = \pi/400$ s the generator voltage by (15.6) is

$$v_g = 100 \sin\frac{\pi}{2} = 100 \text{ V}$$

and at $t = 30$ ms, it is

$$v_g = 100 \sin 6 = -27.94 \text{ V}$$

In the first case we have

$$i = \frac{v_g}{25} = \frac{100}{25} = 4 \text{ A}$$

which checks (15.8), and in the second case we have

$$i = \frac{-27.94}{25} = -1.118 \text{ A}$$

which checks (15.9). These examples illustrate the usefulness of the polarity marks on the generator of Fig. 15.3, since they indicate that by Ohm's law the current is $+v_g/25$ rather than $-v_g/25$.

PRACTICE EXERCISES

15-1.1 Convert to radians the angles (a) 90°, (b) 180°, and (c) 114.59°.

Ans. (a) $\pi/2$, (b) π, (c) 2

15-1.2 Convert to degrees the angles (a) $3\pi/2$ rad, (b) 5π rad, and (c) 1.7453 rad.

Ans. (a) 270, (b) 900, (c) 100

15-1.3 Find the current i in Fig. 15.3 for (a) all time, (b) $t = \pi/800$ s, and (c) $t = 20$ ms, if the generator voltage is changed to

$$v_g = 50 \sin 200t \text{ V}$$

Ans. (a) 2 sin 200t A, (b) 1.41 A, (c) −1.51 A

15.2 THE SINE WAVE

The voltage v given in (15.2) is a special case of the general sinusoid, or sine wave,

$$v = V_m \sin \omega t \qquad (15.10)$$

where V_m is the *peak* voltage, or *maximum* voltage, or *amplitude,* in volts and ω is the *radian frequency* in rad/s. A graph of (15.10) is shown for one cycle ($0 \le \omega t \le 2\pi$ rad) in Fig. 15.4, where it may be seen that the maximum positive value $v = V_m$ is attained at $\omega t = \pi/2$ rad (90°) and the maximum negative value $v = -V_m$ is attained at $\omega t = 3\pi/2$ rad (270°).

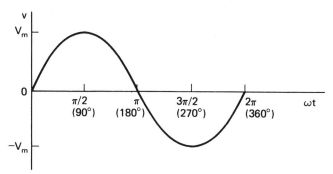

FIGURE 15.4 Sine wave.

Trigonometric Functions: The sine function (15.10) is a function in trigonometry which may be used to relate one side to another in a right triangle. For example, let us consider the right triangle in Fig. 15.5. The angle at C is a *right* angle (90°) and the angles at A and B are denoted by θ (the lowercase Greek letter *theta*) and $90° - \theta$. The sides x and y are the *legs* of the triangle and r is the *hypotenuse.* Side x is the *adjacent* side of θ and side y is its *opposite* side. Using this terminology, the sine of θ is defined by

$$\sin \theta = \frac{\text{opposite side}}{\text{hypotenuse}} \qquad (15.11)$$

or

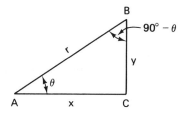

FIGURE 15.5 Right triangle.

$$\sin \theta = \frac{y}{r} \tag{15.12}$$

Two other useful trigonometric functions are the *cosine* of θ (abbreviated cos θ) and the *tangent* of θ (abbreviated tan θ). Referring to Fig. 15.5, these are defined by

$$\cos \theta = \frac{\text{adjacent side}}{\text{hypotenuse}} \tag{15.13}$$

or

$$\cos \theta = \frac{x}{r} \tag{15.14}$$

and

$$\tan \theta = \frac{\text{opposite side}}{\text{adjacent side}} \tag{15.15}$$

or

$$\tan \theta = \frac{y}{x} \tag{15.16}$$

Since by the Pythagorean theorem we have in Fig. 15.5,

$$x^2 + y^2 = r^2 \tag{15.17}$$

we may find any of the three trigonometric functions we have discussed if we know any two of the sides x, y, and r.

Example 15.3: Find the sine, cosine, and tangent of (a) 45°, (b) 30°, and (c) 60°.

Solution: It may be shown that a *45° right triangle,* shown in Fig. 15.6(a), has its two legs equal (chosen as $x = y = 1$ in this case), and a *30°–60° right triangle,* shown in Fig. 15.6(b), has the side opposite 30° equal to half the hypotenuse (chosen as $y = 1$ and $r = 2$). Using (15.17) we may find r in Fig. 15.6(a) as

$$r = \sqrt{x^2 + y^2} = \sqrt{1^2 + 1^2} = \sqrt{2}$$

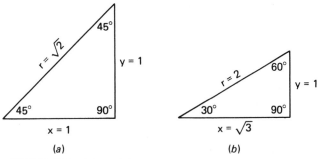

FIGURE 15.6 (a) 45° right angle and (b) 30°–60° right triangle.

and x in Fig. 15.6(b) as

$$x = \sqrt{r^2 - y^2} = \sqrt{2^2 - 1^2} = \sqrt{3}$$

Therefore, we have from Fig. 15.6(a)

$$\sin 45° = \frac{y}{r} = \frac{1}{\sqrt{2}} = 0.707$$

$$\cos 45° = \frac{x}{r} = \frac{1}{\sqrt{2}} = 0.707$$

$$\tan 45° = \frac{y}{x} = \frac{1}{1} = 1$$

and from Fig. 15.6(b)

$$\sin 30° = \frac{y}{r} = \frac{1}{2} = 0.5$$

$$\cos 30° = \frac{x}{r} = \frac{\sqrt{3}}{2} = 0.866$$

$$\tan 30° = \frac{y}{x} = \frac{1}{\sqrt{3}} = 0.577$$

$$\sin 60° = \frac{\text{opposite}}{\text{hypotenuse}} = \frac{\sqrt{3}}{2} = 0.866$$

$$\cos 60° = \frac{\text{adjacent}}{\text{hypotenuse}} = \frac{1}{2} = 0.5$$

$$\tan 60° = \frac{\text{opposite}}{\text{adjacent}} = \frac{\sqrt{3}}{1} = 1.732$$

Use of the Hand Calculator: All the values determined in Example 15.3, as well as any other trigonometric function values, may be found almost effortlessly with any hand calculator with trigonometric function buttons. Most scientific calculators can provide the answers if the angles are specified in degrees or in radians. If a hand calculator is not available, we may look up the values of the trigonometric functions in a table of such functions found in any elementary trigonometry book or any mathematics handbook.

As an example using a number of function values, various points on the sine wave given by (15.10) are listed in Table 15.1. It is seen that v rises from 0 to a peak V_m, drops to 0 and to $-V_m$, and then returns to 0 as ωt varies through one cycle from 0 to 2π rad.

TABLE 15.1   Values of a sinusoidal voltage.

Angle ωt		$\sin \omega t$	$v = V_m \sin \omega t$
Degrees	Radians		
0	0	0	0
30	$\pi/6$	0.500	$0.5 V_m$
45	$\pi/4$	0.707	$0.707 V_m$
60	$\pi/3$	0.866	$0.866 V_m$
90	$\pi/2$	1.000	V_m
180	π	0	0
270	$3\pi/2$	-1.000	$-V_m$
360	2π	0	0

The angle values considered in Table 15.1 range from 0 to 360° and their sine can be found readily using a hand calculator. This is true also of higher angles such as 700°. However, we cannot relate angles greater than 90°, some of which have negative sines and cosines, to the triangle of Fig. 15.5. We will give a more general definition using triangles that is applicable to any angle in Chapter 16.

PRACTICE EXERCISES

15-2.1   Find the amplitude and radian frequency of the function

$$v = 100 \sin 4t \text{ V}$$

Ans. 100 V, 4 rad/s

15-2.2   Find v in Exercise 15.2.1 if (a) $t = 0.1$ s, (b) $t = 0.4$ s, and (c) $t = 1$ s.
Ans. (a) 38.9 V, (b) 99.96 V, (c) -75.7 V

15-2.3 Find the smallest positive time t when v in Exercise 15.2.1 reaches its peak value of 100 V. (*Suggestion:* Note from Fig. 15.4 that this occurs when $\omega t = 4t = \pi/2$ rad.)

Ans. $\pi/8 = 0.393$ s

15-2.4 Find (a) sin 36°, (b) cos 125°, and (c) tan 4 rad.

Ans. (a) 0.588, (b) −0.574, (c) 1.158

15.3   FREQUENCY

In this section we will consider in more detail the amplitude and radian frequency of a sine wave, and we will also discuss another type of frequency, referred to simply as *frequency*. These quantities, as we will see, are of great importance in the study of ac voltages and currents.

Period of a Sine Wave: The sine wave, such as (15.10), repeats the cycle shown in Fig. 15.4 *periodically* as time increases. At $\omega t = 360°$ the wave is passing through zero and rising, exactly as it was at $t = 0$. The time required for one cycle to be executed is thus determined from $\omega t = 360°$ or $\omega t = 2\pi$ rad. This time, denoted by T, is therefore given by

$$T = \frac{2\pi}{\omega} \quad \text{seconds} \tag{15.18}$$

and is called the *period* of the sinusoid. This is the time required for one cycle or the *time per cycle*. A function that repeats itself periodically, such as the sine wave, is called a *periodic* function.

As an example, the sinusoid

$$i = I_m \sin \omega t$$

is sketched for three periods in Fig. 15.7. From the sketch we see that for all t,

$$\sin (\omega t + 2\pi) = \sin \omega t$$

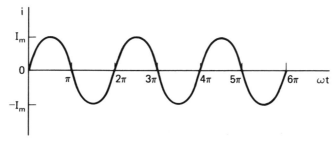

FIGURE 15.7   *Three periods of a sinusoid.*

Frequency: If T is the time per cycle, or seconds per cycle, its reciprocal $1/T$ is the *cycles per second*, or the number of cycles, or periods, executed in 1 s. This quantity, denoted by f, is called the *frequency* of the sine wave. Its units, cycles per second, are called *hertz* (abbreviated Hz), in honor of the German physicist Heinrich R. Hertz (1857–1894). The frequency f is therefore

$$f = \frac{1}{T} \tag{15.19}$$

which by (15.18) is

$$f = \frac{\omega}{2\pi} \quad \text{Hz} \tag{15.20}$$

The frequency f is related to ω by

$$\omega = 2\pi f \tag{15.21}$$

as may be seen from (15.20). There should be no confusion between the two since f is simply called frequency and ω is usually referred to as radian frequency or *angular frequency*. The use of the units Hz in the case of f or rad/s in the case of ω should help distinguish the two. A very common example of frequency is $f = 60$ Hz, which is the frequency of ac voltages generated all over the United States. In this case the radian frequency is

$$\omega = 2\pi(60) = 377 \text{ rad/s}$$

Example 15.4: Find the frequency f and the period T of the sinusoidal ac voltage given by

$$v = 50 \sin 7540t \text{ V}$$

Solution: The radian frequency is $\omega = 7540$ rad/s, so that by (15.20) we have

$$f = \frac{\omega}{2\pi} = \frac{7540}{2\pi} = 1200 \text{ Hz}$$

The period is given by

$$T = \frac{1}{f} = \frac{1}{1200} \text{ s} = 0.833 \text{ ms}$$

Frequency Ranges: The entire frequency range of ac voltages and currents from 1 Hz up may be roughly divided into several groups. The three principal ones are *low frequencies,* ranging from 1 Hz to about 400 Hz, *audio* frequencies, which are those of sound waves that can be heard by the human ear and which range from about 20 Hz to 20,000 Hz (20 kHz), and *radio frequencies,* ranging from about 500 kHz to 100 million hertz, or 100 megahertz (100 MHz).

The range 535 to 1605 kHz is the standard AM radio broadcasting band, and 88 to 108 MHz is the FM radio band. Television broadcast bands range from 54 to 72 MHz for channels 2 and 4 to 174 to 216 MHz for channels 7 through 13. These television bands fall in the classification of *very high frequencies* (VHF), which range from 30 MHz to 300 MHz. *Ultrahigh frequency* (UHF) television channels 14 to 83 broadcast at 300 MHz to 3 gigahertz (3 GHz), or 3 billion hertz. Other frequency ranges are *super-high frequencies* (SHF), ranging from 3 GHz to 30 GHz, and *extra-high frequencies* (EHF), ranging from 30 GHz to 300 GHz. These last two groups are used for amateur broadcasting and government applications, such as satellite communications.

PRACTICE EXERCISES

15-3.1 Find the amplitude, frequency, and period of the sinusoid

$$v = 100 \sin 200\pi t \text{ V}$$

Ans. 100 V, 100 Hz, 10 ms

15-3.2 Find v in Exercise 15.3.1 when (a) $t = 1$ ms, (b) $t = 10$ ms, and (c) $t = 1/400$ s.
Ans. (a) 58.8 V, (b) 0, (c) 100 V

15.4 PHASE

In this section we will consider a more general sinusoidal function, which has the properties of the sinusoids of the previous sections, as well as an additional, very important property known as *phase.*

Phase Angle: A more general expression for the sinusoidal voltage (15.10) is

$$v = V_m \sin (\omega t + \phi) \tag{15.22}$$

where V_m is the amplitude and $\omega = 2\pi f$ is the radian frequency, as before, and ϕ

is the *phase angle,* or simply the *phase.* The sinusoid (15.10) is the special case $\phi = 0$, sketched in Fig. 15.4.

We may note from (15.22) that when we have

$$\omega t + \phi = 0$$

or $\omega t = -\phi$, the voltage is $v = V_m \sin 0 = 0$. Therefore, (15.22) is like (15.10) except that it begins at $v = 0$ when $\omega t = -\phi$. In other words, it is a sine wave like (15.10) shifted to the left by $\omega t = \phi$ rad (or degrees, depending on the units used). Its graph is shown in Fig. 15.8, where we note that it reaches its maximum and minimum values of $\pm V_m$ at $90° - \phi$ and $270° - \phi$, respectively.

For mathematical consistency, ωt and ϕ should each have the same units, such as radians or degrees. However, it is often convenient to give ω in rad/s, t in seconds, and ϕ in degrees, so that we will write expressions like

$$v = 10 \sin (4t + 30°) \tag{15.23}$$

with the understanding that $4t$ is in radians and must be converted to degrees (or 30° must be converted to radians) in order to evaluate v.

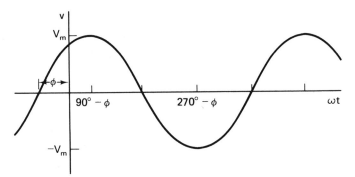

FIGURE 15.8 *Sine wave with phase angle ϕ.*

Example 15.5: Find v in (15.23) if $t = 0.5$ s.

Solution: Converting $4t = 4 \times 0.5 = 2$ rad to degrees, we have

$$2 \text{ rad} = 2 \times \frac{180}{\pi} = 114.6°$$

Therefore, the voltage is

$$v = 10 \sin (114.6° + 30°)$$
$$= 10 \sin 144.6°$$
$$= 5.79 \text{ V}$$

If the phase angle is negative, such as in

$$v = 10 \sin (4t - 30°)$$

the graph is like that of Fig. 15.4 except that it is shifted to the right by 30° (or $30\pi/180$ rad).

From Fig. 15.5 we may see that

$$\cos (90° - \theta) = \frac{y}{r} = \sin \theta \qquad (15.24)$$

and

$$\sin (90° - \theta) = \frac{x}{r} = \cos \theta \qquad (15.25)$$

It may also be shown from trigonometry that

$$\cos (-\alpha) = \cos \alpha \qquad (15.26)$$

and

$$\sin (-\alpha) = -\sin \alpha \qquad (15.27)$$

That is, changing the sign of the angle changes the sign of the sine function but does not change the sign of the cosine function. Therefore, we may write (15.24) as

$$\sin \theta = \cos (\theta - 90°) \qquad (15.28)$$

and replacing θ by $-\theta$ in (15.25) yields

$$\cos \theta = \sin (\theta + 90°) \qquad (15.29)$$

These results indicate that a sine wave is a cosine wave with a phase angle of $-90°$ and a cosine wave is a sine wave with a phase angle of $+90°$. Therefore, both sine and cosine waves are sinusoids with specified phase angles. A plot of $v = \sin \omega t$ and $v = \cos \omega t$ are shown in Fig. 15.9. As expected, they differ only by the phase angle 90°.

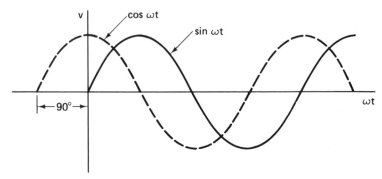

FIGURE 15.9 *Plots of sine and cosine waves.*

Example 15.6: Change the function

$$v = 10 \sin (2t + 15°)$$

to an equivalent cosine function.

Solution: By (15.28) we see that the sine function becomes the cosine function if 90° is subtracted from its angle. Therefore, we have

$$v = 10 \cos (2t - 75°)$$

PRACTICE EXERCISES

15-4.1 Find the amplitude, frequency, period, and phase of the sinusoid

$$v = 20 \sin (100\pi t + 60°) \text{ V}$$

Ans. 20 V, 50 Hz, 20 ms, 60°

15-4.2 Find v in Exercise 15.4.1 when (a) $t = 1$ ms, (b) $t = 15$ ms, and (c) $t = 1/600$ s. (*Suggestion:* Note that 60° = $\pi/3$ rad.)   *Ans.* (a) 19.56 V, (b) −10 V, (c) 20 V

15-4.3 Change the function in Exercise 15.4.1 to a cosine function.

Ans. 20 cos (100πt − 30°)

15.5 AVERAGE VALUES

As we have seen in the case of resistive circuits, a dc voltage v applied to a resistor R causes a dc current $i = v/R$ to flow in the resistor. The power delivered by the source to the resistor is

$$p = vi = Ri^2 \tag{15.30}$$

which is a constant if i is a constant (in which case v is a constant).

Instantaneous Power: The power given in (15.30) depends on the current, or equivalently the voltage, at the instant of time at which we are interested. For this reason it is called the *instantaneous power*, and it is, of course, a function of time if i is a function of t.

If the voltage across the resistor is an alternating voltage,

$$v = V_m \sin \omega t \text{ V} \tag{15.31}$$

as in the circuit of Fig. 15.10, the current by Ohm's law is

$$i = \frac{v}{R} = \frac{V_m}{R} \sin \omega t \text{ A}$$

or

$$i = I_m \sin \omega t \text{ A} \tag{15.32}$$

with a peak value

$$I_m = \frac{V_m}{R} \tag{15.33}$$

In this case the instantaneous power is

$$p = Ri^2$$
$$= RI_m^2(\sin \omega t)^2$$

which we will write in the form

$$p = RI_m^2 \sin^2 \omega t \tag{15.34}$$

Thus the instantaneous power is in general a function of time.

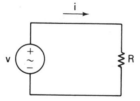

FIGURE 15.10 Alternating voltage applied to a resistor.

Average Value of a Function: In electric circuit theory we are concerned with instantaneous values of quantities such as power, but we are often more interested in their *average* values over some period of time. For example, the power involved in our electric bills is the average power for the period.

The average value of a function over some specified time is defined to be the algebraic area under the curve represented by the function during the given time divided by the given time. For example, $f(t)$ shown in Fig. 15.11 has an average value, which we denote by F_{av}, given by

$$F_{av} = \frac{A_1}{T_1}$$

where A_1 is the algebraic area covered during the time T_1 of interest. In the case shown, if $T_1 = 4$ s, then A_1 is the shaded area

$$A_1 = (8 \times 2) - 2(2) = 12$$

and therefore

$$F_{av} = \frac{12}{4} = 3 \tag{15.35}$$

We note that the area from $t = 2$ to $t = 4$ s is a negative area, whereas that from 0 to 2 s is positive.

If a function $f(t)$ is periodic with period T, its average value over any time T_1, which is a multiple of T, is its average value over T. To see this, we note that each cycle of $f(t)$ has the same area, say A, so that if $T_1 = kT$, the area for T_1 is $A_1 = kA$. The average value is then

$$F_{av} = \frac{A_1}{T_1} = \frac{kA}{kT} = \frac{A}{T}$$

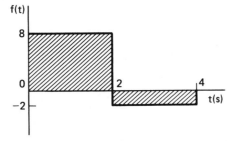

FIGURE 15.11 *A curve with its area shown shaded.*

which is the average over one period T. As an example, if the function shown in Fig. 15.11 is one cycle of a periodic function $f(t)$, then by (15.35) its average value over any number of periods is $F_{av} = 3$.

Average Value of a Sinusoid: The average value over a period of a sinusoid, such as the current of Fig. 15.12, is zero. This is because the area from $\omega t = 0$ to π is exactly like that from π to 2π, except that one is positive and one is negative. Thus we have over one period $T = 2\pi/\omega$ $(\omega T = 2\pi)$, an average value given by

$$I_{av} = 0 \qquad (15.36)$$

It was noted in Chapter 9, in connection with meters, that the d'Arsonval movement was actuated by a current. In the dc case the force on the needle of the meter is constant, but with a changing current the needle will move back and forth. If the current is changing rapidly, the needle will try to follow the current and will therefore rest on a position corresponding to the average value of the current. By (15.36) this average value is zero for an ac current, which, of course, would not be very useful measurement information.

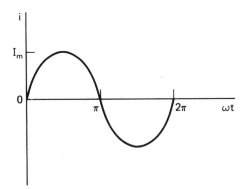

FIGURE 15.12 *Sinusoidal current.*

As was pointed out in Chapter 9, the ac current is *rectified* before it reaches the d'Arsonval meter coil so that the average reading is not zero. A *fully rectified* sine wave, as we observed in Chapter 9, is one whose negative areas are changed to positive areas, as shown for the current in Fig. 15.13. This rectified wave, as we may see from the figure, has a period T defined by $\omega T = \pi$, or

$$T = \frac{\pi}{\omega} \qquad (15.37)$$

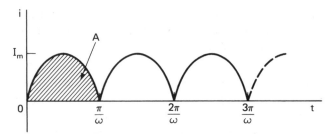

FIGURE 15.13 *Fully rectified sine wave.*

Its average value over T is therefore the shaded area A divided by T, or

$$I_{av} = \frac{A}{T} \qquad (15.38)$$

It may be shown by calculus that the shaded area A in Fig. 15.13 is given by

$$A = \frac{2I_m}{\omega} \qquad (15.39)$$

Therefore, from (15.37) to (15.39) we have

$$I_{av} = \frac{A}{T} = \frac{2I_m/\omega}{\pi/\omega}$$

or

$$I_{av} = \frac{2}{\pi} I_m = 0.637 I_m \qquad (15.40)$$

A justification that (15.39) is true is given in Practice Exercise 15.5.2.

Example 15.7: Find the average value of a full rectification of the sine wave

$$i = 10 \sin 2t \text{ A}$$

Solution: The peak value is $I_m = 10$ and thus by (15.40) we have

$$I_{av} = 0.637 \times 10 = 6.37 \text{ A}$$

The average value, as we see in (15.40), depends only on the amplitude I_m and not on the frequency or phase. Therefore, (15.40) applies for a cosine wave as well as a sine wave and for a sinusoid with any phase angle. A similar result holds for voltages or for any other sinusoidal quantity.

PRACTICE EXERCISES

15-5.1 Find the average value over the time interval 0 to 6 s of the function $f(t)$ shown.

Ans. 4

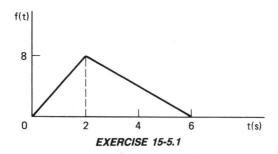

EXERCISE 15-5.1

15-5.2 Verify (15.39) approximately by finding the area under the sine wave shown as being very nearly

$$A = A_1 + A_2 + A_3$$

(Note that the horizontal axis is time.)   *Ans.* $A = \dfrac{2\pi I_m}{3\omega} = \dfrac{2.1 I_m}{\omega}$

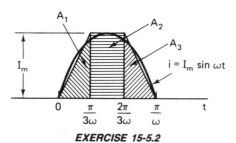

EXERCISE 15-5.2

15-5.3 Find the average value over $t = 0$ to $t = \pi/4$ s of the sinusoid

$$i = 20 \sin 4t \text{ A}$$

Ans. $\dfrac{2}{\pi}(20) = 12.7$ A

15.6 RMS VALUES

As we noted in Section 15.5, the instantaneous power delivered to a resistor R by an ac current

$$i = I_m \sin \omega t \text{ A} \qquad (15.41)$$

is

$$p = RI_m^2 \sin^2 \omega t \text{ W} \qquad (15.42)$$

A typical ac wattmeter, which measures the power delivered by an ac source, reads the average value of the instantaneous power over a period. Thus in the case of (15.42) the reading would be the average power, denoted by P, of p over the period $t = 0$ to $t = T = 2\pi/\omega$.

Average Power: The sketch of the instantaneous power p is shown in Fig. 15.14, where it is seen that p is periodic of period $T = \pi/\omega$, which is half the period $2\pi/\omega$ of the current i. Thus the average power is the average over $t = 0$ to $t = \pi/\omega$. The area A under the curve for one period is shown shaded in Fig. 15.14 and may be obtained by calculus as

$$A = \frac{RI_m^2 \pi}{2\omega} \qquad (15.43)$$

Therefore, the average power delivered to a resistor R by the sinusoidal current (15.41) is

$$P = \frac{A}{T} = \frac{RI_m^2 \pi/2\omega}{\pi/\omega}$$

or

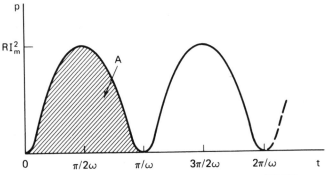

FIGURE 15.14 Plot of the instantaneous power (15.42).

360

$$P = \frac{RI_m^2}{2} \tag{15.44}$$

Root-Mean-Square Value of a Sinusoid: The *root-mean-square (rms) value* of a periodic current or voltage, such as a sinusoid, is defined to be a constant that is equal to the dc current or voltage that would deliver the same average power to a resistor R. In the case of the sinusoidal current of (15.41), its rms value, which we denote as I_{rms}, is equal to the dc current I_{dc} that will deliver the power P of (15.44) to a resistor R. Since the power delivered by I_{dc} is RI_{dc}^2, we must have

$$RI_{dc}^2 = \frac{RI_m^2}{2}$$

Canceling the factor R from both members, we have

$$I_{rms}^2 = I_{dc}^2 = \frac{I_m^2}{2}$$

which is equivalent to

$$I_{rms} = I_m/\sqrt{2} \tag{15.45}$$

or

$$I_{rms} = 0.707 I_m \tag{15.46}$$

Since the rms value does not depend on the frequency or phase, but only on the amplitude, we may say that for the general sinusoid,

$$f = K \sin(\omega t + \phi) \tag{15.47}$$

the rms value is

$$F_{rms} = K/\sqrt{2} \tag{15.48}$$

The function f may be a voltage or a current or any other sinusoidal quantity, and since the phase angle ϕ may be any value, (15.48) holds for cosine functions as well as sine functions.

The term *root-mean-square* comes from the fact that I_{rms} is the square *root* of an average, or *mean*, value of the *square* of the current. The rms value is also called the *effective value*, since the effective value $I_{eff} = I_{rms}$ has the same effect, as far as power is concerned, as a dc current with equal magnitude.

Rms values are very important in ac circuit analysis because, as we may see from (15.44) and (15.45), the average power delivered to a resistor R by the sinusoidal current is

$$P = RI_{rms}^2 \qquad\qquad (15.49)$$

For this reason typical ac ammeters and voltmeters are calibrated to read rms values.

The result given in (15.49) should be easy to remember since it has the identical form of its counterpart in the dc circuit case.

Example 15.8: Find the rms value of (a) the dc current

$$i = 10 \text{ A}$$

and (b) the voltage

$$v = 170 \cos(377t + 30°) \text{ V}$$

Solution: In case (a), since $I_{dc} = I_{rms}$ in general, we have

$$I_{rms} = 10 \text{ A}$$

In (b) the rms value depends only on the amplitude 170 and thus is given by

$$V_{rms} = \frac{V_m}{\sqrt{2}} = \frac{170}{\sqrt{2}} = 120 \text{ V}$$

This is the case of ordinary household ac voltage.

PRACTICE EXERCISES

15-6.1 Find the average power delivered to a 2-kΩ resistor by a sinusoidal current

$$i = 60 \sin 1000t \text{ mA}$$

[*Suggestion:* Use (15.44).]   *Ans.* 3.6 W

15-6.2 Find the voltage across the resistor in Exercise 15.6.1 and the rms values of the current and voltage.   *Ans.* 120 sin 1000t V, 42.4 mA, 84.9 V

15-6.3 Find the rms values of (a) 10 sin 3t, (b) 20 cos 100t, and (c) 5 sin (2t + 15°).
                                    Ans. (a) 7.07, (b) 14.14, (c) 3.54

An ac circuit is one whose sources are ac sinusoids, such as the example of Fig. 15.15. This case is an *RLC* series circuit driven by the ac generator with voltage v_g. As in the case of the *RC* and *RL* circuits considered earlier, the output current *i* will contain a transient component and a steady-state component. The transient will die very soon after the elements are connected and the steady-state output will be a sinusoid that is very similar to the source sinusoid, as we will see in this section.

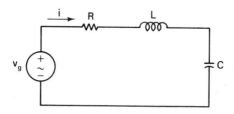

FIGURE 15.15 *RLC circuit.*

Resistor Voltage: As we have seen by Ohm's law, a sinusoidal voltage v_R across a resistor produces a sinusoidal resistor current i_R, and conversely, a sinusoidal current causes a sinusoidal resistor voltage. The frequency and phase of the voltage and current are the same, but their amplitudes will differ unless $R = 1\ \Omega$. This may be seen from

$$v_R = Ri_R$$

where if

$$i_R = I_m \sin \omega t \tag{15.50}$$

then

$$v_R = RI_m \sin \omega t \tag{15.51}$$

Inductor Voltage: To see what form an inductor voltage v takes when its current *i* is the sinusoid

$$i = I_m \sin \omega t \tag{15.52}$$

let us consider Fig. 15.16. The top curve is a plot of one cycle of *i* and the middle curve is a plot of its rate of change di/dt. From $\omega t = 0$ to $\pi/2$, the rate of change, which is the slope of the top curve, is positive. This is shown by the case at point *a*. At $\omega t = 0$ the slope is highest and it gradually gets less until it reaches zero at $\omega t = \pi/2$, the peak of the *i* curve. This is shown in the middle curve, where di/dt starts at its peak at $\omega t = 0$ and drops gradually to 0 at $\omega t = \pi/2$.

363

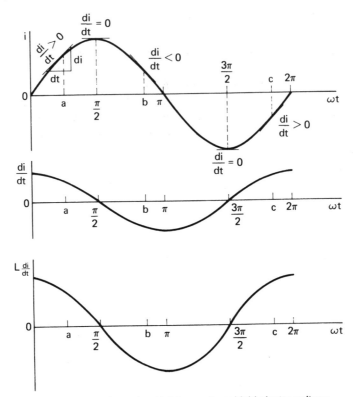

FIGURE 15.16 Steps in obtaining a sinusoidal inductor voltage.

From $\omega t = \pi/2$ to π, di/dt is negative, as shown at point b, and it becomes more negative as t increases, reaching a minimum at $\omega t = \pi$. From that point to $\omega t = 3\pi/2$, di/dt becomes less negative and finally reaches zero, as shown at $3\pi/2$. From $3\pi/2$ to 2π, the cycle is completed with di/dt positive, as shown at point c.

Comparing the graph of di/dt, shown in the middle of Fig. 15.16, with the cosine wave of Fig. 15.9, we see that di/dt appears to be a cosine wave. This is actually the case, as may be shown using calculus. That is, if i is a sine function, then di/dt is a cosine function. The bottom curve of Fig. 15.16 is simply the middle curve multiplied by L, so it also is a cosine wave. Since $L\, di/dt$ is the inductor voltage, we may say then that the inductor voltage is a cosine function. Also, from Fig. 15.16 we may see that the sine wave i and the cosine wave di/dt have the same period and therefore the same frequency.

Since a sine wave and a cosine wave are both sinusoids, from the foregoing discussion we can say that a sinusoidal inductor current causes a sinusoidal inductor voltage of the same frequency. Unlike the case for the resistor, however, the phase angles are different.

AC Currents and Voltages: Using a similar argument with a sinusoidal capacitor voltage v, we may conclude that the capacitor current

$$i = C\frac{dv}{dt}$$

is also a sinusoid of the same frequency, but with a different amplitude and phase. Therefore, all the steady-state voltages and currents in an ac circuit are sinusoids like the source, with the same frequencies but with generally different amplitudes and phases.

As an example, if in Fig. 15.15 we have $R = 4\ \Omega$, $L = 2$ H, $C = 0.5$ F, and the voltage source is

$$v_g = 10 \sin 2t \text{ V} \tag{15.53}$$

then as we will see in Chapter 16, the steady-state current is

$$i = 2 \sin (2t - 36.9°) \text{ A} \tag{15.54}$$

Thus, as expected, the current is a sinusoid like the source, but with a different amplitude and phase.

PRACTICE EXERCISES

15-7.1 It may be shown using calculus that if

$$i = I_m \sin \omega t \text{ A}$$

is the current entering the positive voltage terminal of an inductor with inductance L, the inductor voltage is

$$v = V_m \cos \omega t \text{ V}$$

where

$$V_m = \omega L I_m$$

Find the inductor current if $L = 2$ H and the voltage is

$$v = 40 \cos 4t \text{ V}$$

(Note that this result is consistent with the conclusion drawn from Fig. 15.16.)

Ans. 5 sin 4t A

15-7.2   Find the inductor voltage in Exercise 15.7.1 if $L = 0.2$ H and

$$i = 60 \sin 1000t \text{ mA} \qquad\qquad \textit{Ans.} \; 12 \cos 1000t \text{ V}$$

15.8 SUMMARY

An ac generator produces a sinusoidal voltage in accordance with Faraday's law of electromagnetic induction, which may be expressed in general as

$$v = V_m \sin (\omega t + \phi) \qquad\qquad (15.55)$$

The quantity V_m is the amplitude, ω is the radian frequency, and ϕ is the phase. The sinusoid is periodic, with a period of $T = 2\pi/\omega$, which is the time required for the execution of each cycle. The reciprocal $1/T$ is the frequency f, which is the number of cycles/second, or hertz, executed by the sinusoid.

     The average value of a fully rectified sinusoid with amplitude V_m is $2V_m/\pi$ or $0.637V_m$. The root-mean-square (rms) value of (15.55) is $V_m/\sqrt{2}$ or $0.707V_m$. The rms value is the current most ac ammeters read, and is equal to the value of a dc current that would deliver the same average power to a resistor.

     If an electric circuit has an ac sinusoidal source, all its steady-state voltages and currents will be sinusoids of the same frequency as the source. A resistor current and voltage will have the same phase, but in the general case the amplitudes and phases of the various voltages and currents will be different.

PROBLEMS

15.1   Convert to radians the angles (a) 60°, (b) 720°, and (c) 82°.

15.2   Convert to degrees the angles (a) $\pi/12$ rad, (b) $3\pi/4$ rad, and (c) 6 rad.

15.3   Find the sine of the angles in Problem 15.1.

15.4   Find the cosine of the angles in Problem 15.2.

15.5   Find the current for all time through a 10-kΩ resistor if the voltage is (a) 40 sin 100*t* V and (b) 20 cos 200*t* V.

15.6   The current through a 100-Ω resistor is 2 sin 30*t* A. Find the values of the current and voltage at (a) $t = \pi/90$ s, (b) $t = \pi/60$ s, and (c) $t = 0.2$ s.

15.7   Find the amplitude, radian frequency, frequency in hertz, and period of the voltage

$$v = 50 \cos 400\pi t \text{ V}$$

15.8   Repeat Problem 15.7 for the function

$$v = 25 \sin 40t \text{ V}$$

15.9   Find the period of a sinusoid with frequency (a) 2 kHz, (b) 1 MHz, and (c) 20 Hz.

15.10   Find the smallest positive time when the function

$$i = 20 \sin 50\pi t \text{ A}$$

reaches its peak of 20 A.

15.11   Find the radian frequency of a sinusoid with a period of (a) 1 ms, (b) 20 s, and (c) 1 μs.

15.12   Find v given by

$$v = 20 \cos (6t + 45°) \text{ V}$$

at (a) $t = 0$, (b) $t = \pi/24$ s, and (c) $t = 2$ s.

15.13   Find the average values of the voltages in Problems 15.8 and 15.12 if the sinusoids are fully rectified.

15.14   Find the rms values of the voltages of Problems 15.8 and 15.12.

15.15   Find the average value over one cycle of the *sawtooth* wave shown.

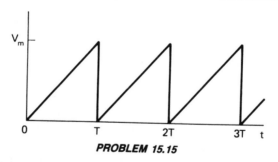

PROBLEM 15.15

15.16   If the function of Fig. 15.11 is periodic with period 4 s, find its rms value. (*Suggestion:* The rms value is the square root of the average of the area under the square of the function.)

15.17   Find the voltage across a 0.1-H inductor if the current is

$$i = 20 \sin 6t \text{ A}$$

(*Suggestion:* See Practice Exercise 15.7.1.)

15.18   Find the inductor current in Problem 15.17 if the voltage is

$$v = 10 \cos 200t \text{ V}$$

16

PHASORS

In Chapter 15 we saw that an ac sinusoidal source applied to an electric circuit of resistors, inductors, and capacitors produced steady-state ac sinusoidal currents and voltages throughout the circuit. These sinusoidal responses have the same frequency as the source, but in general their amplitudes and phases are different.

The equations describing ac circuits involve derivatives, and their rigorous solution requires a knowledge of calculus. However, there is a very elegant method of solution using quantities called *phasors,* which can be applied to ac circuits in exactly the same way that Ohm's law was applied to resistive circuits in the earlier chapters. The phasor method, which is the subject of this chapter, is generally credited to Charles Proteus Steinmetz (1865–1923), a famous electrical engineer with the General Electric Company in the early part of this century.

As we will see, the application of the phasor method requires only a knowledge of algebra and the basic trigonometric functions. The numbers used in the phasor method, however, are not real numbers like 12 V or 6 Ω, but are so-called *complex numbers.* Therefore, we will begin the chapter with a discussion of complex numbers, proceed from there to a discussion of phasors, and complete the chapter by applying the phasor technique to simple circuits. In Chapter 17 we will consider the phasor method in greater detail and apply it to more general circuits.

16.1 IMAGINARY NUMBERS

Numbers like +2 and −5, with which we are familiar, are called *real* numbers, and may be plotted on a *real* axis, as shown in Fig. 16.1. Positive real numbers are measured from 0 (the origin) to the right in the *positive* direction and negative numbers are measured from 0 to the left in the *negative* direction.

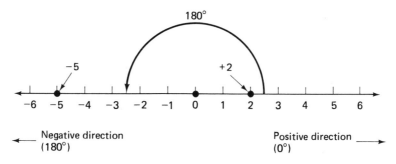

FIGURE 16.1 *Real axis.*

Angular Direction: We may also think of the positive and negative directions in terms of angles. If the positive axis (from 0 to the right) is considered as occupying the angular position of 0°, then if it is rotated 180° in the counterclockwise direction as shown, it will coincide with the negative axis (from 0 to the left). Thus we may think of the positive axis as pointing in the direction 0° and the negative axis as pointing in the direction 180°.

Considering the idea of angular direction, we may think of a negative number as simply a positive number rotated 180° counterclockwise. For example, −5 is +5 rotated 180°, whereas +5 is +5 rotated 0°. The notation we will use is

$$+5 = 5 \,\underline{|0°}$$

and

$$-5 = 5 \,\underline{|180°}$$

which we will read "+5 equals 5 *at* 0° and −5 equals 5 *at* 180°."

Imaginary Numbers: We are familiar with the fact that the square root of a number N is a number which when multiplied by itself equals N. For example,

$$\sqrt{9} = 3$$

because

$$3 \times 3 = 9$$

369

The square root of a negative real number, such as -9, has no real number as its square root, however, since no real number times itself is negative. When the question of square roots of negative numbers was first considered in ancient times, it was decided that since they are not real, they must be *imaginary*. This was an unfortunate name for such numbers, but it has persisted for centuries, and is standard terminology today.

We define the *imaginary unit j* by

$$j = \sqrt{-1} \tag{16.1}$$

and denote an imaginary number by jN, where N is real. For example, $j3$, $j15$, and $-j7$ are imaginary numbers. Also, from (16.1) we see that

$$j^2 = -1 \tag{16.2}$$

so that, as an example,

$$(j3)^2 = j3 \times j3$$
$$= j^2 \times 9$$
$$= -9$$

Thus the square root of -9 is

$$\sqrt{-9} = j3$$

because

$$j3 \times j3 = -9$$

In mathematics books the imaginary unit is denoted by i, which stands for imaginary. However, in circuit theory we cannot use i because this would be confused with the electric current.

Example 16.1: Find (a) $\sqrt{-16}$, (b) $\sqrt{-25}$, and (c) $\sqrt{-2}$.

Solution: In (a) we have

$$\sqrt{-16} = \sqrt{-1}\sqrt{16}$$
$$= j4$$

Similarly, in (b) and (c) we may write

$$\sqrt{-25} = \sqrt{-1}\sqrt{25}$$
$$= j5$$

and

$$\sqrt{-2} = \sqrt{-1}\sqrt{2}$$
$$= j1.414$$

The last answer is rounded off, of course, to three decimal places.

Powers of _j_: From (16.1) we see that the first power of j is $\sqrt{-1}$ and from (16.2) the second power of j is -1. Other powers may be obtained from these, such as

$$j^3 = j^2 \times j = -j \tag{16.3}$$

and

$$j^4 = j^2 \times j^2 = (-1) \times (-1) = 1 \tag{16.4}$$

Thus any power of j can be simplified to ±1 or $\pm j$, because higher powers than 4 reduce to one of these.

Example 16.2: Simplify j^5 and j^{35}.

Solution: Using (16.4), we have

$$j^5 = j^4 \times j = 1 \times j = j$$

and by (16.3) and (16.4) we have

$$j^{35} = j^{32} \times j^3$$
$$= (j^4)^8 \times j^3$$
$$= -j$$

The _j_ Operator: If we multiply the real number 5 by j twice we have

$$5 \times j \times j = 5j^2 = -5 = 5\underline{|180°}$$

Therefore, the operation of multiplying by j twice is equivalent to rotating the real positive number 5 by 180°. It follows that multiplying a real positive number once by j would rotate it one-half of 180° or 90°. For example, we could write

$$j5 = 5\underline{\big|90°}$$

We may think of j as an *operator,* or j *operator,* which applied to any number N (that is, multiplying N by j) results in a 90° rotation of N. In general, j "operating on" N is

$$jN = N\underline{\big|90°} \tag{16.5}$$

The *j* Axis: If N is a real number, then as we have seen, jN is an imaginary number. Also, if N is positive, jN is "at" 90°, or is N rotated 90° counterclockwise. The number jN would then lie on an axis at right angles to the real axis. This axis is called the *imaginary axis,* or the *j axis,* and is vertical if the real axis is horizontal. All imaginary numbers would be plotted on the j axis in the same manner that real numbers are plotted on the real axis. Real and imaginary axes are shown in Fig. 16.2, and as illustrations, the real numbers 2 and −3 and the imaginary numbers $j5$ and $−j4$ are plotted.

The number $−j4$ shown in Fig. 16.2 may be written

$$-j4 = j^3 \times 4$$

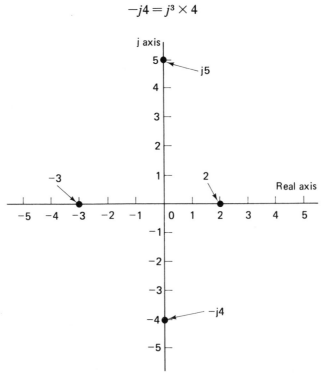

FIGURE 16.2 *Real and imaginary axes.*

Thus it represents three applications of the j operator on the real number 4. Since each operation results in a 90° counterclockwise rotation, we may write $-j4$ as

$$-j4 = 4 \underline{|270°}$$

From Fig. 16.2 we may see that the point $-j4$ may be obtained by rotating $+4$ in the *clockwise* direction by 90°. This is the opposite direction of rotation of the counterclockwise direction, which resulted in a positive angle (positive rotation). Therefore, we will denote the clockwise direction as the negative direction and write

$$-j4 = 4 \underline{|270°} = 4 \underline{|-90°}$$

In the general case, we may write

$$-jN = N \underline{|270°} = N \underline{|-90°} \tag{16.6}$$

where N is real.

PRACTICE EXERCISES

16-1.1   Find (a) $\sqrt{-36}$, (b) $\sqrt{-9}\sqrt{-16}$, and (c) $4\sqrt{-81}$.   *Ans.* (a) $j6$, (b) -12, (c) $j36$

16-1.2   Simplify (a) $-j^6$, (b) j^{17}, and (c) $j^3 \times j^8$.   *Ans.* (a) 1, (b) j, (c) $-j$

16-1.3   Write in real or imaginary form the numbers (a) $6\underline{|0°}$, (b) $4\underline{|180°}$, (c) $17\underline{|90°}$, and
       (d) $2\underline{|-90°}$.   *Ans.* (a) 6, (b) -4, (c) $j17$, (d) $-j2$

16.2   COMPLEX NUMBERS

If we add a real number such as 3 to an imaginary number such as $j4$, the result is the number

$$N = 3 + j4$$

which is called a *complex number*. The real number 3 is called its *real part* and the real number 4, which is the coefficient of j, is called its *imaginary part*. In general, a complex number has the form

$$N = a + jb \tag{16.7}$$

where a and b are real, and a is the real part and b is the imaginary part of N.

Rectangular Form: Complex numbers may be plotted in the plane formed by the real and imaginary axes, as shown for the example $3 + j4$ in Fig. 16.3. The horizontal distance from the j axis is the real part 3 and the vertical distance from the real axis is the imaginary part 4. Because the number $3 + j4$ is plotted on a *rectangular* coordinate system (the axes are at right angles to each other) like the point (3, 4), the number $3 + j4$ and the general case in (16.7) are said to be in *rectangular* form.

Other examples illustrated in Fig. 16.3 are the rectangular-form numbers $-3 + j2$, $-5 - j3$, and $4 - j5$. These illustrate how negative real and/or imaginary parts affect the plotting of the numbers. If we number the *quadrants* in which the axes divide the plane as I, II, III, and IV, as shown in Fig. 16.3, we see that in quadrant I both the real and imaginary parts are positive, in II the real part is negative and the imaginary part is positive, in III both parts are negative, and in IV the real part is positive and the imaginary part is negative.

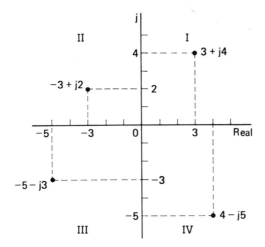

FIGURE 16.3 Plots of complex numbers.

Polar Form: The complex number $N = a + jb$ may also be considered as a rotation of a real number from the real axis in the same way that imaginary numbers were considered in the previous section as real numbers rotated 90°. Such a representation, shown in Fig. 16.4, is called the *polar form* of the complex number, and is expressed

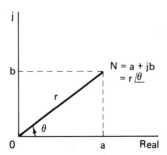

FIGURE 16.4 Complex number in polar form.

as the length r from the origin to the point representing the number, at an angle θ made with the positive real axis. That is,

$$N = a + jb = r\underline{|\theta} \qquad (16.8)$$

where $a + jb$ is the rectangular form and $r\underline{|\theta}$ is the polar form. The length r is also called the *magnitude* of the complex number.

From the triangle formed by sides a, b, and r, we may relate the polar components to the rectangular components. The tangent of θ is

$$\tan \theta = \frac{b}{a}$$

so that θ is "the angle whose tangent is b/a." This statement is written

$$\theta = \arctan \frac{b}{a} \qquad (16.9)$$

or equivalently

$$\theta = \tan^{-1} \frac{b}{a} \qquad (16.10)$$

and sometimes read "θ equals the arctangent of b/a."

By the Pythagorean theorem the radius r of the complex number N of Fig. 16.4 is

$$r = \sqrt{a^2 + b^2} \qquad (16.11)$$

Therefore, to convert the rectangular form $a + jb$ to polar form we use

$$r = \sqrt{a^2 + b^2}$$
$$\theta = \arctan \frac{b}{a} \qquad (16.12)$$

Conversely, since by Fig. 16.4 we have

$$\cos \theta = \frac{a}{r}$$

and

$$\sin \theta = \frac{b}{r}$$

we may convert the polar form to rectangular form by using

$$a = r \cos \theta$$
$$b = r \sin \theta$$

(16.13)

Example 16.3: Convert the rectangular form

$$N = 4 + j3$$

to polar form.

Solution: By (16.12) we have

$$r = \sqrt{4^2 + 3^2} = \sqrt{25} = 5$$

and

$$\theta = \arctan \frac{3}{4}$$
$$= \arctan 0.75$$
$$= 36.9°$$

Therefore, we have

$$4 + j3 = 5 \underline{|36.9°}$$

(16.14)

as shown in Fig. 16.5.

The operations of arctan b/a may be performed on most scientific hand calcula-tors, and may always be done with trigonometric function tables.

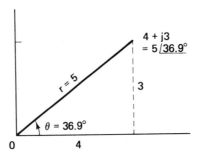

FIGURE 16.5 *Conversion of rectangular to polar form.*

Example 16.4: Convert to rectangular form the polar number

$$N = 10 \underline{|53.1°}$$

Solution: By (16.13) we have

$$a = 10 \cos 53.1° = 10 \ (0.6) = 6$$
$$b = 10 \sin 53.1° = 10 \ (0.8) = 8$$

so that

$$10 \underline{|53.1°} = 6 + j8 \tag{16.15}$$

Example 16.5: Convert to polar form the rectangular number

$$N = -6 + j6$$

Solution: Plotting $-6 + j6$ places N in the second quadrant, as shown in Fig. 16.6. Thus the angle is between 90° and 180°. The angle with the negative real axis is

$$\arctan \frac{6}{6} = \arctan 1 = 45°$$

as we see in Fig. 16.6. Therefore, the angle of N is $180° - 45° = 135°$. The magnitude of N is given by (16.12) as

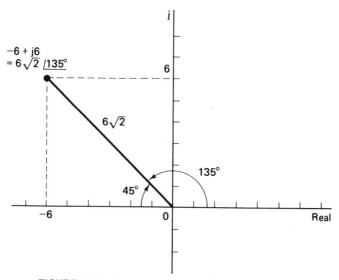

FIGURE 16.6 *Complex number in the second quadrant.*

$$r = \sqrt{6^2 + 6^2} = \sqrt{2(6^2)} = 6\sqrt{2}$$

$$= 8.485$$

Therefore, in polar form the number is

$$-6 + j6 = 6\sqrt{2}\lfloor 135° \qquad (16.16)$$

Example 16.6: Convert to polar form the numbers (a) $-4\sqrt{3} - j4$ and (b) $2.5 - j6$.

Solution: Case (a) is plotted in Fig. 16.7(a), and lies in the third quadrant. The magnitude is

$$r = \sqrt{(4\sqrt{3})^2 + (4)^2} = 8$$

and the angle with the negative real axis is

$$\arctan \frac{4}{4\sqrt{3}} = \arctan 0.577 = 30°$$

Therefore, the angle of the complex number is $180 + 30 = 210°$, as shown, so that we have

$$-4\sqrt{3} - j4 = 8\lfloor 210° \qquad (16.17)$$

We may also note from the figure that its angle is $150°$ in the clockwise direction, which is the negative direction. Therefore, (16.17) may also be written

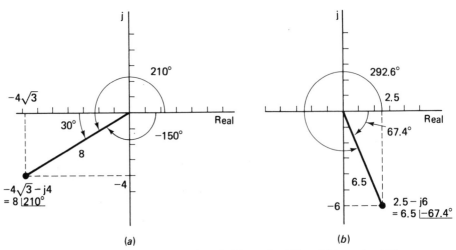

(a)   (b)

FIGURE 16.7 *Complex numbers in (a) quadrant III and (b) quadrant IV.*

$$8 \underline{|210°} = 8 \underline{|-150°}$$

Case (b) is in the fourth quadrant, as we see in Fig. 16.7(b). Its magnitude is

$$r = \sqrt{(2.5)^2 + (6)^2} = 6.5$$

and the angle it makes with the positive real axis is

$$\arctan \frac{6}{2.5} = \arctan 2.4 = 67.4°$$

as shown. Therefore, we have

$$2.5 - j6 = 6.5 \underline{|-67.4°}$$

An alternative representation is

$$6.5 \underline{|-67.4°} = 6.5 \underline{|360° - 67.4°}$$
$$= 6.5 \underline{|292.6°}$$

as shown.

The last three examples illustrate how to use the triangle method, discussed in Chapter 15, to find trigonometric functions of angles outside quadrant I. For example, in Fig. 16.6, the horizontal coordinate is $x = -6$, the vertical coordinate is $y = 6$, and the hypotenuse is $r = 6\sqrt{2}$. Therefore, we have

$$\sin 135° = \frac{y}{r} = \frac{6}{6\sqrt{2}} = 0.707$$

$$\cos 135° = \frac{x}{r} = \frac{-6}{6\sqrt{2}} = -0.707$$

and

$$\tan 135° = \frac{y}{x} = \frac{6}{-6} = -1$$

Example 16.7: Find the sine, cosine, and tangent of (a) 210° and (b) 292.6°.

Solution: For 210° we have from Fig. 16.7(a),

$$x = -4\sqrt{3}$$
$$y = -4$$
$$r = 8$$

so that

$$\sin 210° = \frac{y}{r} = \frac{-4}{8} = -0.5$$

$$\cos 210° = \frac{x}{r} = \frac{-4\sqrt{3}}{8} = -0.866$$

$$\tan 210° = \frac{y}{x} = \frac{-4}{-4\sqrt{3}} = 0.577$$

For 292.6° we have from Fig. 16.7(b),

$$x = 2.5$$

$$y = -6$$

$$r = 6.5$$

so that

$$\sin 292.6° = \frac{y}{r} = \frac{-6}{6.5} = -0.923$$

$$\cos 292.6° = \frac{x}{r} = \frac{2.5}{6.5} = 0.385$$

$$\tan 292.6° = \frac{y}{x} = \frac{-6}{2.5} = -2.4$$

Summarizing, we may note from Figs. 16.5 and 16.7(a), (b) that the sine is positive in quadrants I and II and is negative in quadrants III and IV, the cosine is positive in I and IV and negative in II and III, and the tangent is positive in I and III and negative in II and IV. In all cases the magnitudes of the functions are those of the functions of the angles made with the real axis.

PRACTICE EXERCISES

16-2.1 Convert to polar form the numbers (a) $9 + j12$, (b) $-1 + j2$, (c) $-5 - j12$, and (d) $8 - j15$.   *Ans.* (a) $15\lfloor 53.1°$, (b) $\sqrt{5}\lfloor 116.6°$, (c) $13\lfloor 247.4°$, (d) $17\lfloor -61.9°$

16-2.2 Convert to rectangular form the numbers (a) $10\lfloor 60°$, (b) $5\lfloor 126.9°$, (c) $20\lfloor -45°$, and (d) $10\lfloor 225°$.
          Ans. (a) $5 + j8.66$, (b) $-3 + j4$, (c) $14.14 - j14.14$, (d) $-7.07 - j7.07$

16-2.3 Find (a) $\sin 150°$, (b) $\cos 225°$, (c) $\tan 315°$, and (d) $\sin (-30°)$.
          Ans. (a) 0.5, (b) -0.707, (c) -1, (d) -0.5

which we have taken as a voltage but which may be a current as well. If we denote the rms value of v as $V_{rms} = V$, then we know that

$$V = \frac{V_m}{\sqrt{2}} \tag{16.38}$$

or

$$V_m = \sqrt{2}\, V \tag{16.39}$$

Therefore, in terms of the rms value we may write (16.37) as

$$v = \sqrt{2}\, V \sin(\omega t + \phi) \tag{16.40}$$

The *phasor*, or *phasor* representation, of v in (16.40) is defined to be the complex number

$$\mathbf{V} = V \underline{|\phi} \tag{16.41}$$

The phasor thus depends only on the rms value V, which may be obtained from the amplitude V_m, and the phase ϕ, and can be written down by inspection of the sinusoidal function (16.40). To distinguish them from other complex numbers, phasors are printed in boldface type, as shown.

Example 16.17: Find the phasors of the sinusoidal voltage and current

$$v = 170 \sin(377t + 15°)$$

and

$$i = 17 \sin(377t - 10°)$$

Solution: Since $V_m = 170$, we have

$$V = \frac{V_m}{\sqrt{2}} = \frac{170}{\sqrt{2}} = 120$$

and similarly $I_m = 17$, so that $I = 12$. Therefore, the phasor of v is

$$\mathbf{V} = 120 \underline{|15°}$$

and that of i is

$$\mathbf{I} = 12 \underline{|-10°}$$

In the general case, (16.41) may be written, by (16.38),

$$\mathbf{V} = \frac{V_m}{\sqrt{2}} \underline{\lfloor \phi}\qquad\qquad (16.42)$$

The phasor could be defined in terms of peak values, in which case the factor $\sqrt{2}$ would be omitted in (16.42). However, as we have seen, rms values are usually measured by ac ammeters and voltmeters, and are used in calculating average power. Therefore, we will use rms values, as in (16.41), in defining phasors.

In steady-state ac sinusoidal circuits, all the currents and voltages are sinusoids with the same frequency ω, but with different amplitudes V_m (or I_m) and phase angles ϕ. Therefore, if we know the frequency ω, as we usually will, the specification of the phasor completely describes the sinusoidal wave. That is, if we are given the phasor $\mathbf{V}$ in (16.41), we know V and ϕ. Therefore, we may find $V_m = \sqrt{2}\ V$ and write the sinusoidal function (16.37).

Example 16.18: Given the phasor

$$\mathbf{V} = 70.7\underline{\lfloor 30°}\ \text{V}$$

and the frequency $f = 60$ Hz, find the sinusoidal voltage v.

Solution: Since $V = 70.7$ V, we have

$$V_m = \sqrt{2}\ V = \sqrt{2}\ (70.7) = 100\ \text{V}$$

Also, the angular frequency is

$$\omega = 2\pi f = 2\pi(60) = 377\ \text{rad/s}$$

Therefore, by (16.37) we have

$$v = 100 \sin (377t + 30°)\ \text{V}$$

If the sinusoid is a cosine wave, we may use the result

$$\cos \theta = \sin (\theta + 90°)\qquad\qquad (16.43)$$

previously considered in (15.29), to obtain the phasor representation. All we have to do is add 90° to the phase of the cosine function and it becomes a sine function. This is illustrated in the following example.

Example 16.19: Find the phasor representation of the sinusoids

$$i = 20 \cos 6t\ \text{A}$$

and

$$v = 20 \cos (6t - 30°) \text{ V}$$

Solution: The functions may be converted to sine functions by adding 90° to their phase angles. That is, we may write

$$i = 20 \sin (6t + 90°) \text{ A}$$

and

$$v = 20 \sin (6t - 30° + 90°) \text{ V}$$

or

$$v = 20 \sin (6t + 60°) \text{ V}$$

The phase, therefore, is 90° in the first case and 60° in the second case. Since in both cases the rms value is $20/\sqrt{2} = 14.14$, we have the phasor of i given by

$$\mathbf{I} = 14.14 \underline{|90°} \text{ A}$$

and that of v given by

$$\mathbf{V} = 14.14 \underline{|60°} \text{ V}$$

PRACTICE EXERCISES

16.4.1 Find the phasor of the sinusoids (a) $10 \sqrt{2} \sin 2t$, (b) $20 \sqrt{2} \sin (3t + 15°)$, (c) $40 \sin (7t - 20°)$, and (d) $50 \sqrt{2} \cos (50t + 10°)$

　　　　　　　　　　Ans. (a) $10 \underline{|0°}$, (b) $20 \underline{|15°}$, (c) $28.28 \underline{|-20°}$, (d) $50 \underline{|100°}$

16.4.2 Find the sinusoids whose phasors are (a) $10 \underline{|0°}$, (b) $20/\sqrt{2} \underline{|45°}$, (c) $100 \underline{|-10°}$, and (d) $5 \underline{|90°}$. The frequency is $\omega = 6$ rad/s in each case.

　　　　　　　　　　　　Ans. (a) $14.14 \sin 6t$, (b) $20 \sin (6t + 45°)$,

　　　　　　(c) $141.42 \sin (6t - 10°)$, (d) $7.07 \sin (6t + 90°) = 7.07 \cos 6t$.

16.5　*IMPEDANCE AND ADMITTANCE*

In the case of a resistor, the ratio v/i is the resistance R, which is a measure of the opposition to current. We may define a similar opposition $\mathbf{V}/\mathbf{I}$ to current in the case of the phasors $\mathbf{V}$ and $\mathbf{I}$ of the ac sinusoidal voltage v and current i associated with an element in an ac circuit. Specifically, we may have a circuit element with sinusoidal functions v and i, as shown in Fig. 16.10(a), and the corresponding element

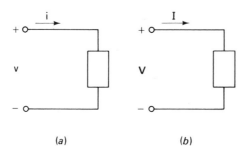

FIGURE 16.10 *Element with (a) sinu-soidal voltage and current and (b) phasor voltage and current.*

(a)   (b)

with *phasor voltage* **V** and *phasor current* **I,** as shown in Fig. 16.10(b). The quantities **V** and **I** are the phasors of *v* and *i,* and the opposition to current is denoted by **Z** and defined as

$$\mathbf{Z} = \frac{\mathbf{V}}{\mathbf{I}} \tag{16.44}$$

Since *v* and *i* are functions of time (sinusoids in this case), and **V** and **I** are phasors, we may think of Fig. 16.10(a) as the *time-domain representation* of the element and of Fig. 16.10(b) as its *phasor-domain* representation. Thus **Z** in (16.44) is independent of time, and, as we will see in this section, is generally a function of frequency.

Impedance: The quantity **Z** in (16.44) is called the *impedance* of the element of Fig. 16.10(b), and is generally a complex number because both **V** and **I** are complex numbers. Thus **Z** plays the same role as *R* in the case of a resistor. Indeed, (16.44) has the exact form of Ohm's law. The chief difference is that *R* is a real positive number where **Z** is complex.

Also like resistance, the impedance is measured in ohms, as may be seen in (16.44). The ratio of **V** to **I** has units of V/A or Ω.

Example 16.20: Find **Z** if in the time domain of Fig. 16.10(a), the voltage is

$$v = 100 \sin (2t + 60°) \text{ V}$$

and the current is

$$i = 5 \sin (2t + 15°) \text{ A}$$

Solution: The phasors are $\mathbf{V} = 100/\sqrt{2} \,\underline{|60°}$ and $\mathbf{I} = 5/\sqrt{2} \,\underline{|15°}$. Therefore, by (16.44) we have

$$\mathbf{Z} = \frac{\mathbf{V}}{\mathbf{I}} = \frac{100/\sqrt{2} \,\underline{|60°}}{5/\sqrt{2} \,\underline{|15°}} = 20 \,\underline{|45°} \; \Omega$$

Impedance of a Resistor: If the element of Fig. 16.10(a) is a resistor with resistance R and current

$$i = I_m \sin \omega t$$

the voltage is

$$v = Ri = RI_m \sin \omega t$$

Therefore, the phasor current and voltage are

$$\mathbf{I} = \frac{I_m}{\sqrt{2}} \underline{|0°}$$

and

$$\mathbf{V} = \frac{RI_m}{\sqrt{2}} \underline{|0°}$$

Consequently, the impedance by (16.44) is

$$\mathbf{Z} = \frac{\mathbf{V}}{\mathbf{I}} = \frac{RI_m/\sqrt{2} \underline{|0°}}{I_m/\sqrt{2} \underline{|0°}} = R \underline{|0°} = R$$

That is, the impedance of a resistor is its resistance R.

In summary, if $\mathbf{Z}_R$ is the impedance of a resistor with resistance R, then

$$Z_R = R \tag{16.45}$$

Impedance of an Inductor: If the element of Fig. 16.10(a) is an inductor with inductance L and current i given by

$$i = I_m \sin \omega t \tag{16.46}$$

then as we saw in Fig. 15.16 and in Practice Exercise 15.7.1, the voltage is a cosine wave given by

$$v = \omega L I_m \cos \omega t$$

By (16.43) the voltage may be written

$$v = \omega L I_m \sin (\omega t + 90°) \tag{16.47}$$

so that the phasor quantities of Fig. 16.10(b) are

$$\mathbf{I} = \frac{I_m}{\sqrt{2}} \underline{|0^\circ}$$

and

$$\mathbf{V} = \frac{\omega L I_m}{\sqrt{2}} \underline{|90^\circ}$$

Therefore, the inductor impedance, which we denote by $\mathbf{Z}_L$, is given by

$$\mathbf{Z}_L = \frac{\omega L I_m/\sqrt{2} \underline{|90^\circ}}{I_m/\sqrt{2} \underline{|0^\circ}} = \omega L \underline{|90^\circ}$$

or, in terms of the j operator,

$$\mathbf{Z}_L = j\omega L \qquad\qquad (16.48)$$

Example 16.21: Find the impedance of a 10-H inductor if the frequency is $\omega = 5$ rad/s.

Solution: Since $L = 10$, we have by (16.48)

$$\mathbf{Z}_L = j\omega L = j(5)(10) = j50 \ \Omega$$

Impedance of a Capacitor: For an inductor we have

$$v = L\frac{di}{dt}$$

and for a capacitor we have

$$i = C\frac{dv}{dt}$$

Therefore, we may obtain the capacitor equation by replacing, in the inductor equation, L by C, v by i, and i by v. If we extend this replacement to (16.46) and (16.47) for the inductor, we have the analogous capacitor voltage

$$v = V_m \sin \omega t$$

and current

$$i = \omega C V_m \sin(\omega t + 90°)$$

The capacitor phasor voltage and current therefore are

$$\mathbf{V} = \frac{V_m}{\sqrt{2}} \underline{|0°}$$

and

$$\mathbf{I} = \frac{\omega C V_m}{\sqrt{2}} \underline{|90°}$$

Thus the capacitor impedance, which we denote by $\mathbf{Z}_C$ is

$$\mathbf{Z}_C = \frac{V_m/\sqrt{2}\,\underline{|0°}}{\omega C V_m/\sqrt{2}\,\underline{|90°}} = \frac{1}{\omega C}\,\underline{|-90°}$$

or in terms of j,

$$\mathbf{Z}_C = -j\frac{1}{\omega C} \tag{16.49}$$

Since $-j = 1/j$, this may be simplified to

$$\mathbf{Z}_C = \frac{1}{j\omega C} \tag{16.50}$$

Example 16.22: Find the impedance of a $10\text{-}\mu\text{F}$ capacitor if the frequency is $\omega = 10{,}000$ rad/s.

Solution: We have $C = 10\ \mu\text{F} = 10 \times 10^{-6}\ \text{F} = 10^{-5}\ \text{F}$, so that by (16.49) the impedance is

$$\mathbf{Z}_C = -j\frac{1}{(10{,}000)(10^{-5})} = -j10\ \Omega$$

The impedances $\mathbf{Z}_R$, $\mathbf{Z}_L$, and $\mathbf{Z}_C$ have been obtained with one or the other, or both, of the phasors $\mathbf{V}$ and $\mathbf{I}$ with a zero phase angle. This was done for mathematical convenience, but the results hold in the general case. Any added phase angle in $\mathbf{V}$ will also appear in $\mathbf{I}$ and will cancel in the ratio $\mathbf{V/I}$.

Another point that should be made is that $\mathbf{Z}_R$ is constant, but $\mathbf{Z}_L$ and $\mathbf{Z}_C$ both depend on frequency. The impedance $\mathbf{Z}_R$ is a positive real number on the positive real axis, $\mathbf{Z}_L$ is on the positive j axis (at an angle of $90°$), and $\mathbf{Z}_C$ is on the negative j axis (at an angle of $270°$ or $-90°$). For low frequencies (small ω), $\mathbf{Z}_L$ is very small but $\mathbf{Z}_C$ is very large. In the dc case, $\omega = 0$, $\mathbf{Z}_L = 0$ and thus the inductor is a short circuit, but $\mathbf{Z}_C$ is infinite, and thus the capacitor is an open circuit. For high frequencies, the reverse is true: $\mathbf{Z}_L$ is large and $\mathbf{Z}_C$ is small.

Admittance: We may think of (16.44) as an "Ohm's law" for ac circuits and write it in the form

$$\mathbf{V} = \mathbf{Z}\mathbf{I} \tag{16.51}$$

This is, of course, the same form as Ohm's law,

$$v = Ri$$

for the resistor, where $\mathbf{Z}$ plays the role of R in providing opposition to the current. In the case of the resistor, the conductance is $G = 1/R$, which is a measure of the *ease* of conducting current. The analogy to conductance in the phasor case is the reciprocal of $\mathbf{Z}$, denoted by $\mathbf{Y}$ and called the *admittance* of the element. That is, the admittance is

$$\mathbf{Y} = \frac{1}{\mathbf{Z}} \tag{16.52}$$

and its SI unit is the mho.

As examples, the admittance of a resistor is

$$\mathbf{Y}_R = \frac{1}{R} = G \tag{16.53}$$

that of an inductor is

$$\mathbf{Y}_L = \frac{1}{\mathbf{Z}_L} = \frac{1}{j\omega L} = -j\frac{1}{\omega L} \tag{16.54}$$

and that of a capacitor is

$$Y_C = \frac{1}{\mathbf{Z}_C} = j\omega C \tag{16.55}$$

Ohm's law in terms of admittance is

$$\mathbf{I} = \mathbf{Y}\mathbf{V} \tag{16.56}$$

Example 16.23: Find the phasor current and the steady-state sinusoidal current of a 2-μF capacitor with voltage

$$v = 10\sqrt{2}\sin(1000t + 15°)\ \text{V}$$

Solution: We see from v that $\omega = 1000$ rad/s and that the phasor voltage is

$$\mathbf{V} = 10\underline{|15°}\ \text{V}$$

Also, by (16.55) we have

$$\mathbf{Y} = j\omega C$$
$$= j(1000)(2 \times 10^{-6}) = j2 \times 10^{-3}$$
$$= 2 \times 10^{-3}\underline{|90°}\ \mho$$

Therefore, by (16.56) the phasor current is

$$\mathbf{I} = \mathbf{YV} = (2 \times 10^{-3}\underline{|90°})(10\underline{|15°})$$
$$= 0.02\underline{|105°}\ \text{A}$$
$$= 20\underline{|105°}\ \text{mA}$$

Therefore, the sinusoidal current is

$$i = 20\sqrt{2}\sin(1000t + 105°)\ \text{mA}$$
$$= 28.28\sin(1000t + 105°)\ \text{mA}$$

PRACTICE EXERCISES

16-5.1 Find the impedance $\mathbf{Z}$ of an element having

$$v = 10\sin 2t\ \text{V}$$

and current

$$i = 2\sin(2t - 30°)\ \text{A} \qquad\qquad Ans.\ 5\underline{|30°}\ \Omega$$

16-5.2 If an element has an impedance

$$\mathbf{Z} = 10\underline{|45°}\ \Omega$$

and a voltage

$$v = 40\sin(6t + 10°)\ \text{V}$$

find its phasor current and its time-domain current.

Ans. $4/\sqrt{2}\lfloor -35° $ A, $4 \sin (6t - 35°)$ A

16-5.3 If the frequency is $\omega = 2000$ rad/s, find the impedance of (a) a 1-kΩ resistor, (b) a 3-H inductor, and (c) a 0.1-μF capacitor.   *Ans.* (a) 1 kΩ, (b) $j6$ kΩ, $-j5$ kΩ

16-5.4 If a 2-H inductor has an impedance of $j1000$ Ω, find the frequency ω.

Ans. 500 rad/s

16.6 KIRCHHOFF'S LAWS AND PHASOR CIRCUITS

In an ac circuit in steady state, as we have noted earlier, all the currents and voltages are sinusoids with the same frequency as the sinusoidal source. Since Kirchhoff's voltage and current laws hold, it follows that the algebraic sum of sinusoids of a certain frequency is a sinusoid of the same frequency. For example, the sum of the sinusoidal voltage drops around a loop equals the sinusoidal voltage rise of the source in the loop. Since the sinusoids may be represented by their phasors, this means that we may replace the sinusoidal voltages and currents in the circuit by their phasors and Kirchhoff's voltage and current laws will still hold.

Phasor Circuits: As an example, let us consider the circuit of Fig. 16.11, which is an ac circuit with all the voltages and the current replaced by their phasor representations. By KCL the same current $\mathbf{I}$ flows through every element and by KVL we have

$$\mathbf{V}_1 + \mathbf{V}_2 = \mathbf{V}_g \qquad (16.57)$$

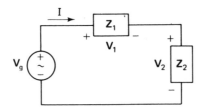

FIGURE 16.11 *Circuit of phasor quantities.*

The quantities $\mathbf{Z}_1$ and $\mathbf{Z}_2$ are the impedances of the elements shown as rectangles, and may be used to relate the current $\mathbf{I}$ to the voltages $\mathbf{V}_1$ and $\mathbf{V}_2$.

We will call the circuit of Fig. 16.11 a *phasor circuit,* since it is obtained from the time-domain circuit by replacing all the sinusoidal voltages and currents by their phasors and labeling all the elements other than sources with their impedances. Since KCL and KVL hold for phasors and the analogies to Ohm's law are valid, such as

$$\mathbf{V}_1 = \mathbf{Z}_1 \mathbf{I}$$

$$\mathbf{V}_2 = \mathbf{Z}_2 \mathbf{I} \qquad (16.58)$$

for the circuit of Fig. 16.11, phasor circuits may be analyzed exactly like resistive circuits. The only difference is that the impedances are complex numbers and not real numbers like the resistances. The current or voltages that are found in the analysis are phasors, which may be converted immediately to the time-domain sinusoidal answers. If we are interested in rms values that the ac meters read, it is not even necessary to make the conversion.

As an example, the circuit of Fig. 16.12(a) is a time-domain series circuit with a sinusoidal source v_g and elements R, L, and C. Its phasor circuit, shown in Fig. 16.12(b), is obtained by replacing v_g, v_R, v_L, v_C, and i by their phasors $\mathbf{V}_g$, $\mathbf{V}_R$, $\mathbf{V}_L$, $\mathbf{V}_C$, and $\mathbf{I}$, and labeling R, L, and C with their impedances.

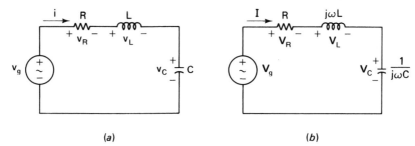

(a)   (b)

FIGURE 16.12 (a) Time-domain circuit and (b) its corresponding phasor circuit.

Suppose that in Fig. 16.12(a), v_g, R, L, and C are given and we are to find the steady-state current i. From the phasor circuit we have KVL,

$$\mathbf{V}_R + \mathbf{V}_L + \mathbf{V}_C = \mathbf{V}_g \qquad (16.59)$$

with

$$\mathbf{V}_R = R\mathbf{I}$$

$$\mathbf{V}_L = j\omega L\mathbf{I}$$

$$\mathbf{V}_C = \frac{1}{j\omega C}\mathbf{I}$$

Therefore, (16.59) becomes

$$R\mathbf{I} + j\omega L\mathbf{I} + \frac{1}{j\omega C}\mathbf{I} = \mathbf{V}_g$$

or

$$\left(R + j\omega L + \frac{1}{j\omega C}\right)\mathbf{I} = \mathbf{V}_g$$

and hence the phasor current is

$$\mathbf{I} = \frac{\mathbf{V}_g}{R + j\omega L + 1/j\omega C} \tag{16.60}$$

The sinusoidal current i follows from $\mathbf{I}$.

Example 16.24: Find i in Fig. 16.12(a) if $R = 4\ \Omega$, $L = 2$ H, $C = 0.5$ F, and

$$v_g = 10 \sin 2t \text{ V}$$

Solution: The source phasor is $\mathbf{V}_g = 10/\sqrt{2}\underline{|0°}$ and $\omega = 2$ rad/s, so that by (16.60) we have

$$\mathbf{I} = \frac{10/\sqrt{2}\underline{|0°}}{4 + j(2)(2) - j1/(2)(0.5)}$$

$$= \frac{10/\sqrt{2}\underline{|0°}}{4 + j3}$$

$$= \frac{10/\sqrt{2}\underline{|0°}}{5\underline{|36.9°}}$$

$$= \frac{2}{\sqrt{2}}\underline{|-36.9°}$$

Therefore the time-domain current is

$$i = 2 \sin (2t - 36.9°) \text{ A}$$

This checks the result in (15.54) given when this circuit was considered in Chapter 15.

PRACTICE EXERCISES

16-6.1 Find the steady-state value of i in Fig. 16.12 if $R = 3\ \Omega$, $L = 1$ H, $C = \frac{1}{4}$ F, and $v_g = 18\sqrt{2} \sin 4t$ V.   *Ans.* 6 sin (4t − 45°) A

16-6.2 Find $\mathbf{I}$ in Fig. 16.11 if $\mathbf{Z}_1 = R = 5\ \Omega$, $\mathbf{Z}_2 = j\omega L = j12\ \Omega$, and $\mathbf{V}_g = 26\ \underline{|0°}$ V. (This is the phasor circuit of an *RL* time-domain circuit.)

$$\text{\textit{Ans.} } \frac{26}{5 + j12} = 2\underline{|-67.4°} \text{ A}$$

16-6.3 Find i in Exercise 16.6.2 if $\omega = 3$ rad/s.   *Ans.* $2\sqrt{2} \sin (3t - 67.4°)$ A

16.7 SUMMARY

Complex numbers are of the rectangular form $N = a + jb$, where a and b are real and $j = \sqrt{-1}$. If $b = 0$, then $N = a$ and is a real number, whereas if $a = 0$, then $N = jb$ and is an imaginary number. Real numbers are plotted on the real axis, which is normally horizontal, and imaginary numbers are plotted on the j axis, which is at 90° to the real axis, and thus is usually vertical. Complex numbers are in one of the four quadrants formed by the two axes, and may also be expressed in the polar form $N = r\underline{|\theta}$, where r is the length of the line representing the number and θ is the angle the line makes with the real axis.

A sinusoid such as

$$v = \sqrt{2}\ V \sin (\omega t + \phi)$$

may be represented by its phasor,

$$\mathbf{V} = V\underline{|\phi}$$

which is a complex number and may be combined with other complex numbers using the rules of addition, subtraction, multiplication, and division. The ratio $\mathbf{Z} = \mathbf{V}/\mathbf{I}$ of the phasor of an element voltage to the phasor of its current is the impedance of the element. This relation,

$$\mathbf{V} = \mathbf{ZI}$$

is like Ohm's law in resistive circuits. Thus ac steady-state circuits may be solved exactly like resistive circuits by replacing all the voltages and currents by their phasors and labeling all the elements except sources with their impedance. The resulting phasor circuit may be solved for the current or voltage phasor of interest and the result used to find the sinusoidal answer.

PROBLEMS

16.1 Convert from rectangular to polar form the numbers (a) $j6$, (b) $6 + j8$, (c) $4 - j4$, (d) $-12 + j9$, and (e) $-1 - j2$.

16.2 Repeat Problem 16.1 for the numbers (a) -4, (b) $150 + j80$, (c) $-5 - j12$, (d) $1 - j\sqrt{3}$, and (e) $-6 + j6$.

16.3 Convert from polar to rectangular form the numbers (a) $8\underline{|90°}$, (b) $100\underline{|-53.1°}$, (c) $8\sqrt{2}\underline{|225°}$, (d) $17\underline{|-28.1°}$, and (e) $10\underline{|150°}$.

16.4 Obtain the following sums in rectangular form:

(a) $(6 + j7) + (3 + j1)$

(b) $(-3 + j2) + (4 - j5)$

(c) $(1 - j2) + (5 + j6)$

(d) $10\underline{|53.1°} + \sqrt{2}\ \underline{|45°}$

(e) $100\underline{|60°} + 100\ \underline{|-60°}$

16.5 Obtain the following differences in rectangular form:

(a) $(3 - j6) - (4 + j1)$

(b) $(-7 - j11) - (-1 + j6)$

(c) $5\underline{|36.9°} - 10\underline{|-53.1°}$

(d) $100\underline{|30°} - 100\underline{|-30°}$

16.6. Find the following products in rectangular form:

(a) $(3 + j4)(3 - j5)$

(b) $(3 + j4)(3 - j4)$

(c) $(1 - j2)(-2 + j6)$

(d) $(-1 - j3)(-2 - j4)$

(e) $(2 + j3)(-j1)$

16.7 Find the following products in polar form:

(a) $(-3 + j4)(8 - j6)$

(b) $(5 + j12)(-j2)$

(c) $(\sqrt{2} + j\sqrt{2})(-3 - j4)$

(d) $(12\underline{|56°})(3\underline{|-12°})$

(e) $(4\underline{|225°})(10\underline{|135°})$

16.8 Find the following quotients in rectangular form:

(a) $(-3 + j4) \div (1 + j2)$

(b) $50 \div (4 + j3)$

(c) $(4 + j4) \div (1 + j1)$

(d) $(20\underline{|10°}) \div (4\underline{|63.1°})$

(e) $[(4 + j5) + (16 - j35)] \div [(6 + j3) + (-1 - j3)]$

16.9 Find the following quotients in polar form:

(a) $(25 + j60) \div (3 - j4)$

(b) $289(1 + j1) \div (-8 + j15)$

(c) $j8 \div (2 + j2)$

(d) $(12\underline{|-10°}) \div (4\underline{|60°})$

(e) $(50\underline{|225°}) \div (10\underline{|-160°})$

16.10 Find the magnitudes of the numbers (a) $-j25$, (b) $-5 - j12$, (c) $-80 - j150$, (d) $\sqrt{2}\ (4 - j4)$, and (e) $20/(4 - j3)$.

16.11 Find the phasors of the following sinusoids:

(a) $60\sqrt{2} \sin(300t - 36°)$

(b) $10\sqrt{2} \sin(5t + 12°)$

(c) $100 \sin 1000t$

(d)   50 cos (20t − 40°)

(e)   20 √2 cos (10t − 90°)

16.12   Find the sinusoid of frequency $\omega = 3$ rad/s whose phasor is (a) $10/\sqrt{2} \,\underline{/40°}$, (b) $7.07\,\underline{/-12°}$, (c) $j4\sqrt{2}$, (d) $(6 + j8)/\sqrt{2}$, and (e) 28.28.

16.13   Find the impedance Z of an ac circuit element if the current is

$$i = 10 \sin (2t + 10°) \text{ A}$$

and the voltage is

$$v = 50 \sin (2t − 30°) \text{ V}$$

16.14   Repeat Problem 16.13 if the voltage is (a) 40 sin (2t + 10°) V and (b) 30 cos 2t V.

16.15   If $\omega = 100$ rad/s, find the impedance of (a) a 1-kΩ resistor, (b) a 0.1-H inductor, and (c) a 10-μF capacitor.

16.16   Repeat Problem 16.15 if $\omega = 10,000$ rad/s.

16.17   Find the impedance of a 2-H inductor if the frequency is (a) 0, (b) 5 rad/s, and (c) 50,000 rad/s.

16.18   Find the impedance of a 2-μF capacitor for the frequencies of Problem 16.17.

16.19   Find the admittance of the 2-H inductor of Problem 16.17.

16.20   Find the admittance of the 2-μF capacitor of Problem 16.18.

16.21   Find the steady-state current i if

$$v_g = 20 \sin (4t + 30°) \text{ V}$$

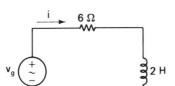

PROBLEM 16.21

16.22   Solve Problem 16.21 if the 2–H inductor is replaced by a 1.5-H inductor and

$$v_g = 24 \sqrt{2} \sin (4t + 15°) \text{ V}$$

16.23   Solve Problem 16.21 if the inductor is replaced by a capacitor with $C = \frac{1}{32}$ F.

16.24   Find the steady-state value of i if $R = 4$ Ω, $L = 1$ H, $C = \frac{1}{4}$ F, and

$$v_g = 10 \sin 4t \text{ V}$$

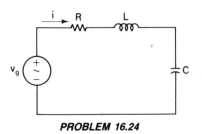

PROBLEM 16.24

16.25 Solve Problem 16.24 if

$$v_g = 20 \sin 2t \text{ V}$$

(Note that in this case the impedance seen by the source appears to be only the resistance R.)

16.26 Solve Problem 16.24 for the values $R = 4 \text{ k}\Omega$, $L = 0.5$ H, $C = 0.05$ μF, and

$$v_g = 20 \sin (10,000t + 60°) \text{ V}$$

17

AC STEADY-STATE ANALYSIS

In Chapter 16 we saw that in the case of circuits with sinusoidal sources, the steady-state currents and voltages may be found by analyzing relatively simple phasor circuits. The phasor circuits are obtained from the time-domain circuits by replacing all the currents and voltages by their phasors and all the resistors, inductors, and capacitors by their impedances. The phasor circuits are then solved for the phasor currents and/or voltages in exactly the same way that resistive circuits are solved. The only difference is that the currents and voltages are phasors, such as **I** and **V**, rather than the time-domain values i and v, and the impedances are complex numbers, whereas the resistances are real numbers.

Once the phasor answers are obtained, they may be converted directly to the sinusoidal time-domain answers. If we are only interested in the rms values of the currents or voltages, the phasor answer is all we need, and thus no conversion is necessary.

The circuits considered in Chapter 16 were relatively simple series circuits. Because of the close kinship between phasor circuits and resistive circuits, however, the phasor method applies to any circuit as long as there is only one sinusoidal source, or in the case of two or more, all the sources have the same frequency. In this chapter we will extend the phasor method to these more general circuits, using resistive circuit techniques such as equivalent impedances, current and voltage division, and loop and nodal analysis. In all cases we will be concerned with the steady-state currents and voltages.

17.1 IMPEDANCE RELATIONSHIPS

As we have seen in Chapter 16, phasor currents and voltages of an element, or of a network of elements, satisfy a kind of Ohm's law, given by

$$\mathbf{V} = \mathbf{ZI} \tag{17.1}$$

where $\mathbf{V}$ is the phasor voltage of the element, $\mathbf{I}$ is its phasor current, and $\mathbf{Z}$ is its impedance, as shown in Fig. 17.1. The network may be a single element, such as a resistor, an inductor, or a capacitor, or it may be a phasor circuit of elements.

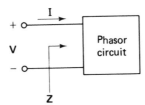

FIGURE 17.1 General phasor circuit.

Impedance: The quantities $\mathbf{V}$ and $\mathbf{I}$ in Fig. 17.1 are phasors of a sinusoidal voltage and current, which we denote by

$$\begin{aligned} v &= V_m \sin{(\omega t + \phi_v)} \text{ V} \\ &= \sqrt{2}\, V \sin{(\omega t + \phi_v)} \text{ V} \end{aligned} \tag{17.2}$$

and

$$\begin{aligned} i &= I_m \sin{(\omega t + \phi_i)} \text{ A} \\ &= \sqrt{2}\, I \sin{(\omega t + \phi_i)} \text{ A} \end{aligned} \tag{17.3}$$

where ϕ_v and ϕ_i are the phase angles of v and i. Thus in the frequency domain the phasors are

$$\mathbf{V} = V \underline{|\phi_v} \text{ V} \tag{17.4}$$

and

$$\mathbf{I} = I \underline{|\phi_i} \text{ A} \tag{17.5}$$

and the impedance is

$$\mathbf{Z} = \frac{\mathbf{V}}{\mathbf{I}} = \frac{V}{I} \underline{|\phi_v - \phi_i} \ \Omega$$

$$= \frac{V_m}{I_m} \underline{|\phi_v - \phi_i|} \ \Omega \qquad (17.6)$$

In polar form the impedance is

$$\mathbf{Z} = |\mathbf{Z}| \underline{|\theta|} \qquad (17.7)$$

where $|\mathbf{Z}|$ is the magnitude and θ is the angle of $\mathbf{Z}$, denoted by

$$\theta = \text{ang } \mathbf{Z} \qquad (17.8)$$

From (17.6) these quantities are

$$|\mathbf{Z}| = \frac{V}{I} = \frac{V_m}{I_m} \qquad (17.9)$$

and

$$\begin{aligned} \text{ang } \mathbf{Z} = \theta &= \phi_v - \phi_i \\ &= \text{ang } \mathbf{V} - \text{ang } \mathbf{I} \end{aligned} \qquad (17.10)$$

Reactance: In the general case of (17.7), $\mathbf{Z}$ is a complex number with real and imaginary parts. If we denote the real part by R and the imaginary part by X, then in rectangular form the impedance is

$$\mathbf{Z} = R + jX \qquad (17.11)$$

The real part R is sometimes called the *resistive* part, or simply the *resistance,* and the imaginary part X is the *reactive part,* or the *reactance.* Both R and X have the unit ohm. As we will see, R is sometimes a constant value of resistance, but in general it is a combination of terms that are functions of frequency. Reactance is always a function of frequency.

Graphical Representation: We may represent the impedance $\mathbf{Z} = R + jX$ graphically, as shown in Fig. 17.2, where it may be seen that the resistive and reactive components are given by

$$R = |\mathbf{Z}| \cos \theta \qquad (17.12)$$
$$X = |\mathbf{Z}| \sin \theta$$

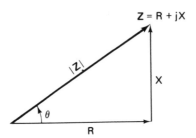

FIGURE 17.2 *Graphical representation of impedance.*

In terms of R and X the polar components are

$$|Z| = \sqrt{R^2 + X^2} \tag{17.13}$$

$$\theta = \arctan \frac{X}{R}$$

The impedance is a complex number, and Fig. 17.2 is simply a graphical representation similar to those we used throughout the previous chapter. In general, the reactance X may be negative as well as positive, in which case we will need to show the real and imaginary axes.

Example 17.1: Find the impedance **Z**, in both polar and rectangular form, which is seen at the source terminals of the phasor circuit of Fig. 17.3. In the time domain the source voltage is

$$v_g = 10\sqrt{2} \sin 5t \text{ V}$$

and the element values are $R = 4 \ \Omega$, $L = 1$ H, and $C = 0.1$ F.

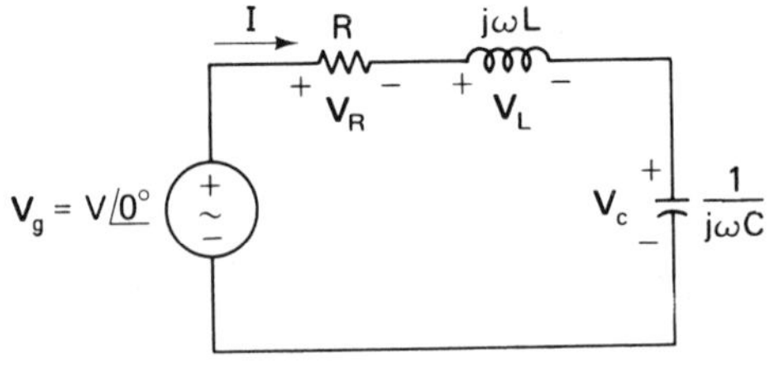

FIGURE 17.3 *RLC circuit example.*

Solution: From the expression for v_g we see that $\omega = 5$ rad/s and $\mathbf{V}_g = 10 \underline{|0°} $ V. Also, by Kirchhoff's voltage law we have in Fig. 17.3,

$$\mathbf{V}_R + \mathbf{V}_L + \mathbf{V}_C = \mathbf{V}_g$$

or

$$R\mathbf{I} + j\omega L\mathbf{I} + \frac{1}{j\omega C}\mathbf{I} = \mathbf{V}_g$$

Substituting in the known values we have

$$\left[4 + j(5)(1) - j\frac{1}{(5)(0.1)}\right] \mathbf{I} = 10\underline{|0°}$$

which simplifies to

$$(4 + j3)\,\mathbf{I} = 10 = \mathbf{V}_g$$

From this result we may find $\mathbf{Z}$ seen at the source terminals, which is given by

$$\mathbf{Z} = \frac{\mathbf{V}_g}{\mathbf{I}}$$

or

$$\mathbf{Z} = 4 + j3 \; \Omega \tag{17.14}$$

This is the rectangular form, from which we have

$$R = 4\,\Omega \qquad X = 3\,\Omega$$

From (17.13) the magnitude and phase are

$$|\mathbf{Z}| = \sqrt{4^2 + 3^2} = 5$$

$$\theta = \arctan \tfrac{3}{4} = 36.9°$$

Therefore, the polar form is

$$\mathbf{Z} = 5\underline{|36.9°} \; \Omega \tag{17.15}$$

Conductance and Susceptance: Since the admittance $\mathbf{Y}$ is the reciprocal $1/\mathbf{Z}$ of the impedance, it also is a complex number and may be written in rectangular or polar form. In rectangular form the general case is denoted by

$$\mathbf{Y} = G + jB \tag{17.16}$$

where the real part G is called the *conductance* and the imaginary part B is called the *susceptance*. The standard unit in each case is mho.

Example 17.2: Find the admittance $\mathbf{Y}$ seen at the source terminals in Example 17.1 in both rectangular and polar form.

Solution: In rectangular form we have, by (17.14),

$$\mathbf{Y} = \frac{1}{\mathbf{Z}} = \frac{1}{4 + j3}$$

which is rationalized to

$$\mathbf{Y} = \frac{1}{4+j3} \cdot \frac{4-j3}{4-j3}$$

$$= \frac{4-j3}{25}$$

$$= \frac{4}{25} + j\left(-\frac{3}{25}\right)$$

Therefore, comparing this result with (17.16), we have the conductance

$$G = \frac{4}{25} \; \mho$$

and the susceptance

$$B = -\frac{3}{25} \; \mho$$

The polar form may be found from the rectangular form, but an easier method is to obtain it directly from the polar form of $\mathbf{Z}$, given in (17.15). In this case we have

$$\mathbf{Y} = \frac{1}{5 \lfloor 36.9°} = 0.2 \lfloor -36.9° \; \mho$$

Special Cases: In the special cases of resistances, inductances and capacitances, considered in Section 16.5, the impedances are

$$\mathbf{Z}_R = R$$
$$\mathbf{Z}_L = j\omega L = \omega L \lfloor 90° \qquad\qquad (17.17)$$
$$\mathbf{Z}_C = \frac{1}{j\omega C} = -j\left(\frac{1}{\omega C}\right) = \frac{1}{\omega C} \lfloor -90°$$

These are indicated in the phasor circuits for a resistor, an inductor, and a capacitor, shown in Fig. 17.4(a), (b), and (c).

Comparing these results with the general impedance

$$\mathbf{Z} = R + jX$$

we see that in the case of a resistor, the reactance is zero, and thus the impedance is purely resistive. Impedances of inductors and capacitors have no resistive component

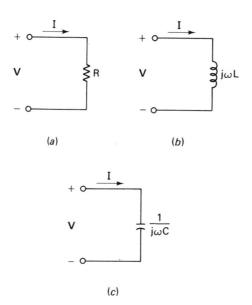

FIGURE 17.4 *Phasor circuits for (a) a resistor, (b) an inductor, and (c) a capacitor.*

and thus are purely reactive. The reactance of an inductor is called *inductive reactance,* and is denoted by X_L. Its value, by (17.17), is

$$X_L = \omega L = 2\pi f L \tag{17.18}$$

so that

$$\mathbf{Z}_L = jX_L = X_L \underline{|90°} \tag{17.19}$$

The reactance of a capacitor is called *capacitive reactance,* and is denoted by X_C, which we define by

$$X_C = \frac{1}{\omega C} = \frac{1}{2\pi f C} \tag{17.20}$$

Therefore, by (17.17) we see that the impedance of a capacitor is

$$\mathbf{Z}_C = -jX_C = X_C \underline{|-90°} \tag{17.21}$$

PRACTICE EXERCISES

17-1.1 Find the impedance seen at the terminals of the source in the phasor circuit shown. Give the answer in both rectangular and polar form.

Ans. $6 - j6 = 6\sqrt{2}\underline{|-45°}$ Ω

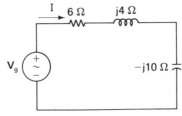

EXERCISE 17-1.1

17-1.2 Find the admittance seen at the terminals of the source in Exercise 17.1.1. Give the answer in both rectangular and polar form.

$$Ans. \ \frac{1}{12} + j\frac{1}{12} = \frac{1}{6\sqrt{2}} \underline{|45°} \ \mho$$

17-1.3 If the time-domain voltage corresponding to V_g in Exercise 17.1.1 is

$$v_g = 24 \sin (2t - 15°) \ V$$

find the phasor current I and its corresponding time-domain current.

$$Ans. \ 2 \underline{|30°} \ A, \ 2\sqrt{2} \sin (2t + 30°) \ A$$

17-1.4 Find the inductive reactance of a 100-mH inductor and the capacitive reactance of a 1-μF capacitor if the frequency is (a) 10 rad/s and (b) 10,000 rad/s.

$$Ans. \ (a) \ 1 \ \Omega, \ 100 \ k\Omega, \ (b) \ 1 \ k\Omega, \ 100 \ \Omega$$

17.2 PHASE RELATIONSHIPS

Two sinusoids with the same frequency but with different phases will resemble each other, but one will be displaced in time from the other. For example, the sine waves

$$v_1(t) = V_m \sin \omega t$$

and

$$v_2(t) = V_m \sin (\omega t + \phi)$$

are sketched in Fig. 17.5, with v_1 shown dashed and v_2 shown solid. The solid curve

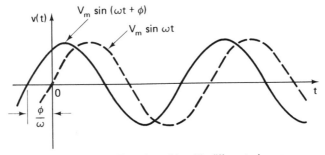

FIGURE 17.5 *Two sinusoids with different phases.*

is simply the dashed curve displaced $\omega t = \phi$ radians, or $t = \phi/\omega$ seconds, to the left.

Leading and Lagging: Points on the solid curve $V_m \sin (\omega t + \phi)$, such as its peaks, occur ϕ rad, or ϕ/ω s, earlier than the corresponding points on the dashed curve $V_m \sin \omega t$. Accordingly, we will say that $V_m \sin (\omega t + \phi)$ *leads* $V_m \sin \omega t$ by ϕ rad (or degrees). In the general case, the sinusoid

$$v_1 = V_1 \sin (\omega t + \alpha)$$

leads the sinusoid

$$v_2 = V_2 \sin (\omega t + \beta)$$

by $\alpha - \beta$. An equivalent expression to v_1 leads v_2 by $\alpha - \beta$ is that v_2 *lags* v_1 by $\alpha - \beta$. If $\alpha - \beta$ is negative, we would normally say that v_2 leads v_1 by $\beta - \alpha$ or v_1 lags v_2 by $\beta - \alpha$.

As an example, if

$$v_1 = 2 \sin (3t + 30°)$$

and

$$v_2 = 6 \sin (3t + 5°)$$

then v_1 leads v_2 by $30° - 5° = 25°$, or v_2 lags v_1 by $25°$.

Resistor Phase Relationship: In the case of a resistance R, if the voltage is

$$v = V_m \sin (\omega t + \phi) \tag{17.22}$$

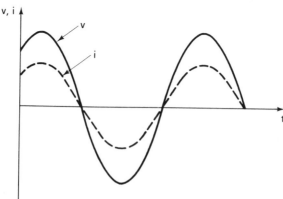

FIGURE 17.6 *Voltage and current waveforms for a resistor.*

then the current is $i = v/R$, or

$$i = I_m \sin (\omega t + \phi) \tag{17.23}$$

where $I_m = V_m/R$. Thus the sinusoidal voltage and current for a resistor have the same phase (ϕ in this case), and are said to be *in phase*. This phase relationship is shown in Fig. 17.6, where the resistor voltage is the solid line and the current is the dashed line. The peaks and dips of both curves occur at the same time.

Inductor Phase Relationship: In the case of the inductor, if the voltage phasor is

$$\mathbf{V} = V\underline{\phi}$$

then the current phasor is

$$\mathbf{I} = \frac{\mathbf{V}}{\mathbf{Z}_L} = \frac{V\underline{\phi}}{\omega L \underline{90°}}$$

or

$$\mathbf{I} = I\underline{\phi - 90°}$$

where $I = V/\omega L$. Therefore, the time-domain voltage and current are

$$v = \sqrt{2}\, V \sin (\omega t + \phi)$$

and

$$i = \sqrt{2}\, I \sin (\omega t + \phi - 90°)$$

and the voltage leads the current, or the current lags the voltage, by $\phi - (\phi - 90°) = 90°$, which of course, is the angle of $\mathbf{Z}_L$. Another expression is that the current and voltage are 90° *out of phase*. This is shown graphically in Fig. 17.7,

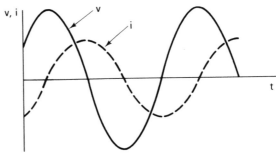

FIGURE 17.7   *Voltage and current waveforms for an inductor.*

where it may be seen that the current lags, its peaks or dips occurring after those of v.

Capacitor Phase Relationship: For a capacitor, if the voltage phasor is

$$\mathbf{V} = V\underline{|\phi}$$

then the current phasor is

$$\mathbf{I} = \frac{\mathbf{V}}{\mathbf{Z}_C} = \frac{V\underline{|\phi}}{1/\omega C\underline{|-90°}}$$

or

$$\mathbf{I} = I\underline{|\phi + 90°}$$

where $I = \omega CV$. Therefore, the time-domain functions are

$$v = \sqrt{2}\,V\sin\,(\omega t + \phi)$$

and

$$i = \sqrt{2}\,I\sin\,(\omega t + \phi + 90°)$$

so that again i and v are 90° out of phase, but i is leading v by $\phi + 90° - \phi = 90°$. This is shown graphically in Fig. 17.8, where it is seen that the peaks of i occur before those of v.

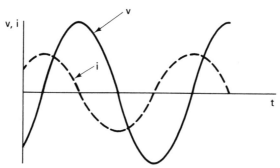

FIGURE 17.8   *Voltage and current waveforms for a capacitor.*

Example 17.3: A circuit element has a time-domain voltage

$$v = 30\,\sqrt{2}\,\sin\,(6t + 20°)\,\text{V}$$

Find the phase relationship if the element is (a) a 2-H inductor, (b) a $\frac{1}{4}$-F capacitor, and (c) an element with an impedance of $\mathbf{Z} = 4 + j3\ \Omega$.

Solution: The frequency is $\omega = 6$ rad/s and the voltage phasor is

$$\mathbf{V} = 30\,\underline{|20°}$$

so that in (a) the current phasor is

$$\mathbf{I} = \frac{\mathbf{V}}{\mathbf{Z}} = \frac{\mathbf{V}}{\omega L\,\underline{|90°}} = \frac{30\,\underline{|20°}}{6(2)\,\underline{|90°}}$$

or

$$\mathbf{I} = 2.5\,\underline{|-70°}$$

Therefore, the current is

$$i = 2.5\,\sqrt{2}\sin{(6t - 70°)}\ \text{A}$$

and the current lags the voltage by $20° - (-70°) = 90°$.
  In (b) we have

$$\mathbf{I} = \frac{30\,\underline{|20°}}{1/(6)(\frac{1}{4})\,\underline{|-90°}} = 45\,\underline{|110°}$$

and

$$i = 45\,\sqrt{2}\sin{(6t + 110°)}\ \text{A}$$

Thus the current leads the voltage by $110° - 20° = 90°$.
  Finally, in (c) we have

$$\mathbf{I} = \frac{30\,\underline{|20°}}{4 + j3} = \frac{30\,\underline{|20°}}{5\,\underline{|36.9°}} = 6\,\underline{|-16.9°}$$

and

$$i = 6\,\sqrt{2}\sin{(6t - 16.9°)}\ \text{A}$$

Therefore, the voltage leads the current, or the current lags the voltage, by $20° - (-16.9°) = 36.9°$.

General Case: In every case in Example 17.3, the current lags the voltage (or the voltage leads the current) by the angle of the impedance. This is the case in general, since if the voltage phasor is

$$\mathbf{V} = V\underline{|\phi}$$

and the impedance is

$$\mathbf{Z} = |\mathbf{Z}|\underline{|\theta}$$

then the current phasor is

$$\mathbf{I} = \frac{V\underline{|\phi}}{|\mathbf{Z}|\underline{|\theta}} = I\underline{|\phi - \theta}$$

where $I = V/|\mathbf{Z}|$. Therefore, the time-domain quantities are

$$v = \sqrt{2}\ V \sin\ (\omega t + \phi)$$

and

$$i = \sqrt{2}\ I \sin\ (\omega t + \phi - \theta)$$

Thus the voltage leads the current by $\phi - (\phi - \theta) = \theta$, which is the angle of $\mathbf{Z}$.

PRACTICE EXERCISES

17-2.1   Find the angle by which

$$v_1 = 10\ \sin\ (2t + 10°)$$

leads v_2 given by (a) 3 sin $(2t + 10°)$, (b) 4 sin $(2t - 25°)$, and (c) 8 sin $(2t + 30°)$.
                                              Ans. (a) 0°, (b) 35°, (c) −20°

17-2.2   Find the voltage phasor if $|\mathbf{Z}| = 10\ \Omega$, $\mathbf{I} = 4\underline{|0°}$ A, and the current lags the voltage by 30°.   *Ans.* 40$\underline{|30°}$ V

17.3   VOLTAGE AND CURRENT DIVISION

Since phasor circuits obey the same basic rules (Ohm's and Kirchhoff's laws) as resistive circuits, all the techniques we have used for resistive circuits apply to phasor circuits. Equivalent impedances may be found for series and parallel connections, voltage and current division apply, the network theorems are valid, and nodal and loop analysis methods may be used. In the remainder of the chapter and in Chapter 18 we will consider these other topics, beginning with equivalent impedances and voltage and current division in this section.

Series Impedances: In the case of N impedances connected in series, as shown in Fig. 17.9, we see that by KVL

$$V = V_1 + V_2 + \cdots + V_N \tag{17.24}$$

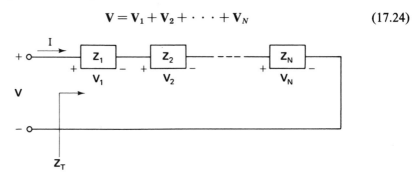

FIGURE 17.9 *Series impedances.*

Also, the same current **I** flows through every element so that by Ohm's law we have

$$V_1 = Z_1 I$$

$$V_2 = Z_2 I$$

$$\vdots$$

$$V_N = Z_N I$$

Substituting these values into (17.24) results in

$$V = Z_1 I + Z_2 I + \cdots + Z_N I$$
$$= (Z_1 + Z_2 + \cdots + Z_N) I$$

If Z_T is the equivalent impedance seen at the terminals of Fig. 17.9, we must have

$$V = Z_T I$$

Comparing these last two results yields

$$Z_T = Z_1 + Z_2 + \cdots + Z_N \tag{17.25}$$

as is the case for series resistors. That is, the equivalent impedance is the sum of the series impedances.

Example 17.4: Find the steady-state current i in the circuit of Fig. 17.10(a) if

$$v_g = 100 \sin 3t \text{ V}$$

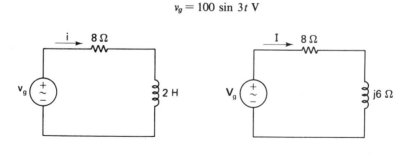

(a)   (b)

FIGURE 17.10   *(a) Time-domain circuit and (b) its phasor circuit.*

Solution: The phasor circuit is shown in Fig. 17.10(b) where $\omega = 3$ rad/s,

$$\mathbf{V}_g = \frac{100}{\sqrt{2}} \underline{|0°} \text{ V}$$

and

$$\mathbf{Z}_L = j\omega L = j(3)(2) = j6 \ \Omega$$

The equivalent impedance seen by the source is the sum of the series impedances, given by

$$\mathbf{Z}_T = 8 + j6 = 10 \underline{|36.9°} \ \Omega$$

Thus the phasor current is

$$\mathbf{I} = \frac{\mathbf{V}_g}{\mathbf{Z}_T} = \frac{100/\sqrt{2} \underline{|0°}}{10 \underline{|36.9°}}$$

or

$$I = \frac{10}{\sqrt{2}} \underline{|-36.9°} \text{ A}$$

Therefore, the sinusoidal steady-state current is

$$i = 10 \sin (3t - 36.9°) \text{ A}$$

Example 17.5: Find the equivalent impedance $\mathbf{Z}$ seen at the input terminals of the circuit of Fig. 17.11.

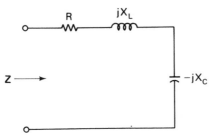

FIGURE 17.11 RLC series phasor circuit.

Solution: The three impedances shown are in series, so that their equivalent impedance is the sum

$$\mathbf{Z} = R + jX_L - jX_C$$

or

$$\mathbf{Z} = R + j(X_L - X_C)$$

Comparing this result with the general case $\mathbf{Z} = R + jX$, we see that the total reactance is

$$X = X_L - X_C$$

or

$$X = \omega L - \frac{1}{\omega C}$$

Thus, in general, the net reactance is the inductive reactance minus the capacitive reactance.

Parallel Admittances: In the case of N parallel admittances, as shown in Fig. 17.12, we have by KCL

$$\mathbf{I} = \mathbf{I}_1 + \mathbf{I}_2 + \cdots + \mathbf{I}_N$$

where

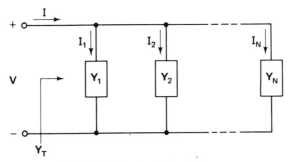

FIGURE 17.12 Parallel admittances.

$$\mathbf{I}_1 = \mathbf{Y}_1\mathbf{V}$$

$$\mathbf{I}_2 = \mathbf{Y}_2\mathbf{V}$$

$$\vdots$$

$$\mathbf{I}_N = Y_N\mathbf{V}$$

Combining these results, we have

$$\mathbf{I} = \mathbf{Y}_1\mathbf{V} + \mathbf{Y}_2\mathbf{V} + \cdots + \mathbf{Y}_N\mathbf{V}$$

$$= (\mathbf{Y}_1 + \mathbf{Y}_2 + \cdots + \mathbf{Y}_N)\mathbf{V}$$

Therefore, if $\mathbf{Y}_T$ is the equivalent admittance seen at the terminals, we have

$$\mathbf{I} = \mathbf{Y}_T\mathbf{V}$$

Comparing these last two results yields

$$\mathbf{Y}_T = \mathbf{Y}_1 + Y_2 + \cdots + \mathbf{Y}_N \tag{17.26}$$

as in the case for parallel conductances.

In the case of two parallel elements ($N = 2$), we have

$$\mathbf{Z}_T = \frac{1}{\mathbf{Y}_T} = \frac{1}{\mathbf{Y}_1 + \mathbf{Y}_2} = \frac{\mathbf{Z}_1\mathbf{Z}_2}{\mathbf{Z}_1 + \mathbf{Z}_2} \tag{17.27}$$

That is, the equivalent impedance is the product over the sum of the parallel impedances, as was true of parallel resistances.

Voltage Division: For Fig. 17.13 we may write

$$\mathbf{I} = \frac{\mathbf{V}}{\mathbf{Z}_1 + \mathbf{Z}_2}$$

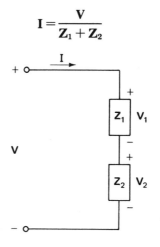

FIGURE 17.13 Circuit illustrating voltage division.

and

$$V_1 = Z_1 I$$

$$V_2 = Z_2 I$$

Substituting for I in the last two equations, we have

$$V_1 = \frac{Z_1}{Z_1 + Z_2} V$$

$$V_2 = \frac{Z_2}{Z_1 + Z_2} V$$

(17.28)

These results illustrate *voltage division* for phasors. They are, of course, identical to their resistive circuit counterparts.

Current Division: As in the case of voltage division, *current division* also is exactly the same for phasor circuits as it is for resistive circuits. This is illustrated by the circuit of Fig. 17.14, where the equivalent admittance at the terminals is

$$Y_T = Y_1 + Y_2$$

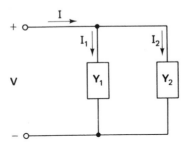

FIGURE 17.14 *Circuit illustrating current division.*

Therefore, we have

$$V = \frac{I}{Y_T} = \frac{I}{Y_1 + Y_2}$$

Also from the figure we have

$$I_1 = Y_1 V$$

$$I_2 = Y_2 V$$

which become, upon substitution for V,

$$\mathbf{I}_1 = \frac{\mathbf{Y}_1}{\mathbf{Y}_1 + \mathbf{Y}_2} \mathbf{I} = \frac{\mathbf{Z}_2}{\mathbf{Z}_1 + \mathbf{Z}_2} \mathbf{I}$$

$$\mathbf{I}_2 = \frac{\mathbf{Y}_2}{\mathbf{Y}_1 + \mathbf{Y}_2} \mathbf{I} = \frac{\mathbf{Z}_1}{\mathbf{Z}_1 + \mathbf{Z}_2} \mathbf{I}$$

(17.29)

Example 17.6: Find the steady-state values of i and v in the series–parallel circuit of Fig. 17.15 if

$$v_g = 10 \sqrt{2} \sin 2t \text{ V}$$

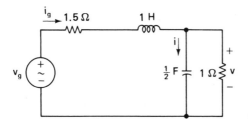

FIGURE 17.15   Series-parallel circuit.

Solution: The source phasor is $\mathbf{V}_g = 10\underline{/0°}$ V and the frequency is $\omega = 2$ rad/s. Therefore, the inductor and capacitor impedances are

$$\mathbf{Z}_L = j\omega L = j(2)(1) = j2 \ \Omega$$

$$\mathbf{Z}_c = -j\frac{1}{\omega C} = -j\frac{1}{(2)(\frac{1}{2})} = -j1 \ \Omega$$

as shown in the phasor circuit of Fig. 17.16.

We may combine the 1.5-Ω resistor and the $j2$-Ω inductor, which are in series, into an equivalent impedance,

$$\mathbf{Z}_1 = 1.5 + j2 \ \Omega$$

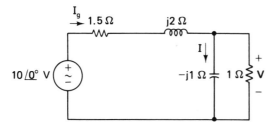

FIGURE 17.16   Phasor circuit of Fig. 17.15.

Similarly, the $-j1$-Ω capacitor and the 1-Ω resistor are in parallel, with the equivalent impedance

$$\mathbf{Z}_2 = \frac{(-j1)(1)}{-j1 + 1} = \frac{-j1}{1 - j1} \cdot \frac{1 + j1}{1 + j1}$$

$$= \frac{1 - j1}{2} \ \Omega$$

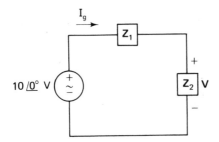

FIGURE 17.17 *Equivalent circuit of Fig. 17.16.*

These are shown in the equivalent circuit of Fig. 17.17. From Fig. 17.17 we have, by voltage division,

$$\mathbf{V} = \frac{\mathbf{Z}_2}{\mathbf{Z}_1 + \mathbf{Z}_2} \, 10 \underline{|0°}$$

$$= \frac{\dfrac{1 - j1}{2}}{1.5 + j2 + \dfrac{1 - j1}{2}} \, 10 \underline{|0°}$$

$$= \frac{10(1 - j1)}{4 + j3} = \frac{10\sqrt{2} \underline{|-45°}}{5 \underline{|36.9°}}$$

or

$$\mathbf{V} = 2\sqrt{2} \underline{|-81.9°} \ = \frac{4}{\sqrt{2}} \underline{|-81.9°} \ \mathbf{V}$$

Therefore, the steady-state sinusoidal voltage is

$$v = 4 \sin(2t - 81.9°) \, \mathbf{V}$$

The current $\mathbf{I}_g$ is, from Fig. 17.17,

$$\mathbf{I}_g = \frac{10 \underline{|0°}}{\mathbf{Z}_1 + \mathbf{Z}_2} = \frac{10 \underline{|0°}}{1.5 + j2 + \dfrac{1 - j1}{2}}$$

which may be simplified to

$$\mathbf{I}_g = 4 \underline{|-36.9°} \ \mathbf{A}$$

Finally, by current division in Fig. 17.16 we have

$$\mathbf{I} = \frac{1}{1 - j1} \mathbf{I}_g = \frac{4 \underline{|-36.9°}}{\sqrt{2} \underline{|-45°}}$$

$$= \frac{4}{\sqrt{2}} \underline{|8.1°} \ \mathbf{A}$$

Therefore, the sinusoidal current is

$$i = 4 \sin (2t + 8.1°) \text{ A}$$

PRACTICE EXERCISES

17-3.1 Find the equivalent impedance of a series connection of a 1-kΩ resistor, a 0.1-H inductor, and a 1-μF capacitor if $\omega = 1000$ rad/s.   *Ans.* $1 - j0.9$ kΩ

17-3.2 Find the equivalent impedance of a parallel connection of a 20-kΩ resistor and a 0.1-μF capacitor if $\omega = 500$ rad/s.   *Ans.* $10 - j10$ kΩ

17-3.3 Find the steady-state values of i_1 and i_2.

Ans. $5 \sin (4t + 53.1°)$ A, $10 \sin (4t + 53.1°)$ A

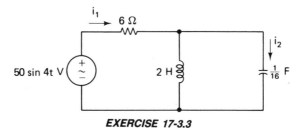

EXERCISE 17-3.3

17-3.4 Find $\mathbf{Z}_T$.

Ans. $\dfrac{R\omega^2 L^2}{R^2 + \omega^2 L^2} + j\dfrac{R^2 \omega L}{R^2 + \omega^2 L^2}$

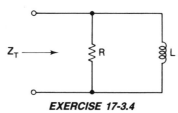

EXERCISE 17-3.4

17.4   NODAL ANALYSIS

Phasor circuits are analyzed using nodal and loop analysis in exactly the same manner that resistive circuits are analyzed. The answers obtained are phasors, and if the time-domain answers are desired, we must make the necessary conversion from phasors to sinusoids. In this section we will consider nodal analysis and in the following section we will discuss loop analysis.

Example 17.7: To illustrate nodal analysis, let us find the steady-state node voltage v in the circuit of Fig. 17.18(a).

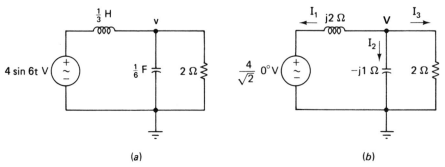

(a)   (b)

FIGURE 17.18 *(a) Time-domain circuit and (b) its corresponding phasor circuit.*

Solution: Noting that $\omega = 6$ rad/s, we may obtain the phasor circuit shown in Fig. 17.18(b). The nodal equation at the node labeled **V** is

$$\mathbf{I_1} + \mathbf{I_2} + \mathbf{I_3} = 0$$

which in terms of **V** may be written, using Ohm's law, as

$$\frac{\mathbf{V} - 4/\sqrt{2}\lfloor 0°}{j2} + \frac{\mathbf{V}}{-j1} + \frac{\mathbf{V}}{2} = 0$$

Multiplying through by $j2$ and collecting terms, we have

$$(1 - 2 + j1)\,\mathbf{V} = \frac{4}{\sqrt{2}}$$

which results in

$$\mathbf{V} = \frac{4/\sqrt{2}}{-1 + j1} = \frac{4/\sqrt{2}}{\sqrt{2}\lfloor 135°}$$

or

$$\mathbf{V} = 2\lfloor -135°\ \text{V}$$

The time-domain voltage is therefore

$$v = 2\sqrt{2}\sin(6t - 135°)\ \text{V}$$

This example is relatively simple since there is only one unknown node voltage and therefore only one node equation is required. In general, we may have several node voltages and thus several equations to be solved simultaneously. The work is exactly like resistive circuit analysis, but, of course, the numbers are complex. We will now illustrate nodal analysis with a circuit of two unknown node voltages.

Example 17.8: Find the node voltage V_2 in the circuit of Fig. 17.19.

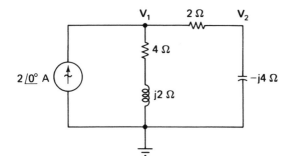

FIGURE 17.19 *Phasor circuit with two unknown node voltages.*

Solution: The node equation at node V_1 is

$$\frac{V_1}{4+j2}+\frac{V_1-V_2}{2}=2\underline{|0°}$$

or

$$\left(\frac{1}{4+j2}+\frac{1}{2}\right)V_1-\frac{1}{2}V_2=2\underline{|0°} \qquad (17.30)$$

At node V_2 we have

$$\frac{V_2-V_1}{2}+\frac{V_2}{-j4}=0$$

or

$$-\frac{1}{2}V_1+\left(\frac{1}{2}+\frac{1}{-j4}\right)V_2=0 \qquad (17.31)$$

From (17.31) we may find V_1 in terms of V_2, given by

$$V_1=\left(1+j\frac{1}{2}\right)V_2$$

which substituted into (17.30) gives

$$\left(\frac{1}{4+j2}+\frac{1}{2}\right)\left(1+j\frac{1}{2}\right)V_2-\frac{1}{2}V_2=2$$

or

$$\left[\left(\frac{4-j2}{20}+\frac{1}{2}\right)\left(1+j\frac{1}{2}\right)-\frac{1}{2}\right]\mathbf{V}_2 = 2$$

Solving for $\mathbf{V}_2$, we have

$$\mathbf{V}_2 = \frac{2}{\left(\dfrac{4-j2}{20}+\dfrac{1}{2}\right)\left(1+j\dfrac{1}{2}\right)-\dfrac{1}{2}}$$

which simplifies to

$$\mathbf{V}_2 = \frac{8}{1+j1} = 4\sqrt{2}\,\underline{|-45°}\ \text{V}$$

We may note that in this example the node equations (17.30) and (17.31) may be obtained directly using the shortcut method for resistive circuits. The coefficient of $\mathbf{V}_1$ in the first equation (17.30) is the sum of the admittances connected to node $\mathbf{V}_1$ and the coefficient of $\mathbf{V}_2$ is the negative of the admittance between the two nodes. In the second equation (17.31) the coefficient of $\mathbf{V}_1$ is the negative of the admittance between the two nodes, and the coefficient of $\mathbf{V}_2$ is the sum of the admittances connected to node $\mathbf{V}_2$. This was true for resistances in the resistive circuits.

Also we may note that in Example 17.8 the node voltage $\mathbf{V}_1$ may be found directly using one node equation by combining the 2-Ω and $-j4$-Ω impedances into a single impedance. Then $\mathbf{V}_2$ may be found from $\mathbf{V}_1$ by voltage division. The node equation of $\mathbf{V}_1$ is

$$\frac{\mathbf{V}_1}{4+j2}+\frac{\mathbf{V}_1}{2-j4} = 2\,\underline{|0°}$$

from which we have

$$\left(\frac{1}{4+j2}\cdot\frac{4-j2}{4-j2}+\frac{1}{2-j4}\cdot\frac{2+j4}{2+j4}\right)\mathbf{V}_1 = 2$$

or

$$\left(\frac{4-j2+2+j4}{20}\right)\mathbf{V}_1 = \left(\frac{3+j1}{10}\right)\mathbf{V}_1 = 2$$

Therefore, we have

$$\mathbf{V}_1 = \frac{2(10)}{3+j1}\cdot\frac{3-j1}{3-j1} = 6-j2\ \text{V}$$

Finally, by voltage division we see from Fig. 17.19 that

$$\mathbf{V}_2 = \frac{-j4}{2-j4}\,\mathbf{V}_1$$

$$= \frac{-j4}{2-j4}\,(6-j2)$$

which simplifies to

$$\mathbf{V}_2 = 4 - j4 = 4\sqrt{2}\,\underline{|-45°}\ \text{V}$$

The method of elimination works very well in Example 17.8 because of the form of (17.31). We could also have used determinants to get $\mathbf{V}_1$ and $\mathbf{V}_2$ directly.

PRACTICE EXERCISES

17-4.1 Find the steady-state node voltage v using nodal analysis in the corresponding phasor circuit.   *Ans.* $6 \sin(3t - 8.1°)$ V

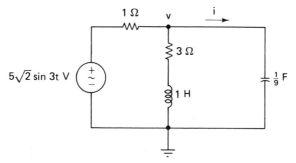

EXERCISE 17-4.1

17-4.2 Find the steady-state node voltage v using nodal analysis in the corresponding phasor circuit.   *Ans.* $4\sqrt{2} \sin(6t - 45°)$ V

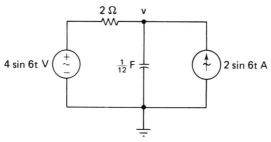

EXERCISE 17-4.2

17.5 LOOP ANALYSIS

Like nodal analysis, loop analysis of steady-state ac phasor circuits is performed exactly as in the case of resistive circuits. In this section we will illustrate the procedure by considering two examples.

Example 17.9: Find the steady-state currents i_1 and i_2 in the circuit of Fig. 17.20.

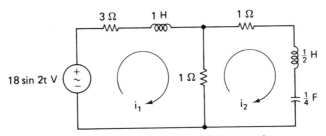

FIGURE 17.20 *Circuit with two loop currents.*

Solution: The frequency is $\omega = 2$ rad/s, which may be used to obtain the impedances and the phasor circuit shown in Fig. 17.21(a). The 3-Ω and $j2$-Ω impedances are in series, as are the 1-Ω, $j1$-Ω, and $-j2$-Ω impedances. Thus we may combine these in equivalent impedances $3 + j2$ and $1 + j1 - j2 = 1 - j1$, as shown in Fig. 17.21(b).

The loop, or mesh, equations are written exactly as in resistive circuit analysis. In Fig. 17.21(b) the equation for the mesh labeled $\mathbf{I}_1$ is

$$(3 + j2)\mathbf{I}_1 + 1(\mathbf{I}_1 - \mathbf{I}_2) = \frac{18}{\sqrt{2}}\underline{|0°}$$

or

$$(4 + j2)\mathbf{I}_1 - \mathbf{I}_2 = \frac{18}{\sqrt{2}}\underline{|0°} \tag{17.32}$$

and that for the mesh labeled $\mathbf{I}_2$ is

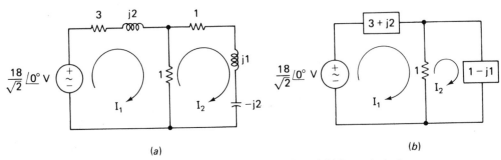

(a)

(b)

FIGURE 17.21 *(a) Phasor circuit and (b) its equivalent.*

$$1(\mathbf{I}_2 - \mathbf{I}_1) + (1 - j1)\mathbf{I}_2 = 0$$

or

$$-\mathbf{I}_1 + (2 - j1)\mathbf{I}_2 = 0 \qquad (17.33)$$

From this last result we have

$$\mathbf{I}_1 = (2 - j1)\mathbf{I}_2 \qquad (17.34)$$

which substituted into (17.32) yields

$$(4 + j2)(2 - j1)\mathbf{I}_2 - \mathbf{I}_2 = \frac{18}{\sqrt{2}}$$

Solving for $\mathbf{I}_2$, we have

$$\mathbf{I}_2 = \frac{18/\sqrt{2}}{(4 + j2)(2 - j1) - 1} = \frac{2}{\sqrt{2}} \underline{|0°} \qquad (17.35)$$

Finally, by (17.34) we have

$$\mathbf{I}_1 = (2 - j1)\left(\frac{2}{\sqrt{2}} \underline{|0°}\right)$$

$$= (\sqrt{5} \underline{|-26.6°})\left(\frac{2}{\sqrt{2}} \underline{|0°}\right)$$

or

$$\mathbf{I}_1 = \frac{2\sqrt{5}}{\sqrt{2}} \underline{|-26.6°} \text{ A} \qquad (17.36)$$

The time-domain mesh currents, obtained from (17.36) and (17.35), are given by

$$i_1 = 2\sqrt{5} \sin(2t - 26.6°) \text{ A}$$

and

$$i_2 = 2 \sin 2t \text{ A}$$

As a final note on this example, we observe that the mesh equations (17.32) and (17.33) may be obtained directly using the shortcut method for resistive circuits. In the first equation the coefficient of $\mathbf{I}_1$ is the sum of the impedances in the first mesh and that of $\mathbf{I}_2$ is the negative of the impedance common to the first and second meshes. In the second equation the coefficient of $\mathbf{I}_2$ is the sum of the impedances in the second mesh and that of $\mathbf{I}_1$ is the negative of the impedance common to the second and first meshes. This was the case for resistances in the resistive circuits.

The circuits we have considered thus far have only one source and consequently one frequency ω. More than one source with different frequencies would pose a problem because we could not determine the impedances that depend on ω. As we will see later, this problem can be resolved using superposition. If all the sources have the same frequency, however, we may perform the analysis exactly as we would for resistive circuits, as the following example indicates.

Example 17.10: Find the steady-state current i in the two-source circuit of Fig. 17.22(a).

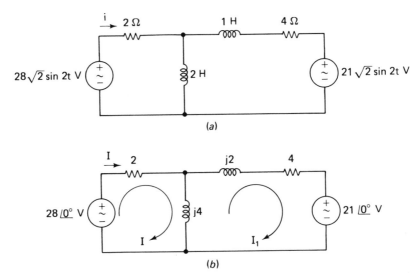

FIGURE 17.22 (a) Two-source circuit and (b) its phasor circuit.

Solution: The phasor circuit is shown in Fig. 17.22(b), for which we have the mesh equations

$$(2 + j4)\mathbf{I} - j4\mathbf{I}_1 = 28$$

$$-j4\mathbf{I} + (4 + j6)\mathbf{I}_1 = -21$$

Multiplying the first of these by $4 + j6$ and the second by $j4$ yields

$$(2 + j4)(4 + j6)\mathbf{I} - j4(4 + j6)\mathbf{I}_1 = 28(4 + j6)$$

$$-j4(j4)\mathbf{I} + j4(4 + j6)\mathbf{I}_1 = -21(j4)$$

Adding these equations eliminates $\mathbf{I}_1$ and results in

$$[(2 + j4)(4 + j6) - j4(j4)]\mathbf{I} = 28(4 + j6) - 21(j4)$$

which simplifies to

$$j28\mathbf{I} = 112 + j84$$

Therefore, we have

$$\mathbf{I} = \frac{112 + j84}{j28} = 3 - j4 = 5\lfloor{-53.1°}\ \text{A}$$

and thus the sinusoidal current is

$$i = 5\sqrt{2}\sin(2t - 53.1°)\ \text{A}$$

PRACTICE EXERCISES

17-5.1 Find the steady-state current i in the circuit of Practice Exercise 17.4.1 using mesh analysis. Check by using the phasor voltage obtained in that exercise.

Ans. $2\sin(3t + 81.9°)$ A

17-5.2 Find the steady-state value of i using mesh analysis in the corresponding phasor circuit.

Ans. $0.5\sin(t + 53.1°)$ A

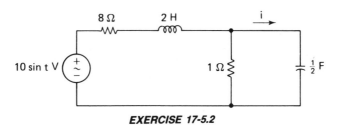

EXERCISE 17-5.2

17.6 PHASOR DIAGRAMS

Since phasors are complex numbers, they may be represented by *vectors* (or lines at angles) sketched in a plane, and the phasor addition and subtraction operations may be carried out graphically. Such a sketch is called a *phasor diagram* and may be quite useful in analyzing steady-state ac circuits.

To illustrate the use of phasor diagrams, let us consider the circuit of Fig. 17.23, which is an *RLC* series phasor circuit. We first observe that the current **I** is common to all the elements, and thus we will take it as our *reference* phasor. That is, we will assign it a phase angle of 0°, and draw it on the positive real axis. Then all the other phasors will be drawn in positions relative to the reference phasor. (The actual phase angles may be found at the end of the process, since the phasor diagram

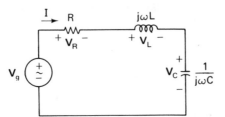

FIGURE 17.23 RLC series phasor circuit.

will determine the relative phase angles, by indicating, for example, whether $\mathbf{V}_L$ leads or lags $\mathbf{V}_g$, and by how much.)

In other words, we will take the reference phasor $\mathbf{I}$ as

$$\mathbf{I} = I\underline{/0°} \tag{17.37}$$

and sketch all the other phasors relative to it.

Resistor Phasor Diagram: In the case of the resistor R in Fig. 17.23, we know that its voltage $\mathbf{V}_R$ is in phase with the current. That is, we have by (17.37)

$$\mathbf{V}_R = R\mathbf{I} = RI\underline{/0°} \tag{17.38}$$

This is shown in the phasor diagram of Fig. 17.24(a), where the scale used for the reference phasor $\mathbf{I}$ is different from that used for $\mathbf{V}_R$.

Inductor Phasor Diagram: For the inductor L of Fig. 17.23, the current $\mathbf{I}$ lags the voltage $\mathbf{V}_L$ by 90°, as we see from (17.37) and the relation

$$\mathbf{V}_L = j\omega L\mathbf{I} = \omega LI\underline{/90°} \tag{17.39}$$

This is shown in the phasor diagram of Fig. 17.24(b), in which the angle of the current phasor is 90° less than that of the voltage phasor.

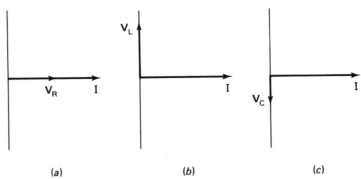

FIGURE 17.24 Phasor diagrams of current and voltage for (a) a resistor, (b) an inductor, and (c) a capacitor.

Capacitor Phasor Diagram: For the capacitor C of Fig. 17.23, we have

$$\mathbf{V}_C = -j\frac{1}{\omega C}\mathbf{I} = \frac{I}{\omega C}\underline{|{-90°}} \tag{17.40}$$

Therefore, as expected, the current leads the voltage by 90°. This is shown in the phasor diagram of Fig. 17.24(c), where the angle of $\mathbf{V}$ is 90° less than that of $\mathbf{I}$.

Phasor Diagram for the Circuit of Fig. 17.23: In the case of the *RLC* series circuit of Fig. 17.23, we may find the current $\mathbf{I}$ graphically by means of a phasor diagram. We first choose $\mathbf{I}$ as in (17.37) with an arbitrary angle of 0° and an arbitrary magnitude $|\mathbf{I}| = I$. Then we may find the phasors $\mathbf{V}_R$, $\mathbf{V}_L$, and $\mathbf{V}_C$, as shown in Fig. 17.24, but we will plot all of them on the same phasor diagram with $\mathbf{I}$, as shown in Fig. 17.25.

By Kirchhoff's voltage law we know from Fig. 17.23 that

$$\mathbf{V}_R + \mathbf{V}_L + \mathbf{V}_C = \mathbf{V}_g$$

We may perform the addition graphically on the phasor diagram and find $\mathbf{V}_g$. For example, $\mathbf{V}_L + \mathbf{V}_C$ will be a number on the j axis since $\mathbf{V}_L$ has a positive imaginary part and $\mathbf{V}_C$ has a negative imaginary part. If $|\mathbf{V}_L| > |\mathbf{V}_C|$, the result will be a positive j-axis number, as shown in Fig. 17.25. Completing the parallelogram of $\mathbf{V}_L + \mathbf{V}_C$ and $\mathbf{V}_R$ gives the sum $\mathbf{V}_g$. At this point we know the magnitude $|\mathbf{V}_g|$ and the angle θ by which $\mathbf{V}_g$ leads or lags $\mathbf{I}$. Since $\mathbf{V}_g$ is known at the beginning, we will know what is necessary to correct the calculated $\mathbf{V}_g$ to get the actual $\mathbf{V}_g$. That is, we will know how to adjust its magnitude and phase to make them correct. These same adjustments will then make $\mathbf{I}$, $\mathbf{V}_R$, $\mathbf{V}_L$, and $\mathbf{V}_C$ correct.

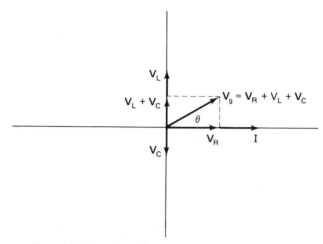

FIGURE 17.25   *Phasor diagram for the circuit of Fig. 17.23.*

Example 17.11: In Fig. 17.23, let $R = 3$ Ω, $j\omega L = j6$ Ω, $1/j\omega C = -j2$ Ω, and $\mathbf{V}_g = 10\underline{|0°}$ V. Use the phasor diagram to find $\mathbf{I}$.

Solution: We take $\mathbf{I}$ as the reference phasor and arbitrarily give it the value

$$\mathbf{I} = 1\underline{|0°} \text{ A} \tag{17.41}$$

Then we have

$$\mathbf{V}_R = 3\mathbf{I} = 3 = 3\underline{|0°} \text{ V}$$
$$\mathbf{V}_L = j6\mathbf{I} = j6 = 6\underline{|90°} \text{ V} \tag{17.42}$$
$$\mathbf{V}_C = -j2\mathbf{I} = -j2 = 2\underline{|-90°} \text{ V}$$

These quantities are sketched in Fig. 17.26, where their sum $\mathbf{V}_g$ is found graphically to be

$$\mathbf{V}_g = \mathbf{V}_R + \mathbf{V}_L + \mathbf{V}_C = 3 + j4 = 5\underline{|53.1°} \text{ V}$$

To correct the calculated value $5\underline{|53.1°}$ V to the actual value $10\underline{|0°}$ V, we must multiply the amplitude by 2 and subtract 53.1° from the phase. These adjustments will also correct the calculated values of $\mathbf{I}$, $\mathbf{V}_R$, $\mathbf{V}_L$, and $\mathbf{V}_C$, given in (17.41) and (17.42), to their actual values, given by

$$\mathbf{I} = 2\underline{|-53.1°} \text{ A}$$
$$\mathbf{V}_R = 6\underline{|-53.1°} \text{ V}$$
$$\mathbf{V}_L = 12\underline{|36.9°} \text{ V}$$
$$\mathbf{V}_C = 4\underline{|-143.1°} \text{ V}$$

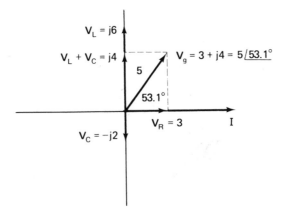

FIGURE 17.26 *Example phasor diagram.*

Other Cases: In the phasor diagram of Fig. 17.25, we have $|\mathbf{V}_L| > |\mathbf{V}_C|$, in which case the current $\mathbf{I}$ lags the voltage $\mathbf{V}_g$. That is, the inductive reactance has a larger magnitude than the capacitive reactance, so that the net reactance is inductive. We may also have the cases $|\mathbf{V}_C| > |\mathbf{V}_L|$ and $|\mathbf{V}_C| = |\mathbf{V}_L|$, shown in Fig. 17.27(a) and

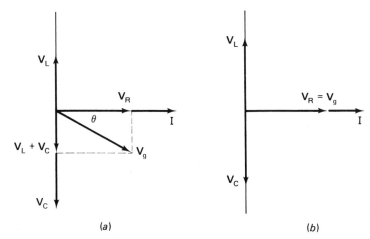

(a)   (b)

FIGURE 17.27 *Phasor diagrams for which (a)* $|\mathbf{V}_\mathrm{C}| > |\mathbf{V}_\mathrm{L}|$ *and (b)* $|\mathbf{V}_\mathrm{c}| = |\mathbf{V}_\mathrm{L}|$.

(b). In the first case, the circuit has a net capacitive reactance and the current $\mathbf{I}$ leads the voltage $\mathbf{V}_g$ by the angle θ, as shown. The second case is the interesting one where the inductive reactance and the capacitive reactance exactly cancel each other, and the circuit behaves as a purely resistive circuit.

Example 17.12: Using a phasor diagram, find $\mathbf{V}$ in the circuit of Fig. 17.28.

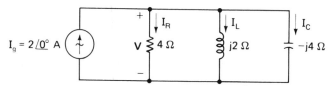

FIGURE 17.28 *RLC parallel phasor circuit.*

Solution: Since the voltage phasor $\mathbf{V}$ is common to all the elements we will take it as the reference phasor, and arbitrarily take it as

$$\mathbf{V} = 1\underline{|0°}\ \text{V}$$

The element currents are then

$$\mathbf{I}_R = \frac{\mathbf{V}}{4} = \frac{1}{4}\,\mathrm{A} = \frac{1}{4}\underline{|0°}\ \mathrm{A}$$

$$\mathbf{I}_L = \frac{\mathbf{V}}{j2} = -j\frac{1}{2}\,\mathrm{A} = \frac{1}{2}\underline{|-90°}\ \mathrm{A}$$

$$\mathbf{I}_C = \frac{\mathbf{V}}{-j4} = j\frac{1}{4}\,\mathrm{A} = \frac{1}{4}\underline{|90°}\ \mathrm{A}$$

These quantities are sketched in the phasor diagram of Fig. 17.29, where their sum is found graphically to be

$$\mathbf{I}_g = \mathbf{I}_R + \mathbf{I}_L + \mathbf{I}_C$$

$$= \frac{1}{4} - j\frac{1}{4}\,\mathrm{A}$$

$$= \frac{\sqrt{2}}{4}\underline{|-45°}\ \mathrm{A}$$

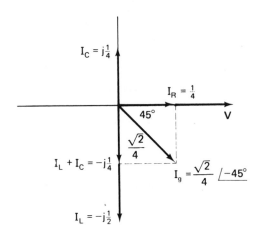

FIGURE 17.29 *Phasor diagram for the circuit of Fig. 17.28.*

We see from the phasor diagram that the current $\mathbf{I}_g$ lags the voltage $\mathbf{V}$ by 45°. The actual phase angle and magnitude of $\mathbf{I}_g$ may be found by comparing the calculated value $\sqrt{2}/4\ \underline{|-45°}$ of $\mathbf{I}_g$ with its actual value of $2\underline{|0°}$. Thus to correct $\mathbf{I}_g$ to its actual value requires that we multiply the calculated magnitude by $8/\sqrt{2}$ ($8/\sqrt{2} \times \sqrt{2}/4 = 2$) and add 45° to the calculated phase. Making these same corrections in the case of the other calculated phasors gives the correct values

$$\mathbf{V} = \frac{8}{\sqrt{2}} \times 1\underline{|0° + 45°} = 4\sqrt{2}\underline{|45°}\ \mathrm{V}$$

$$\mathbf{I}_R = \frac{8}{\sqrt{2}} \times \frac{1}{4}\underline{|0° + 45°} = \sqrt{2}\underline{|45°}\ \mathrm{A}$$

$$\mathbf{I}_L = \frac{8}{\sqrt{2}} \times \frac{1}{2} \underline{|-90° + 45°} = 2\sqrt{2} \underline{|-45°} \ \mathbf{A}$$

$$\mathbf{I}_C = \frac{8}{\sqrt{2}} \times \frac{1}{4} \underline{|90° + 45°} = \sqrt{2} \underline{|135°} \ \mathbf{A}$$

PRACTICE EXERCISES

17-6.1 Use a phasor diagram to find **I**. (Take the reference phasor to be **I**.)

Ans. $2 \underline{|-36.9°} \ \mathbf{A}$

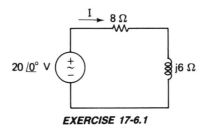

EXERCISE 17-6.1

17-6.2 Use a phasor diagram to find **V**. (Take the reference phasor to be **V**.)

Ans. $12 \underline{|-53.1°} \ \mathbf{V}$

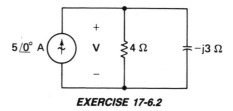

EXERCISE 17-6.2

17.7 SUMMARY

Phasor circuits with ac sinusoidal sources may be analyzed exactly as dc resistive circuits are analyzed. The only difference is that in the ac case the numbers involved are complex, whereas in the dc case the numbers are real. Because of this similarity, we may apply the same resistive circuit techniques we used earlier to the phasor circuits.

We may combine series and parallel impedances into equivalent impedances and use Ohm's law to obtain phasor voltages and currents. The impedances are complex numbers, and thus have a real part, called the resistive component, and an imaginary part, called the reactance.

Resistor currents and voltages are in phase (have the same phase angles), but currents and voltages associated with inductors and capacitors differ in phase by 90°. Inductor currents lag the voltages by 90° (the current phase angle is 90° less than that of the voltage), and capacitor currents lead (have greater phase angles) the voltages by 90°.

Voltage and current division are applied to phasor circuits exactly as they are to resistive circuits. Many relatively simple circuits may be completely analyzed this way without writing any circuit equations. In the general case, however, Kirchhoff's laws are used in writing loop and nodal equations exactly as they are written for resistive circuits.

Finally, phasor diagrams, which are sketches of the phasors of a circuit, may be used to obtain phasor voltages and currents graphically. The phase and magnitude relations between phasors are shown graphically and addition and subtraction of phasors are performed in accordance with Kirchhoff's laws.

PROBLEMS

17.1   Find the impedance **Z**.

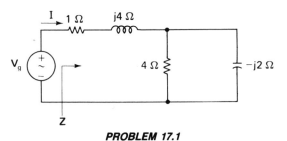

PROBLEM 17.1

17.2   If the time-domain voltage corresponding to V_g in Problem 17.1 is

$$v_g = 15 \sin (2t - 25°) \text{ V}$$

find the phasor current and its corresponding time-domain current.

17.3   Find the inductive reactance of a 10-mH inductor and the capacitive reactance of a 2-μF capacitor if the frequency is (a) 100 rad/s, (b) 100,000 rad/s, and (c) 1000 Hz.

17.4   Find the impedance of a 2-mH inductor and of a 0.1-μF capacitor if the frequency is (a) 10 rad/s, (b) 10,000 rad/s, and (c) 60 Hz.

17.5   Determine if v_1 leads or lags v_2 and by how much, if v_1 is given by

$$v_1 = 40 \sin (30t + 30°)$$

and v_2 is given by (a) 10 sin (30t − 12°), (b) 5 sin 30t, and (c) 8 sin (30t + 40°).

17.6 Find the impedance **Z** seen by the source and use the result to obtain the phasor **I** and its corresponding time-domain value i if the frequency is $\omega = 5$ rad/s.

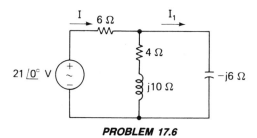

PROBLEM 17.6

17.7 In the corresponding phasor circuit, find the impedance **Z** seen by the source and use the result to get the sinusoidal steady-state current i.

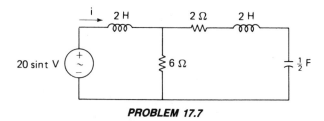

PROBLEM 17.7

17.8 Find the steady-state value of v using phasors.

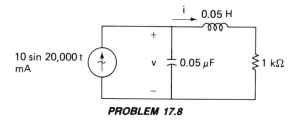

PROBLEM 17.8

17.9 Find the steady-state values of i and v using phasors and current division.

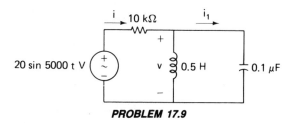

PROBLEM 17.9

17.10 Find the steady-state value of v using phasors and voltage division.

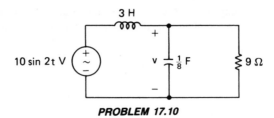

PROBLEM 17.10

17.11   Find the current I_1 and its corresponding time-domain value i_1 in Problem 17.6 using current division.

17.12   Find the steady-state value of i in Problem 17.8 using current division in the corresponding phasor circuit.

17.13   Find the steady-state current i_1 in Problem 17.9 using current division in the corresponding phasor circuit.

17.14   Find the steady-state value of v using nodal analysis.

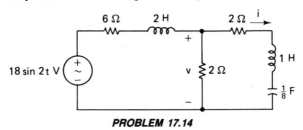

PROBLEM 17.14

17.15   Find the steady-state value of v using nodal analysis.

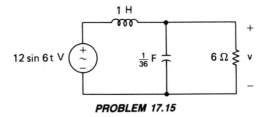

PROBLEM 17.15

17.16   Find the steady-state value of v using nodal analysis.

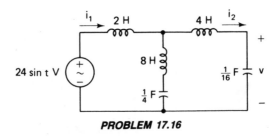

PROBLEM 17.16

17.17   Find the steady-state value of i in Problem 17.14 using mesh analysis.

17.18 Find the steady-state values of i_1 and i_2 in Problem 17.16 using mesh analysis.

17.19 Find the steady-state current i using mesh analysis.

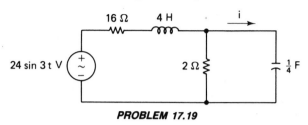

PROBLEM 17.19

17.20 Find the steady-state values of i_1 and i_2 using mesh analysis.

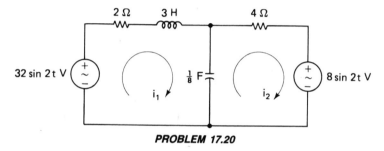

PROBLEM 17.20

17.21 Find **V** using nodal analysis.

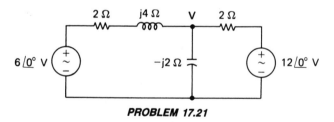

PROBLEM 17.21

17.22 Find the steady-state value of v.

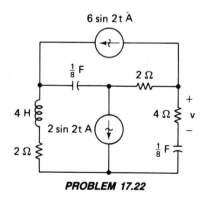

PROBLEM 17.22

17.23   Find the steady-state current i by obtaining the phasor circuit, taking **I**, the phasor of i, as the reference phasor, and constructing the phasor diagram for the cases (a) $\omega = 4$ rad/s and (b) $\omega = 2$ rad/s. (Take **I** $= 1\underline{|0°}$ A as the reference phasor and make the necessary corrections.)

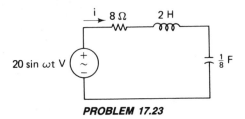

PROBLEM 17.23

17.24   Use the method of Problem 17.23 to find the steady-state voltage v. (Take its phasor as **V** $= 1\underline{|0°}$ V as reference and make the necessary corrections.)

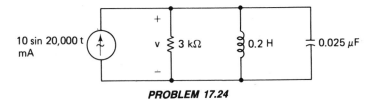

PROBLEM 17.24

18

AC NETWORK THEOREMS

Since the phasor circuits are exactly like dc resistive circuits except that the numbers involved are complex instead of real, all the network theorems that are valid for resistive circuits are also valid for phasor circuits. Every network theorem discussed in Chapter 8 for resistive circuits carries over to phasor circuits if we replace dc currents and voltages by phasor currents and voltages, and resistances by impedances.

In this chapter we will consider superposition, Thévenin's theorem, Norton's theorem, source conversions, and Y–Δ conversions. As we will see, these theorems are almost identical to the corresponding resistive circuit theorems, and may be used in the same way. In the case of superposition, if the circuit contains two or more sources of the same frequency, any current or voltage may be found as the sum of components due to each source acting alone. If the sources have different frequencies, superposition *must* be applied because our definition of impedance allows us to use only one frequency at a time.

18.1 SUPERPOSITION

If an ac circuit has more than one source, the principle of superposition may be applied as in the dc circuit case to find any voltage or current. We find the phasor voltage or current due to each source acting alone (that is, with the other sources killed). If all the sources have the same frequency, we may add the phasor voltages or currents due to each source to get the total phasor voltage or current. If the frequencies are different, we must convert the phasor components to their corresponding time-domain values, which are then added to obtain the total time-domain voltage or current.

Example 18.1: Use superposition to find the steady-state current i in the circuit of Fig. 18.1.

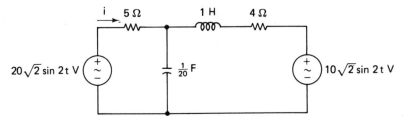

FIGURE 18.1   *Two-source, single frequency circuit.*

Solution: The phasor circuit is shown in Fig. 18.2, where **I** is the phasor of i. By superposition, **I** is given by

$$\mathbf{I} = \mathbf{I_1} + \mathbf{I_2} \tag{18.1}$$

where $\mathbf{I_1}$ is the current due to the 20-V source alone (with the 10-V source replaced by a short circuit) and $\mathbf{I_2}$ is the current due to the 10-V source alone (with the 20-V source

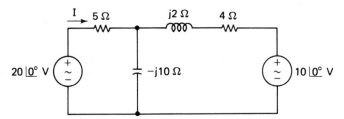

FIGURE 18.2   *Phasor circuit for Fig. 18.1.*

replaced by a short circuit). The phasor circuits for $\mathbf{I_1}$ and $\mathbf{I_2}$ are shown in Fig. 18.3(a) and (b). In (a) the 10-V source is killed and in (b) the 20-V source is killed.

In Fig. 18.3(a) the impedance seen by the source is

$$\mathbf{Z_1} = 5 + \frac{-j10\,(4 + j2)}{-j10 + 4 + j2}$$

$$= 5 + \frac{-j10\,(4 + j2)}{4 - j8}$$

$$= 5 + \frac{5(4 - j8)}{4 - j8}$$

$$= 5 + 5 = 10 \ \Omega$$

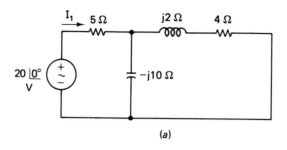

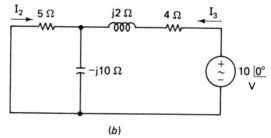

FIGURE 18.3 *Phasor circuit for (a) I₁ and (b) I₂ in Fig. 18.2.*

Therefore, the current I_1 is given by

$$I_1 = \frac{20\,\underline{|0°}}{Z_1} = \frac{20\,\underline{|0°}}{10} = 2\,\underline{|0°}\ A$$

The impedance seen by the source in Fig. 18.3(b) is

$$Z_2 = 4 + j2 + \frac{-j10(5)}{5 - j10}$$

$$= 4 + j2 - \frac{j10}{1 - j2} \cdot \frac{1 + j2}{1 + j2}$$

$$= 4 + j2 - \frac{10(-2 + j1)}{5}$$

$$= 4 + j2 + 4 - j2 = 8$$

Therefore, the current I_3 is

$$I_3 = \frac{10\,\underline{|0°}}{8} = \frac{5}{4}\,\underline{|0°}\ A$$

By current division the current I_2 is given by

$$I_2 = -\frac{-j10}{5 - j10} \cdot I_3$$

where we note that the negative sign is required because of the polarity of I_2. Substituting for I_3, we have

$$I_2 = \frac{j2}{1 - j2} \cdot \frac{5}{4}$$

$$= \frac{j5}{2(1 - j2)} \cdot \frac{1 + j2}{1 + j2}$$

$$= \frac{5(-2 + j1)}{2(5)}$$

or

$$I_2 = \frac{-2 + j1}{2} = \frac{\sqrt{5}}{2} \underline{|153.4°} \text{ A}$$

The phasor current I in Fig. 18.2 is therefore

$$I = I_1 + I_2$$

$$= 2\underline{|0°} + \frac{-2 + j1}{2} \text{ A}$$

or

$$I = 2 + \frac{-2 + j1}{2}$$

$$= \frac{2 + j1}{2} = \frac{\sqrt{5}}{2} \underline{|26.6°} \text{ A}$$

The time-domain current is therefore

$$i = \frac{\sqrt{5}\sqrt{2}}{2} \sin(2t + 26.6°) \text{ A}$$

or

$$i = 1.58 \sin(2t + 26.6°) \text{ A}$$

Circuit with Sources of Different Frequencies: If a circuit has two or more sources with different frequencies, we cannot find the steady-state currents or voltages directly because phasor circuits and impedances cannot be considered for more than one frequency. We may, however, use superposition to consider two or more circuits, each with a single source and therefore a single frequency. We will illustrate the procedure with the following example.

Example 18.2: Find the steady-state current i in the circuit of Fig. 18.4.

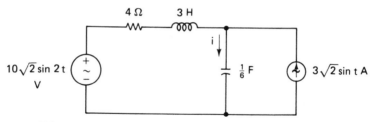

FIGURE 18.4 *Circuit with two sources of different frequencies.*

Solution: By superposition the current is given by

$$i = i_1 + i_2$$

where i_1 is due to the voltage source alone (with the current source killed) and i_2 is due to the current source alone (with the voltage source killed). The phasor circuits are shown for these cases in Fig. 18.5(a) and (b). In (a) $\mathbf{I}_1$ is the phasor of i_1, $\omega = 2$ rad/s (that of the voltage source), and the current source is replaced by an open circuit. In (b) $\mathbf{I}_2$ is the phasor of i_2, $\omega = 1$ rad/s (that of the current source), and the voltage source is replaced by a short circuit.

In Fig. 18.5(a) the impedance seen by the source is

$$\mathbf{Z} = 4 + j6 - j3 = 4 + j3 = 5 \underline{|36.9°} \ \Omega$$

so that the current is

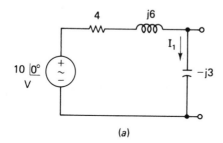

(a)

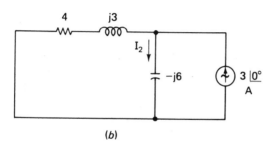

(b)

FIGURE 18.5 *Phasor circuit for finding (a) i_1 and (b) i_2.*

$$\mathbf{I_1} = \frac{10\underline{|0°}}{\mathbf{Z}} = \frac{10\underline{|0°}}{5\underline{|36.9°}} = 2\underline{|-36.9°} \text{ A}$$

Therefore, the sinusoidal current is

$$i_1 = 2\sqrt{2} \sin (2t - 36.9°) \text{ A}$$

In Fig. 18.5(b) we have by current division,

$$\mathbf{I_2} = \frac{4+j3}{4+j3-j6} \cdot 3\underline{|0°} = \frac{3(4+j3)}{4-j3}$$

$$= \frac{3(5\underline{|36.9°})}{5\underline{|-36.9°}}$$

or

$$\mathbf{I_2} = 3\underline{|73.8°} \text{ A}$$

Therefore, the sinusoidal current is

$$i_2 = 3\sqrt{2} \sin (t + 73.8°) \text{ A}$$

By superposition the total steady-state current i is

$$i = i_1 + i_2$$

or

$$i = 2\sqrt{2} \sin (2t - 36.9°) + 3\sqrt{2} \sin (t + 73.8°) \text{ A}$$

Superposition may also be applied if one or more of the sources is a dc source. In this case the component due to the source is obtained from a circuit with all other sources killed and the inductors and capacitors replaced by short circuits and open circuits, respectively. A problem of this type is considered in Practice Exercise 18.1.1.

PRACTICE EXERCISES

18.1.1 Find the steady-state current i. (*Suggestion:* Use superposition and note that the component due to the dc source is obtained with the inductors and the ac source replaced by a short circuit and the capacitor replaced by an open circuit.)

Ans. $3 \sin (2t - 53.1°) - 4$ A

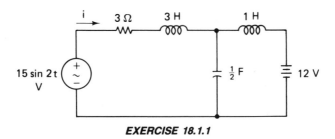

EXERCISE 18.1.1

18.1.2 Find the steady-state node voltage v.

Ans. $8 \sin (2t + 36.9°) + 2 \sqrt{5} \sin (3t + 26.6°)$ V

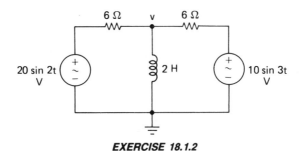

EXERCISE 18.1.2

18.2 THÉVENIN'S AND NORTON'S THEOREMS

In the case of phasor circuits, Thévenin's and Norton's theorems are identical to the resistive case, except that the open-circuit voltage $\mathbf{V}_{oc}$ and the short-circuit current $\mathbf{I}_{sc}$ are phasors, and instead of the Thévenin resistance R_{th} of the dead circuit we have the *Thévenin impedance* $\mathbf{Z}_{th}$.

Thévenin's Theorem: In the case of a phasor circuit, such as that of Fig. 18.6, Thévenin's theorem says that the circuit is equivalent at the terminals *a-b* to the *Thévenin equivalent circuit* of Fig. 18.7(a), consisting of a voltage source $\mathbf{V}_{oc}$ in series with an impedance $\mathbf{Z}_{th}$. The source is the *open-circuit phasor voltage* across the open terminals *a-b* in Fig. 18.6, and the impedance is the Thévenin impedance seen at terminals *a-b* with all the sources killed in the phasor circuit. This is, of course, identical in form to Thévenin's theorem for resistive circuits given in Chapter 8.

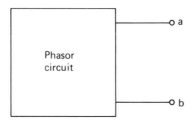

FIGURE 18.6 *Phasor circuit to be replaced by its Thévenin or Norton equivalent.*

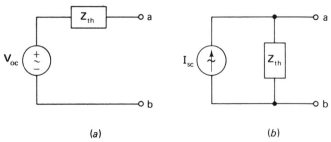

FIGURE 18.7 *(a) Thévenin equivalent circuit and (b) Norton equivalent circuit in the phasor domain.*

Norton's Theorem: Again referring to the phasor circuit of Fig. 18.6, Norton's theorem says that an equivalent circuit, called the *Norton equivalent circuit,* is that of Fig. 18.7(b). The impedance $\mathbf{Z}_{th}$ is the same as in the Thévenin equivalent circuit, and $\mathbf{I}_{sc}$ is the *short-circuit phasor current* that would flow if terminals *a* and *b* were connected by a short circuit.

Also, as in the resistive circuit case, the open-circuit voltage and the short-circuit current are related to each other. In the phasor case the relationship is

$$\mathbf{V}_{oc} = \mathbf{Z}_{th}\mathbf{I}_{sc} \tag{18.2}$$

Thus we may find any two of the quantities $\mathbf{V}_{oc}$, $\mathbf{I}_{sc}$, and $\mathbf{Z}_{th}$, and use the result to find the third quantity. We will illustrate the procedure with an example.

Example 18.3: Replace the phasor circuit to the left of terminals *a-b* in Fig. 18.8 by its Thévenin equivalent and find the phasor current **I**.

Solution: The open-circuit voltage $\mathbf{V}_{oc}$ is found by opening terminals *a-b*, as shown in Fig. 18.9(a). Since there is no current in the inductor, $\mathbf{V}_{oc}$ is the voltage across the capacitor. Therefore, by voltage division we have

$$\mathbf{V}_{oc} = \frac{-j4}{4 - j4} \cdot 6\underline{|0°} = 3 - j3$$

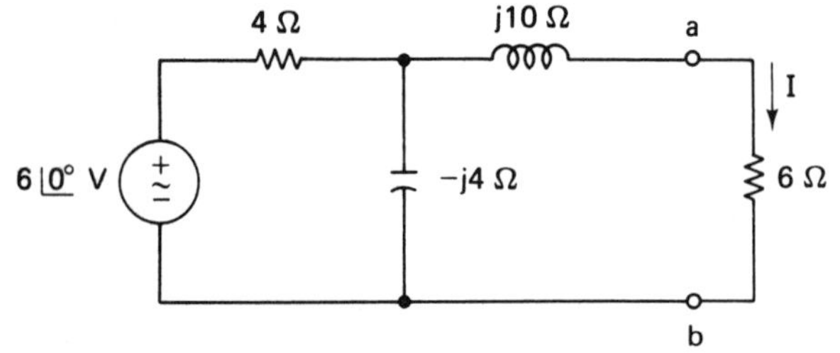

FIGURE 18.8 *Circuit to be replaced by its Thévenin equivalent.*

or

$$\mathbf{V}_{oc} = 3\sqrt{2}\underline{|-45°}\ \text{V}$$

To find $\mathbf{Z}_{th}$ we kill the source, resulting in the circuit of Fig. 18.9(b). Thus $\mathbf{Z}_{th}$ is the $j10$-Ω impedance in series with the parallel combination of 4 Ω, and $-j4$ Ω, and is given by

$$\mathbf{Z}_{th} = j10 + \frac{4(-j4)}{4 - j4}$$

$$= 2 + j8\ \Omega = 8.246\underline{|75.96°}\ \Omega$$

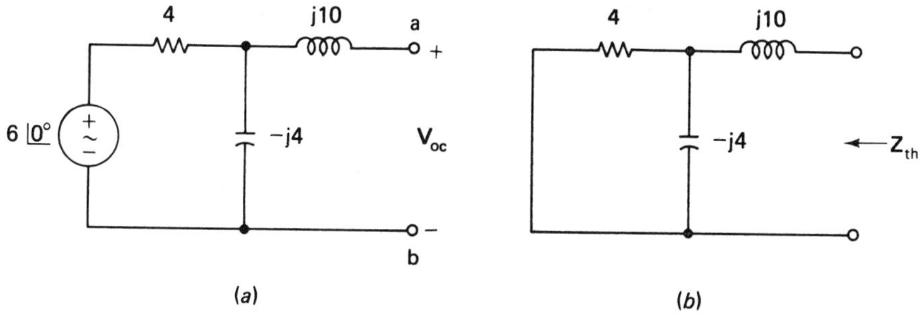

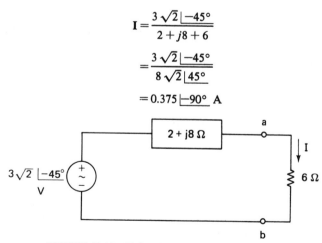

FIGURE 18.9   Circuits for obtaining (a) V_{oc} and (b) Z_{th}.

Therefore, the Thévenin equivalent circuit is that of Fig. 18.10, where we have added the 6-Ω load resistor.

The current $\mathbf{I}$ in Fig. 18.8 may now be found using Fig. 18.10. The result is

$$\mathbf{I} = \frac{3\sqrt{2}\underline{|-45°}}{2 + j8 + 6}$$

$$= \frac{3\sqrt{2}\underline{|-45°}}{8\sqrt{2}\underline{|45°}}$$

$$= 0.375\underline{|-90°}\ \text{A}$$

FIGURE 18.10   Thévenin equivalent of Fig. 18.8.

Example 18.4: Replace the circuit to the left of terminals *a-b* in Fig. 18.8 by its Norton equivalent and find $\mathbf{I}$.

Solution: We have already found $\mathbf{V}_{oc}$ and $\mathbf{Z}_{th}$ in Example 18.3, so that by (18.2) we have

$$\mathbf{I}_{sc} = \frac{\mathbf{V}_{oc}}{\mathbf{Z}_{th}} = \frac{3\sqrt{2}\lfloor -45°}{8.246\lfloor 75.96°}$$

or

$$\mathbf{I}_{sc} = 0.515\lfloor -120.96° \text{ A}$$

Therefore, the Norton equivalent circuit with the 6-Ω load is that of Fig. 18.11. From this circuit we have by current division,

$$\mathbf{I} = \frac{2 + j8}{2 + j8 + 6}(0.515\lfloor -120.96°)$$

$$= \frac{(\sqrt{68}\lfloor 75.96°)(0.515\lfloor -120.96°)}{\sqrt{128}\lfloor 45°}$$

$$= 0.375\lfloor -90° \text{ A}$$

which checks the result obtained in Example 18.3.

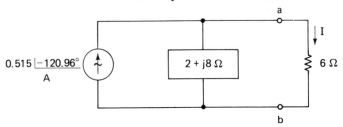

FIGURE 18.11 *Norton equivalent of Fig. 18.8.*

PRACTICE EXERCISES

18.2.1 Find the Thévenin equivalent of the circuit to the left of terminals *a-b*.

Ans. $\mathbf{V}_{oc} = 5\lfloor -90° \text{ V}$, $\mathbf{Z}_{th} = \dfrac{1 + j2}{2}\ \Omega$

EXERCISE 18.2.1

18.2.2 Find the Norton equivalent of the circuit to the left of terminals *a-b* in Exercise 18.2.1.

$$Ans. \ \mathbf{I}_{sc} = 2\sqrt{5}\,\underline{/-153.4°}\ \text{A}, \ \mathbf{Z}_{th} = \frac{1+j2}{2}\ \Omega$$

18.3 VOLTAGE AND CURRENT PHASOR SOURCE CONVERSIONS

As in the case of resistive circuits in Section 8.4, phasor voltage sources may be converted to equivalent phasor current sources, and vice versa, by using Thévenin's and Norton's theorems. The Thévenin and Norton phasor circuits, shown earlier in Fig. 18.7(a) and (b) are equivalents of the same phasor circuit and therefore are equivalents of each other. The impedance $\mathbf{Z}_{th}$ is the same in both circuits, and $\mathbf{V}_{oc}$ and $\mathbf{I}_{sc}$ are related by

$$\mathbf{V}_{oc} = \mathbf{Z}_{th}\mathbf{I}_{sc} \tag{18.3}$$

Thus the *practical* voltage source (an ideal voltage source in series with an impedance) of Fig. 18.7(a) is equivalent at the terminals to the *practical* current source (an ideal current source in parallel with an impedance) of Fig. 18.7(b).

Source Conversions: Using the equivalence of the Thévenin and Norton phasor circuits we may make voltage-to-current source conversions or current-to-voltage source conversions, as we did for resistive circuits. We simply replace the practical voltage source by its Norton equivalent, which is a practical current source, or the practical current source by its Thévenin equivalent, which is a voltage source. This is easily done by using (18.3) and the fact that the impedance of each source is the same.

Example 18.5: Convert the practical voltage source of Fig. 18.12(a) to an equivalent current source.

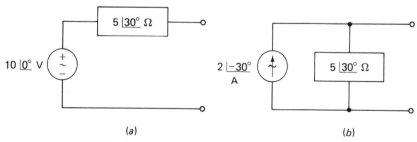

(a)   *(b)*

FIGURE 18.12 *Equivalent voltage and current sources.*

Solution: The internal impedance of the current source is

$$\mathbf{Z} = 5\underline{|30°}\ \Omega$$

which is the same as that of the voltage source. The ideal current source **I** is the short-circuit current obtained in Fig. 18.12(a) and is thus given by

$$\mathbf{I} = \frac{10\underline{|0°}}{5\underline{|30°}} = 2\underline{|-30°}\ \text{A}$$

The equivalent practical current source is therefore that of Fig. 18.12(b), with **I** and **Z** in parallel. This circuit is, of course, the Norton equivalent of Fig. 18.12(a).

Example 18.6: Find the Thévenin and Norton equivalents of the circuit of Fig. 18.13 using successive source transformations.

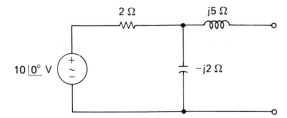

FIGURE 18.13 *Circuit with a voltage source.*

Solution: The practical voltage source consisting of the 10-V source and the 2-Ω resistor may be replaced by an equivalent current source of

$$\mathbf{I}_1 = \frac{10\underline{|0°}}{2} = 5\underline{|0°}\ \text{A}$$

in parallel with a 2-Ω resistor, as shown in Fig. 18.14(a). The 2-Ω resistor and the −*j*2-Ω capacitor are now in parallel and may be replaced by an equivalent impedance of

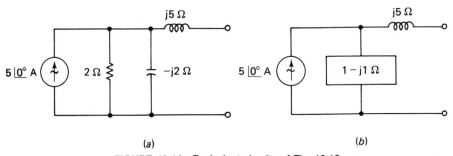

(a)   (b)

FIGURE 18.14 *Equivalent circuits of Fig. 18.13.*

$$\frac{2(-j2)}{2 - j2} = 1 - j1 \ \Omega$$

This is shown in Fig. 18.14(b).

The 5-A source and the $1 - j1$-Ω impedance of Fig. 18.14(b) form a practical current source, which is equivalent to a voltage source of

$$\mathbf{V}_2 = (1 - j1)(5 \underline{/0°})$$

$$= 5 - j5$$

$$= 5 \sqrt{2} \underline{/-45°} \ \text{V}$$

in series with an impedance of $1 - j1 \ \Omega$, as shown in Fig. 18.15(a). We may combine the two series impedances into their equivalent of

$$1 - j1 + j5 = 1 + j4 \ \Omega$$

as shown in Fig. 18.15(b). This last circuit is a voltage source in series with an impedance, and therefore is the Thévenin equivalent of Fig. 18.13.

The Norton equivalent circuit is found by converting Fig. 18.15(b) to an equivalent practical current source. The internal impedance is $1 + j4 \ \Omega$, as in Fig. 18.15(b), and the current source is

$$\mathbf{I}_3 = \frac{5 \sqrt{2} \underline{/-45°}}{1 + j4}$$

$$= \frac{5 \sqrt{2} \underline{/-45°}}{\sqrt{17} \underline{/76°}}$$

$$= 1.715 \underline{/-121°} \ \text{A}$$

The Norton equivalent is shown in Fig. 18.16.

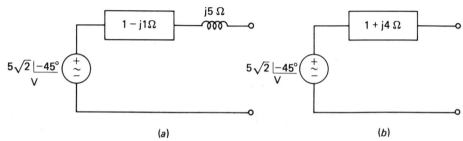

FIGURE 18.15   (a) Equivalent circuit and (b) the Thévenin equivalent circuit of Fig. 18.13.

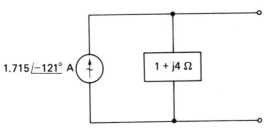

FIGURE 18.16 Norton equivalent of Fig. 18.13.

PRACTICE EXERCISES

18.3.1 Using successive source conversions, replace the circuit to the left of terminals *a-b* by its Thévenin equivalent and find **V**.

Ans. $\mathbf{V}_{oc} = 6 - j6$ V, $\mathbf{Z}_{th} = 1 + j1$ Ω, $\mathbf{V} = -j12$ V

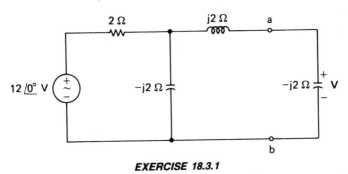

EXERCISE 18.3.1

18.3.2 Replace both voltage sources by equivalent current sources and find **V**.

Ans. $2 - j6$ V

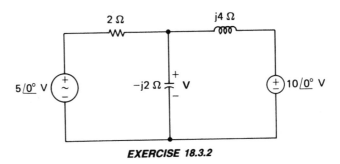

EXERCISE 18.3.2

18.4 Y AND Δ PHASOR NETWORKS

As in the resistive circuits of Section 8.6, we may have Y and Δ connections of impedances as shown in Fig. 18.17(a) and (b). The only difference between these circuits and the corresponding resistive circuits is that we have impedances that are generally complex numbers instead of resistances that are real numbers.

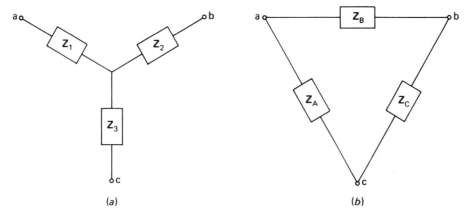

FIGURE 18.17 *(a) Y and (b) Δ (—) impedance connections.*

Y–Δ Conversions: We may convert from a Y connection to a Δ equivalent connection, or from a Δ to an equivalent Y, by formulas identical in form to those for the resistor circuits given in (8.18) and (8.20). In the case of the Y–Δ conversion (conversion from the Y to the Δ of Fig. 18.17), we have

$$Z_A = \frac{Z_1 Z_2 + Z_2 Z_3 + Z_3 Z_1}{Z_2}$$

$$Z_B = \frac{Z_1 Z_2 + Z_2 Z_3 + Z_3 Z_1}{Z_3} \qquad (18.4)$$

$$Z_C = \frac{Z_1 Z_2 + Z_2 Z_3 + Z_3 Z_1}{Z_1}$$

The Y and Δ connections are shown together in Fig. 18.18, so that we may formulate the Y–Δ transformation in words. Referring to (18.4) and Fig. 18.18, we may note that in each case the numerator is the sum of products of the impedances of the Y network taken two at a time, and the denominator is the impedance in the Y that is *opposite* the impedance being computed in the Δ. That is,

$$Z_\Delta = \frac{\text{sum of products in } Y}{\text{opposite } Z \text{ in } Y} \qquad (18.5)$$

This is very much like the corresponding case (8.19) for resistors.

Example 18.7: Find the Δ equivalent of the Y network of Fig. 18.19(a).

Solution: From Fig. 18.19(a) we have

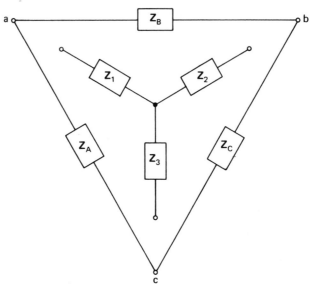

FIGURE 18.18 *Circuit for Y − Δ and Δ − Y conversions.*

$$\mathbf{Z}_1\mathbf{Z}_2 + \mathbf{Z}_2\mathbf{Z}_3 + \mathbf{Z}_3\mathbf{Z}_1 = j2(-j4) - j4(8) + 8(j2)$$

$$= 8 - j16$$

which is the numerator in (18.4) or (18.5) in every case. Thus we may write

$$\mathbf{Z}_A = \frac{8 - j16}{\mathbf{Z}_2} = \frac{8 - j16}{-j4} = 4 + j2 \ \Omega$$

$$\mathbf{Z}_B = \frac{8 - j16}{\mathbf{Z}_3} = \frac{8 - j16}{8} = 1 - j2 \ \Omega$$

$$\mathbf{Z}_C = \frac{8 - j16}{\mathbf{Z}_1} = \frac{8 - j16}{j2} = -8 - j4 \ \Omega$$

The equivalent Δ connection is therefore that of Fig. 18.19(b).

Δ–Y Conversions: The conversion from the Δ connection of Fig. 18.17(b) to the equivalent Y connection is done with the Δ–Y conversion formulas, given by

$$\mathbf{Z}_1 = \frac{\mathbf{Z}_A\mathbf{Z}_B}{\mathbf{Z}_A + \mathbf{Z}_B + \mathbf{Z}_C}$$

$$\mathbf{Z}_2 = \frac{\mathbf{Z}_B\mathbf{Z}_C}{\mathbf{Z}_A + \mathbf{Z}_B + \mathbf{Z}_C} \qquad\qquad (18.6)$$

$$\mathbf{Z}_3 = \frac{\mathbf{Z}_A\mathbf{Z}_C}{\mathbf{Z}_A + \mathbf{Z}_B + \mathbf{Z}_C}$$

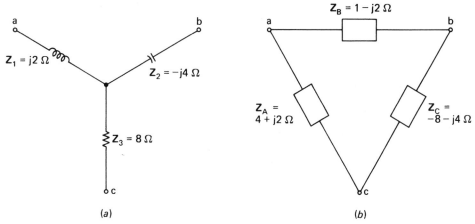

FIGURE 18.19 *(a) Y network and (b) its equivalent Δ network.*

These equations are identical in form to the resistive equations (8.20). In each case the denominator is the sum of the impedances of the Δ, and the numerator, referring to Fig. 18.18, is the product of the two Δ impedances that are *adjacent* to the Y impedance (on each side of the Y impedance). That is,

$$\mathbf{Z_Y} = \frac{\text{product of two adjacent } \mathbf{Z}\text{s in } \Delta}{\text{sum of } \mathbf{Z}\text{s in } \Delta} \qquad (18.7)$$

Example 18.8: Find the equivalent impedance $\mathbf{Z}_T$ of the circuit of Fig. 18.20.

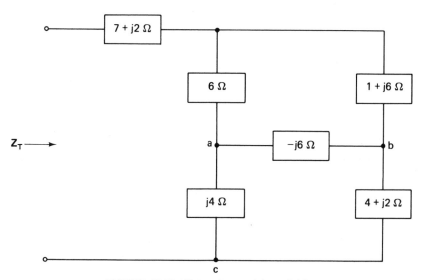

FIGURE 18.20 *Network containing a bridge.*

Solution: No two impedances are in series or parallel, but the network exclusive of the $7 + j2$ –Ω impedance is a bridge network that can be simplified by means of Y–Δ or Δ–Y conversions. We will use a Δ–Y conversion to change the Δ with terminals *a, b,* and *c* to an equivalent Y.

The equivalent Y that is to replace the Δ at terminals *a, b,* and *c* is shown dashed in Fig. 18.21. Using (18.7), we have for the Y impedances

$$Z_1 = \frac{j4\,(-j6)}{(j4) + (-j6) + (4 + j2)} = \frac{24}{4} = 6\ \Omega$$

$$Z_2 = \frac{-j6(4 + j2)}{4} = 3 - j6\ \Omega$$

$$Z_3 = \frac{j4(4 + j2)}{4} = -2 + j4\ \Omega$$

Replacing the Δ by the Y yields the circuit of Fig. 18.22.

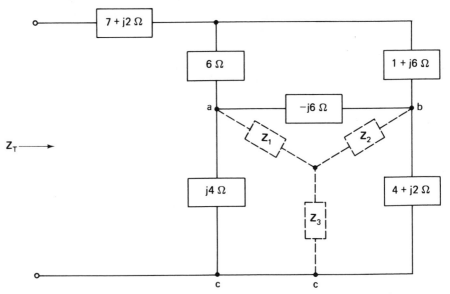

FIGURE 18.21 *Network of Fig. 18.20 showing equivalent Y to be used.*

The two 6-Ω resistors are in series and may be replaced by their equivalent impedance of $6 + 6 = 12\ \Omega$. Similarly, the $1 + j6$– and $3 - j6$-Ω impedances are in series and equivalent to $1 + j6 + 3 - j6 = 4\ \Omega$. The resulting circuit is that of Fig. 18.23. From this circuit we may see that the 12- and 4-Ω resistors are in parallel, and this parallel combination is in series with the impedances of $7 + j2\ \Omega$ and $-2 + j4\ \Omega$. Therefore, the equivalent impedance is

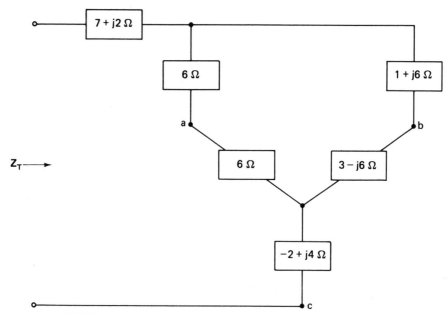

FIGURE 18.22   *Network of Fig. 18.21 with the Δ replaced by the Y.*

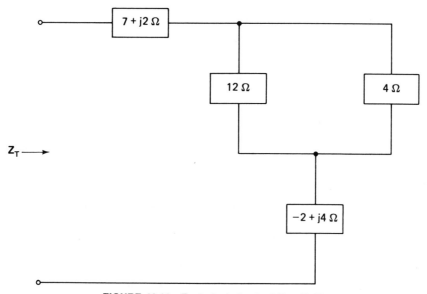

FIGURE 18.23   *Equivalent circuit of Fig. 18.22.*

$$\mathbf{Z}_T = (7+j2) + \frac{12(4)}{12+4} + (-2+j4)$$

$$= (7+3-2) + j(2+4)$$

$$= 8 + j6 \ \Omega$$

Case of Equal Impedances: If all impedances in a Y connection are alike and equal to $\mathbf{Z}_y$, all impedances in the equivalent Δ are also equal. To see this, let $\mathbf{Z}_\Delta$ be the value of any Δ impedance. Then by (18.4) we have for $\mathbf{Z}_A = \mathbf{Z}_\Delta$,

$$\mathbf{Z}_\Delta = \frac{\mathbf{Z}_y\mathbf{Z}_y + \mathbf{Z}_y\mathbf{Z}_y + \mathbf{Z}_y\mathbf{Z}_y}{\mathbf{Z}_y} = \frac{3\mathbf{Z}_y^2}{\mathbf{Z}_y}$$

or

$$\mathbf{Z}_\Delta = 3\mathbf{Z}_y \qquad\qquad (18.8)$$

Since this result holds for the other two Δ impedances, we see that they are all equal to $\mathbf{Z}_\Delta$. From (18.8) we have the corresponding conversion from Δ to Y given by

$$\mathbf{Z}_y = \frac{\mathbf{Z}_\Delta}{3} \qquad\qquad (18.9)$$

Example 18.9: If all the Δ impedances are alike and given by $\mathbf{Z}_\Delta = 30\,\underline{|60°}\ \Omega$, find the equivalent Y connection.

Solution: All the Y impedances are alike also and are given by (18.9) as

$$\mathbf{Z}_y = \frac{\mathbf{Z}_\Delta}{3} = \frac{30\,\underline{|60°}}{3}$$

$$= 10\,\underline{|60°}\ \Omega$$

PRACTICE EXERCISES

18.4.1 In Fig. 18.17(a) the impedances of the Y network are

$$\mathbf{Z}_1 = j10 \ \Omega$$

$$\mathbf{Z}_2 = 10 + j10 \ \Omega$$

$$\mathbf{Z}_3 = -j4 \ \Omega$$

Find the impedances Z_A, Z_B, and Z_C of the equivalent Δ network.

Ans. $2 + j4\ \Omega$, $-15 - j5\ \Omega$, $6 + j2\ \Omega$

18.4.2 Find the impedances Z_1, Z_2, and Z_3 of the Y network that is equivalent to the Δ network of Fig. 18.17(b) with

$$Z_A = 2 - j4\ \Omega$$

$$Z_B = j6\ \Omega$$

$$Z_C = -j2\ \Omega$$

Ans. $12 + j6\ \Omega$, $6\ \Omega$, $-4 - j2\ \Omega$

18.4.3 Replace the Y connection of Z_2, Z_3, and Z_4 by its equivalent Δ and find Z_T, if

$$Z_1 = \frac{2 + j1}{5}\ \Omega$$

$$Z_2 = Z_3 = Z_4 = 1 - j1\ \Omega$$

$$Z_5 = Z_6 = j3\ \Omega$$

Ans. $4 - j1\ \Omega$

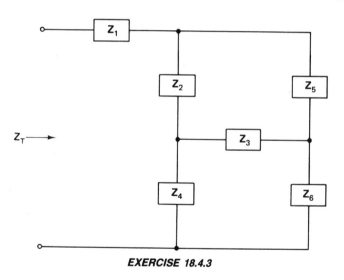

EXERCISE 18.4.3

18.5 SUMMARY

Network theorems, such as superposition, Thévenin's theorem, Norton's theorem, source conversions, and Y–Δ conversions, are almost identical in the ac steady-state case to the corresponding resistive circuit theorems. The only difference is that the currents and voltages of the resistive circuits are replaced by phasor currents and voltages, and resistances are replaced by impedances. In the case of Thévenin's and Norton's theorems, the open-circuit voltage and short-circuit current are the phasors V_{oc} and I_{sc} and the impedance seen at the terminals of the dead circuit is Z_{th}. Source conversions and Y–Δ conversion rules are identical in form to their corresponding resistive circuit rules.

Superposition is used to find phasor voltages or currents by killing all the sources except one and finding the current or voltage it contributes to the total. The total is then found by algebraically adding the contributions of each source acting alone. In the case of circuits with two or more sources with different frequencies, we must use superposition because we can use only one frequency at a time in the phasor circuits. If all the sources have the same frequency, superposition can be carried out in the phasor domain. However, phasors corresponding to different frequencies must be first converted to the time domain, and the time-domain quantities are then added to get the total.

PROBLEMS

18.1 Use superposition to find the phasor currents I_1, I_2, and I, where $I = I_1 + I_2$ is the current shown and I_1 and I_2 are due to the 11-V source and the 7-V source acting alone.

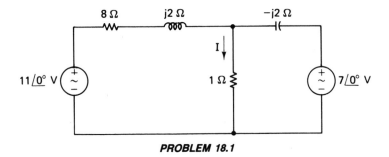

PROBLEM 18.1

18.2 Use superposition to find the phasor voltages V_1, V_2, and V, where $V = V_1 + V_2$ is the voltage shown and V_1 and V_2 are due to the 20-V source and the 2-A source acting alone.

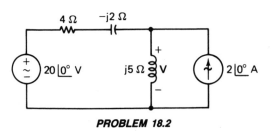

PROBLEM 18.2

18.3 Find the steady-state value of v if i_g is a dc source of 4 A.

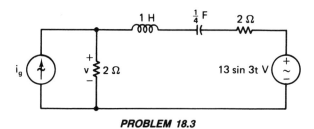

PROBLEM 18.3

18.4 Find the steady-state value of v in Problem 18.3 if

$$i_g = 2 \sin 2t \text{ A}$$

18.5 Solve Problem 17.21 using superposition.

18.6 Solve Problem 17.22 using superposition.

18.7 Replace the network to the left of terminals a-b by its Thévenin equivalent and use the result to find **V**.

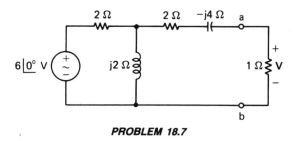

PROBLEM 18.7

18.8 Replace the network to the left of terminals a-b in Problem 18.7 by its Norton equivalent and use the result to find **V**.

18.9 In the corresponding phasor circuit, replace the portion to the left of terminals a-b by its Thévenin equivalent and use the result to find the phasor **I** of i and the steady-state value of i.

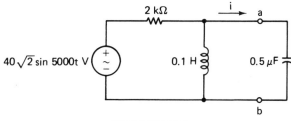

PROBLEM 18.9

18.10 Find **I** by replacing the network to the left of terminals *a-b* by its Thévenin equivalent.

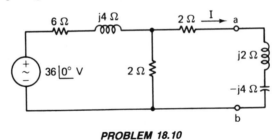

PROBLEM 18.10

18.11 Replace the circuit to the left of terminals *a-b* by its Norton equivalent and find **I**.

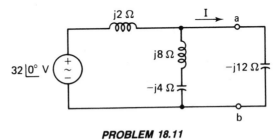

PROBLEM 18.11

18.12 Find **V** in Problem 18.7 by using successive source conversions to obtain the Thévenin equivalent of the network to the left of terminals *a-b*.

18.13 Use source conversions to obtain the Thévenin equivalent of the circuit to the left of terminals *a-b* and use the result to find **V**.

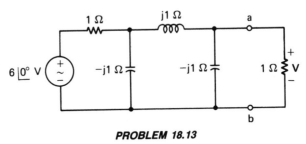

PROBLEM 18.13

18.14 Replace both sources by equivalent current sources and find **I**.

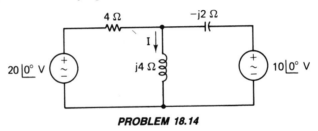

PROBLEM 18.14

18.15 Convert the Y network of Fig. 18.17(a) to the equivalent Δ network of Fig. 18.17(b) if

$$\mathbf{Z}_1 = 1 - j2 \ \Omega$$

$$\mathbf{Z}_2 = j5 \ \Omega$$

$$\mathbf{Z}_3 = 1 + j2 \ \Omega$$

18.16 Convert the Δ network of Fig. 18.17(b) to the equivalent Y network of Fig. 18.17(a) if

$$\mathbf{Z}_A = 2 + j4 \ \Omega$$

$$\mathbf{Z}_B = j2 \ \Omega$$

$$\mathbf{Z}_C = 2 - j6 \ \Omega$$

18.17 Find $\mathbf{Z}_T$ using a Y–Δ or Δ–Y conversion in the circuit of Practice Exercise 18-4.3 if

$$\mathbf{Z}_1 = 4 + j2 \ \Omega$$

$$\mathbf{Z}_2 = j4 \ \Omega$$

$$\mathbf{Z}_3 = 1 + j1 \ \Omega$$

$$\mathbf{Z}_4 = -j2 \ \Omega$$

$$\mathbf{Z}_5 = -j3 \ \Omega$$

$$\mathbf{Z}_6 = j3/2 \ \Omega$$

18.18 Find $\mathbf{Z}_T$ using a Y–Δ or Δ–Y conversion in the circuit of Practice Exercise 18-4.3 if

$$\mathbf{Z}_1 = \frac{3 + j1}{2} \ \Omega$$

$$\mathbf{Z}_2 = \mathbf{Z}_3 = \mathbf{Z}_4 = 2 + j1 \ \Omega$$

$$\mathbf{Z}_5 = \mathbf{Z}_6 = -j3 \ \Omega$$

19

AC STEADY-STATE POWER

In Chapter 15 we noted that the *instantaneous power* delivered to an element is the product of the element voltage and current, and is in general a function of the time, or instant, when it is being considered. We also defined *average power* in the case of periodic voltages and currents, such as ac sinusoids, as the average over a period of the instantaneous power, and we noted that average power is usually the power that is read by an ac wattmeter.

In this chapter we will consider in some detail the average power delivered to ac steady-state circuits, and we will see that average power is closely associated with the phasors of the ac voltages and currents. We will also define the *power factor* of an ac steady-state load, and new types of power called *apparent power, reactive power,* and *complex power.* Finally, we will consider how wattmeters, discussed in Chapter 9, are connected to measure ac steady-state power.

19.1 AVERAGE POWER

As we have seen, the instantaneous power p delivered to a load with voltage v and current i, as shown in Fig. 19.1, is given by

$$p = vi \qquad (19.1)$$

The load may be a single element or it may be a circuit of elements.

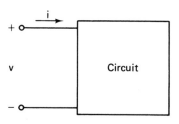

FIGURE 19.1 *Circuit absorbing power.*

AC Circuits: If the circuit of Fig. 19.1 is an ac circuit in steady state, the voltage and current are sinusoids, which we denote by

$$v = V_m \sin \omega t \tag{19.2}$$

and

$$i = I_m \sin (\omega t - \theta) \tag{19.3}$$

We have chosen the voltage to have zero phase angle and the current to be lagging it by the phase angle θ. This is a general case since if the current is actually leading the voltage, the angle θ will be negative. The instantaneous power in this case is

$$p = vi = V_m I_m \sin \omega t \sin (\omega t - \theta) \tag{19.4}$$

To find the average value of p, we need two identities from trigonometry, given by

$$\sin (\omega t - \theta) = \sin \omega t \cos \theta - \cos \omega t \sin \theta \tag{19.5}$$

and

$$\sin \omega t \cos \omega t = \frac{1}{2} \sin 2\omega t \tag{19.6}$$

Using these identities we may write (19.4) in the form

$$p = V_m I_m \sin \omega t \, (\sin \omega t \cos \theta - \cos \omega t \sin \theta)$$
$$= V_m I_m \, (\cos \theta \sin^2 \omega t - \sin \theta \sin \omega t \cos \omega t)$$

or

$$p = V_m I_m \cos \theta \sin^2 \omega t - \frac{1}{2} V_m I_m \sin \theta \sin 2\omega t$$

If we denote p_1 and p_2 by

$$p_1 = (V_m I_m \cos \theta) \sin^2 \omega t \tag{19.7}$$

and

$$p_2 = \left(-\frac{1}{2} V_m I_m \sin \theta\right) \sin 2\omega t \tag{19.8}$$

the instantaneous power may be written

$$p = p_1 + p_2 \tag{19.9}$$

Average AC Steady-State Power: The average value of p in (19.9), which we denote by P, is given by

$$P = P_1 + P_2 \tag{19.10}$$

where P_1 is the average value of p_1 and P_2 is the average value of p_2. The quantity P is the steady-state ac power read by a wattmeter.

As we saw in Section 15.6, the average value of the square of a sine wave, such as

$$f_1(t) = K_1 \sin^2 \omega t$$

is F_1, given by

$$F_1 = \frac{K_1}{2}$$

[This may be seen from (15.42) and (15.44), where $K_1 = R I_m^2$.] That is, the average value is simply half the coefficient of $\sin^2 \omega t$. Applying this result to (19.7), we see that the average value P_1 of p_1 is

$$P_1 = \frac{1}{2} V_m I_m \cos \theta$$

The average value P_2 of p_2 is by (19.8) the average value of a sinusoid, which from Chapter 15 we know to be zero. That is,

$$P_2 = 0$$

Substituting these values of P_1 and P_2 into (19.10), we have the average power delivered to the ac load given by

$$P = \frac{V_m I_m}{2} \cos \theta \qquad (19.11)$$

Phasor Relationships: In the case of the phasor circuit of Fig. 19.2, the phasors **V** and **I** are those of v and i of (19.2) and (19.3). Thus they are given by

$$\mathbf{V} = V \underline{/0°} \text{ V}$$
$$\mathbf{I} = I \underline{/-\theta} \text{ A} \qquad (19.12)$$

where V and I are the rms values,

$$V = V_{rms} = \frac{V_m}{\sqrt{2}}$$
$$I = I_{rms} = \frac{I_m}{\sqrt{2}} \qquad (19.13)$$

which were discussed in Chapter 15.

Using (19.13) we may express the average power in terms of rms values. To see this we note that

$$VI = \frac{V_m}{\sqrt{2}} \cdot \frac{I_m}{\sqrt{2}} = \frac{V_m I_m}{2}$$

Therefore, we have in (19.11),

$$P = VI \cos \theta \qquad (19.14)$$

The standard unit of P is the watt (W), as was the case for instantaneous power. This may be seen in (19.14) since $\cos \theta$ has no dimensions.

The impedance **Z**, shown in Fig. 19.2, is given by

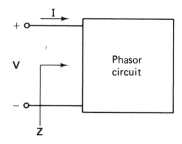

FIGURE 19.2 *Phasor circuit absorbing power.*

$$\mathbf{Z} = \frac{\mathbf{V}}{\mathbf{I}} = \frac{V\underline{|0°}}{I\underline{|-\theta}} = \frac{V}{I}\underline{|\theta}\ \Omega$$

or in other words,

$$\mathbf{Z} = |\mathbf{Z}|\underline{|\theta}\ \Omega \tag{19.15}$$

where

$$|\mathbf{Z}| = \frac{V}{I} = \frac{V_m}{I_m} \tag{19.16}$$

and $\theta = 0 - (-\theta)$, which is the angle of $\mathbf{V}$ minus the angle of $\mathbf{I}$, denoted by

$$\theta = \text{ang } \mathbf{V} - \text{ang } \mathbf{I} \tag{19.17}$$

These features of $\mathbf{Z}$ are shown in the diagram of Fig. 19.3(a), and the relation of θ to $\mathbf{V}$ and $\mathbf{I}$ is shown in Fig. 19.3(b).

   Thus the average power may be completely determined from the phasors, without any need for conversion to the time domain. We need only the magnitudes V and I of the voltage and current phasors and the angle θ between them.

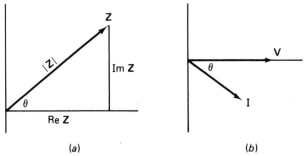

(a)   (b)
FIGURE 19.3   Diagrams of (a) Z and (b) V and I.

Relationship of Power to Impedance: We may eliminate V in (19.14) and obtain an expression for the average power in terms of the current and the impedance. From (19.16) we have

$$V = |\mathbf{Z}|\ I$$

and from Fig. 19.3(a) we see that

$$\cos\theta = \frac{\text{Re } \mathbf{Z}}{|\mathbf{Z}|}$$

Substituting these values into (19.14), we have

$$P = (|\mathbf{Z}|\, I)\left(I \frac{\operatorname{Re} \mathbf{Z}}{|\mathbf{Z}|}\right)$$

or

$$P = I^2 \operatorname{Re} \mathbf{Z} \qquad\qquad (19.18)$$

which is very similar to $p = i^2 R$ in the case of power delivered to a resistor.

Example 19.1: Find the average power delivered to a load with impedance

$$\mathbf{Z} = 4 + j3 \; \Omega$$

if the rms value of the current is

$$I = 2 \text{ A}$$

Solution: The real part of $\mathbf{Z}$ is

$$\operatorname{Re} \mathbf{Z} = 4$$

so that by (19.18) we have

$$P = I^2 \operatorname{Re} \mathbf{Z}$$
$$= (2)^2(4)$$
$$= 16 \text{ W}$$

Example 19.2: Find the ac steady-state power delivered by the source in the circuit of Fig. 19.4, by using (19.14) and by using (19.18).

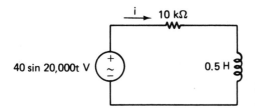

$$\text{40 sin 20,000t V} \qquad \qquad i \quad \text{10 k}\Omega \qquad \qquad \text{0.5 H}$$

FIGURE 19.4 *Circuit in the ac steady-state.*

Solution: The phasor circuit corresponding to Fig. 19.4 has a source voltage

$$\mathbf{V} = \frac{40}{\sqrt{2}}\underline{/0^\circ} \text{ V} \qquad\qquad (19.19)$$

and the impedance seen by the source is

$$\mathbf{Z} = R + j\omega L$$
$$= 10 \times 10^3 + j(20{,}000)(0.5)$$
$$= (10 + j10) \times 10^3 \ \Omega$$

or

$$\mathbf{Z} = 10 + j10 \ \text{k}\Omega \tag{19.20}$$

In polar form this is

$$\mathbf{Z} = 10\sqrt{2} \ \underline{|45°} \ \text{k}\Omega \tag{19.21}$$

From (19.19) we see that

$$V = \frac{40}{\sqrt{2}} \ \text{V}$$

and from (19.21) we have

$$|\mathbf{Z}| = 10\sqrt{2} \ \text{k}\Omega \qquad \theta = 45°$$

Therefore, the rms value of the current is

$$I = \frac{V}{|\mathbf{Z}|} = \frac{40/\sqrt{2}}{10\sqrt{2}} = 2 \ \text{mA}$$

(We note that V/kΩ = mA.) Therefore, by (19.14) we have

$$P = VI \cos \theta$$
$$= \left(\frac{40}{\sqrt{2}}\right)(2) \cos 45°$$
$$= 40 \ \text{mW}$$

The answer is in mW since the factors are volts times milliamperes.
From (19.20) we see that

$$\text{Re} \ \mathbf{Z} = 10 \ \text{k}\Omega$$

so that by (19.18) we have

$$P = I^2 \ \text{Re} \ \mathbf{Z}$$
$$= (2)^2(10)$$
$$= 40 \ \text{mW}$$

PRACTICE EXERCISES

19-1.1 In the time domain the voltage and current of a load are

$$v = 10 \sin (2t + 75°) \text{ V}$$

$$i = 2 \sin (2t + 15°) \text{ A}$$

Find the average power delivered to the load. *Ans.* 5 W

19-1.2 Find the average power delivered to a 2-kΩ resistor by a current

$$i = 4 \sin (100t + 30°) \text{ mA} \qquad Ans. \text{ 16 mW}$$

19-1.3 Find the average power delivered by the source. *Ans.* 30 mW

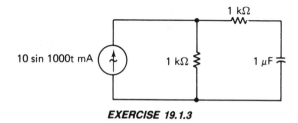

EXERCISE 19.1.3

19-1.4 Find the average power absorbed by the 10-Ω resistor. (*Suggestion:* Find the phasor **I** corresponding to *i.*) *Ans.* 2.5 W

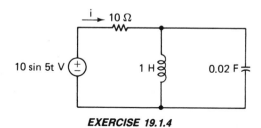

EXERCISE 19.1.4

19.2 POWER FACTOR

As we saw in Section 19.1, the average power delivered to an ac steady-state load is given by

$$P = VI \cos \theta \qquad (19.22)$$

where V and I are the rms voltage and current values and θ is the angle by which the current leads or lags the voltage. In practice, the rms voltage and current values are easily measured and their product VI is called the *apparent power*. This is perhaps because it seems *apparent* that power is a product of voltage and current, as it is in the resistive dc case.

Expressions for Apparent Power: We will denote the apparent power by S and write

$$S = VI \quad \text{VA} \tag{19.23}$$

The units, as noted, are voltamperes (VA), chosen to distinguish apparent power from average power, whose units are watts. In the case of large voltages, such as those in the power industry, apparent power is often given in kilovoltamperes (kVA).

Definition of Power Factor: Since $\cos \theta$ is never greater than 1, we see from (19.22) and (19.23) that the average power P can never exceed the apparent power. When $\cos \theta = 1$, they are equal, but otherwise P is less than S. The ratio of the average power to the apparent power is defined as the *power factor*. Thus if we denote the power factor by F_p, we see from (19.22) and (19.23) that

$$F_p = \frac{P}{S} = \frac{P}{VI} \tag{19.24}$$

or

$$F_p = \cos \theta \tag{19.25}$$

The angle θ, sometimes called the *power-factor angle*, is of course the angle of the impedance $\mathbf{Z}$ of the load.

Example 19.3: Find the power factor of a load consisting of a series connection of a 100-Ω resistor and a 0.1-H inductor if the frequency is $\omega = 500$ rad/s.

Solution: The impedance of the load is

$$\mathbf{Z} = R + j\omega L$$
$$= 100 + j(500)(0.1) = 100 + j50 \ \Omega$$

The angle of $\mathbf{Z}$ is therefore

$$\theta = \tan^{-1} \frac{50}{100} = 26.6°$$

and the power factor is

$$F_p = \cos \theta = \cos 26.6° = 0.894$$

Resistive Load: In the case of a purely resistive load with resistance R, the voltage and current are in phase, and $\theta = 0°$. Thus the power factor is

$$F_p = \cos 0 = 1$$

and the average and apparent powers are equal and given by

$$P = S = VI$$

Other expressions, using Ohm's law, are

$$P = S = RI^2$$

and

$$P = S = \frac{V^2}{R}$$

Reactive Load: If the load is purely reactive, as in the case of an inductor or a capacitor, the power factor is zero. This is because for an inductor we have

$$\mathbf{Z} = \mathbf{Z}_L = jX_L = X_L \underline{|90°}$$

and for a capacitor we have

$$\mathbf{Z} = \mathbf{Z}_C = -jX_C = X_C \underline{|-90°}$$

Thus the power factor angle θ is $+90°$ for an inductor and $-90°$ for a capacitor. In either case we have the power factor given by

$$F_p = \cos (\pm 90°) = 0$$

An inductor or a capacitor, or any network composed of inductors and capacitors, therefore, dissipates zero average power. For this reason, inductors and capacitors are sometimes called *lossless* elements. Physically, lossless elements store energy during part of the period and release it during the other part, so that the average delivered power is zero.

Leading and Lagging Power Factors: A load for which $-90° < \theta < 0$ is equivalent to an RC combination, such as

$$\mathbf{Z} = R - jX_C$$

The impedance diagram for $\mathbf{Z}$ is shown in Fig. 19.5(a), where it may be seen that the

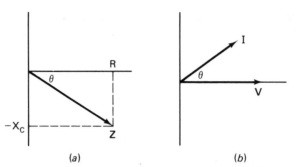

FIGURE 19.5 *Diagrams for (a) the impedance and (b) the voltage and current of an RC load.*

angle of **Z,** which we denote by $-\theta$, is negative and in the fourth quadrant. Since $-\theta$ is negative, the current **I** will *lead* the voltage **V** by θ, as shown in Fig. 19.5(b). The phasor $\mathbf{V} = V\underline{|0°}$ has been taken as the reference, so that

$$\mathbf{I} = \frac{\mathbf{V}}{\mathbf{Z}} = \frac{V\underline{|0°}}{|\mathbf{Z}|\,\underline{|-\theta}}$$

$$= I\underline{|\theta}$$

Thus for an *RC* load the phase angle is between $-90°$ and 0, and the current *leads* the voltage.

For an *RL* load the impedance is

$$\mathbf{Z} = R + jX_L$$

which has the impedance diagram of Fig. 19.6(a). The phase angle θ is positive, lying between 0 and 90°, and the current *lags* the voltage, as shown in Fig. 19.6(b).

Since cos θ is the same for positive or negative θ, we cannot tell from the power factor whether the load is of *RC* or *RL* type (or whether the inductive reactance is greater or less than the capacitive reactance). For example, suppose that we have

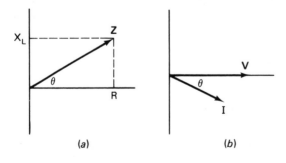

FIGURE 19.6 *Diagrams for (a) the impedance and (b) the voltage and current of an RL load.*

an *RC* load for which $\theta = -15°$ and an *RL* load for which $\theta = +15°$. In both cases the power factor is

$$F_p = \cos(\pm 15°) = 0.966$$

To avoid this difficulty in identifying the loads, we will define the power factor as *leading* or *lagging* according to whether the current is leading or lagging. For instance, in the example just considered, the *RC* load with $\theta = -15°$ and a leading current will be said to have a power factor of 0.966 leading. Similarly, the *RL* load with $\theta = +15°$ and a lagging current will be said to have a power factor of 0.966 lagging.

Example 19.4: Find the power factor seen by the source in Fig. 19.7, and state whether it is leading or lagging.

Solution: The frequency is $\omega = 2$ rad/s, so that the impedance seen by the source is

$$\mathbf{Z} = R + j\omega L - j\frac{1}{\omega C}$$

$$= 10 + j(2)(3) - j\frac{1}{2(\frac{1}{4})}$$

$$= 10 + j4 \ \Omega$$

The power factor angle therefore is

$$\theta = \tan^{-1}\frac{4}{10} = 21.8°$$

and the power factor is

$$F_p = \cos 21.8° = 0.928$$

Since the power-factor angle is positive, the net reactance is inductive. (We may also see this from the fact that $X = \omega L - 1/\omega C = 4$.) Therefore, the current lags the voltage and the power factor is 0.928 lagging.

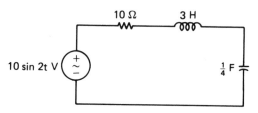

FIGURE 19.7 *RLC circuit.*

PRACTICE EXERCISES

19-2.1 Find the apparent power for (a) a load that draws 20 A from a 230-V line (rms values in each case), and (b) a load consisting of a 100-Ω resistor in parallel with a 10-μF capacitor connected to a source of 100 sin 1000*t* V.

Ans. (a) 4.6 kVA, (b) 50 √2 VA

19-2.2 Find the power factor of a load consisting of a 1-kΩ resistor in parallel with a 2-μF capacitor if the frequency is ω = 500 rad/s. *Ans.* 0.707 leading

19-2.3 Find the power factor of the load seen by the source. *Ans.* 0.894 lagging

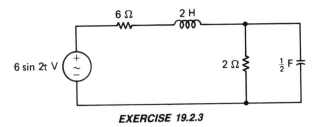

EXERCISE 19.2.3

19.3 POWER TRIANGLE

As we saw in Section 19.2, the average power P and the apparent power $S = VI$ are related by

$$P = S \cos \theta \tag{19.26}$$

or equivalently

$$\cos \theta = \frac{P}{S}$$

Therefore, we may relate P, S, and θ in a right triangle with hypotenuse S and a side P adjacent to θ. This triangle is called the *power triangle*, and is shown for $\theta > 0$ (lagging power factor) in Fig. 19.8(a) and for $\theta < 0$ (leading power factor) in Fig. 19.8(b). The quantity Q, shown as the other leg of the triangle, is also a type of power, as we will see.

Reactive Power: Since the average power P is real, the power triangle is drawn with P along the real axis. Thus Q is parallel to the imaginary axis, like the imaginary part of a complex number. We may relate Q to S and θ by noting from the triangle that

$$\sin \theta = \frac{Q}{S}$$

482

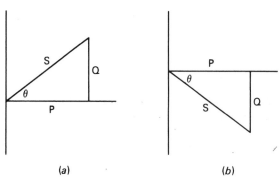

FIGURE 19.8 *Power triangle for the case of (a) lagging power factor and (b) leading power factor.*

and therefore

$$Q = S \sin \theta \qquad (19.27)$$

or

$$Q = VI \sin \theta \qquad (19.28)$$

Since $\cos \theta$ and $\sin \theta$ are dimensionless, we see from (19.26) and (19.27) that P and Q have the same units, and thus that Q is also a type of power. To distinguish the two, however, the unit of Q is defined as the *var*, which stands for *voltampere reactive*. The word "reactive" is appropriate because Q is at right angles to the real quantity P in the same way that the reactance X is at right angles to the real quantity R in the impedance diagram. For this reason, Q is sometimes called the *reactive power.*

Complex Power: From Fig. 19.8 we see that P and Q may be considered as the real and imaginary parts of a complex number represented by the hypotenuse, with magnitude S and angle θ. If we denote this number by **S**, we may write

$$\mathbf{S} = P + jQ \qquad (19.29)$$

In polar form, since $|\mathbf{S}| = S$, we have

$$\mathbf{S} = S \underline{/\theta}$$

or

$$\mathbf{S} = VI \underline{/\theta} \qquad (19.30)$$

Since both its components are power quantities, **S** is sometimes called the *complex power.* It is a very useful quantity because its real part is the *real* power, or average power, and its imaginary part is the reactive power. Also, its magnitude is the apparent power and its angle is the power factor angle.

We will take the standard units of complex power to be VA, as in the case of apparent power. This will distinguish it from P and Q.

Relation of the Complex Power to the Voltage and Current Phasors: We may find the complex power **S** from the voltage and current phasors since they contain V, I, and θ. However, there is an easier way to obtain **S** directly from the voltage and current phasors. To see this, let us choose the voltage phasor as

$$\mathbf{V} = V \underline{|\phi} \tag{19.31}$$

Then if the impedance is

$$\mathbf{Z} = |\mathbf{Z}| \underline{|\theta}$$

the current phasor is

$$\mathbf{I} = \frac{\mathbf{V}}{\mathbf{Z}} = \frac{V \underline{|\phi}}{|\mathbf{Z}| \underline{|\theta}}$$

or

$$\mathbf{I} = I \underline{|\phi - \theta} \tag{19.32}$$

where $I = V/|\mathbf{Z}|$. Now let us consider the conjugate **I*** of the phasor **I**. We recall that in polar form the conjugate is obtained by changing the sign of the angle, so that from (19.32) we have

$$\mathbf{I}^* = I \underline{|\theta - \phi} \tag{19.33}$$

Finally, forming the product **VI***, we have by (19.31) and (19.33)

$$\mathbf{VI}^* = (V \underline{|\phi})(I \underline{|\theta - \phi})$$
$$= VI \underline{|\phi + \theta - \phi}$$

or

$$\mathbf{VI}^* = VI \underline{|\theta}$$

Comparing this result with (19.30), we have

$$\mathbf{S} = \mathbf{VI}^* \tag{19.34}$$

Example 19.5: A load has an impedance of $\mathbf{Z} = 10\lfloor 15°$ Ω and a voltage $\mathbf{V} = 50\lfloor 0°$ V. Find the complex power, the apparent power, the average power, the reactive power, and the power factor associated with the load.

Solution: The current is given by

$$I = \frac{\mathbf{V}}{\mathbf{Z}} = \frac{50\lfloor 0°}{10\lfloor 15°} = 5\lfloor -15°\ \text{A}$$

and therefore its conjugate is

$$I^* = 5\lfloor 15°\ \text{A}$$

The complex power is

$$\mathbf{S} = \mathbf{VI}^* = (50\lfloor 0°\ )(5\lfloor 15°\ ) = 250\lfloor 15°$$

Therefore, the apparent power is

$$S = |\mathbf{S}| = 250\ \text{VA}$$

and the power factor angle is

$$\theta = \text{ang }\mathbf{S} = 15°$$

Thus the power factor is

$$F_p = \cos 15° = 0.966 \text{ lagging}$$

The average power is

$$P = S\cos\theta$$
$$= 250\cos 15°$$
$$= 241.5\ \text{W}$$

and the reactive power is

$$Q = S\sin\theta$$
$$= 250\sin 15°$$
$$= 64.7\ \text{vars}$$

Reactive Power and Power Factor: We may relate the reactive power Q to the current and the impedance by noting that

$$V = |\mathbf{Z}|I$$

and from Fig. 19.3 that

$$\sin \theta = \frac{\text{Im } \mathbf{Z}}{|\mathbf{Z}|}$$

Substituting these values into (19.28), we have

$$Q = (|\mathbf{Z}|I)(I)\left(\frac{\text{Im } \mathbf{Z}}{|\mathbf{Z}|}\right)$$

which simplifies to

$$Q = I^2 \text{ Im } \mathbf{Z} \tag{19.35}$$

Also, since Im $\mathbf{Z} = X$, the reactance, we have

$$Q = I^2 X \tag{19.36}$$

We may also eliminate I rather than V to obtain

$$Q = \frac{V^2 \text{ Im } \mathbf{Z}}{|\mathbf{Z}|^2} \tag{19.37}$$

From this result we see that if the load is inductive, then $X > 0$ and consequently Q is positive. This is also the case of a lagging power factor. If, on the other hand, the load is capacitive, then $X < 0$ ($X = -X_C$, for example), and thus Q is negative. This is the case of a leading power factor. Finally, if the load is purely resistive, then $X = 0$ and consequently $Q = 0$. This is the case of unity power factor ($F_p = 1$).

Summing up, we see that Q is related to the power factor in the following ways. If $Q = 0$, the power factor is unity and the load is resistive. If $Q > 0$, the power factor is lagging and the load is inductive. If $Q < 0$, the power factor is leading and the load is capacitive. Also, large positive or negative Q corresponds to a low power factor, whereas small positive or negative Q indicates a large power factor (near 1).

Example 19.6: A given load absorbs an average power of $P = 10$ kW. Find the power factor if the reactive power is (a) $Q = 0$, (b) $Q = 500$ vars, (c) $Q = 5$ kvars, and (d) $Q = -20$ kvars.

Solution: From Fig. 19.8 we see that the power factor angle is given by

$$\theta = \tan^{-1} \frac{Q}{P} \tag{19.38}$$

Therefore, in (a) we have

$$\theta = \tan^{-1} 0 = 0$$

and a power factor of

$$F_p = \cos 0 = 1$$

In (b) we have

$$\theta = \tan^{-1} \frac{500}{10,000} = 2.862°$$

and a power factor of

$$F_p = \cos 2.862° = 0.999 \text{ lagging}$$

(Note that $\theta > 0$.) In (c) we have

$$\theta = \tan^{-1} \frac{5,000}{10,000} = 26.6°$$

and

$$F_p = \cos 26.6° = 0.894 \text{ lagging}$$

Finally, in (d) we have

$$\theta = \tan^{-1} \frac{-20}{10} = -63.43°$$

and

$$F_p = 0.447 \text{ leading}$$

PRACTICE EXERCISES

19-3.1 A load absorbs a real, or average, power of $P = 4$ kW and a reactive power of $Q = -3$ kvars. If the rms value of the load voltage is 500 V, find the apparent power, the rms value of the current, the power factor, and the load impedance.

Ans. 5 kVA, 10 A, 0.8 leading, $50\underline{/-36.9°}$ Ω

19-3.2 Find the complex power delivered to a load that has a 0.9 lagging power factor and absorbs an average power of 1 kW.

Ans. $1000 + j484$ VA

19-3.3   The voltage and current of a load are

$$\mathbf{V} = 100 \underline{|0°} \ \text{V}$$

and

$$\mathbf{I} = 5 \underline{|30°} \ \text{A}$$

Find the complex power, the apparent power, the average power, the reactive power, and the power factor.

Ans. 500 $\underline{|-30°}$ VA, 500 VA, 433 W, −250 vars, 0.866 leading

19.4   POWER-FACTOR CORRECTION

In practice the power factor of a load is very important. In industrial applications, for example, loads may require many kilowatts to operate, and the power factor greatly affects the electric bill. To see this, let us solve for the rms value of the current in (19.24), resulting in

$$I = \frac{P}{VF_p} \tag{19.39}$$

From this we see that if the voltage V is fixed, as it usually is by the power line, and we wish to draw a fixed power P, then changing the power factor F_p may greatly affect the current that must be supplied by the electric company.

If the power factor is low, the current must be high, and a higher power factor (nearer to 1) requires a lower current. For this reason power companies encourage a high power factor, say 0.9 or higher, and impose a penalty on large industrial users with lower power factors.

Method of Correcting the Power Factor: We may make a low power factor higher, or *correct* the power factor, without changing the average power drawn, by adding a reactive element in parallel with the load. Since the element is reactive, its impedance is

$$\mathbf{Z} = 0 + jX$$

with Re $\mathbf{Z} = 0$. Therefore, it does not absorb any average power, which leaves P of the load unchanged.

If the original power factor is lagging, as it often is in practice because many motors and appliances are inductive, the reactive element that is added should have a leading power factor to make the overall power factor "less lagging," and thus higher. This means that a capacitor is required as the added element. In the rare instances where the power factor to be corrected is a leading power factor, we would add an inductor in parallel with the load.

The corrected case is shown in Fig. 19.9, where the load $\mathbf{Z}$ is the original load and $\mathbf{Z}_1$ is the parallel element added to correct the power factor. The line voltage is $\mathbf{V}$, the load current is $\mathbf{I}$, and the current drawn from the line is $\mathbf{I}_1$. Since the voltage $\mathbf{V}$ is across the load $\mathbf{Z}$, the load will draw the same current $\mathbf{I}$ as it would have without the correction.

The only change caused by the correction is that the current drawn from the line is $\mathbf{I}_1$ rather than $\mathbf{I}$. With the correction, $\mathbf{I}_1$ should be less in magnitude than $\mathbf{I}$.

If the power factor of $\mathbf{Z}$ is lagging, then for correction we must have

$$\mathbf{Z}_1 = -jX_C \tag{19.40}$$

which is the impedance of a capacitor. If the power factor of $\mathbf{Z}$ is leading, the parallel element must be an inductor with

$$\mathbf{Z}_1 = jX_L \tag{19.41}$$

Example 19.7: A load with a power factor of 0.8 lagging absorbs an average power of 500 W from a line having an rms voltage of 100 V. If the power factor is corrected to 0.95 lagging, as in Fig. 19.9, find the rms value of the current drawn from the line before correction and after correction of the power factor.

Solution: Before correction we have by (19.39)

$$I = \frac{P}{VF_p} = \frac{500}{(100)(0.8)} = 6.25 \text{ A}$$

After correction the load still draws 6.25 A, but the current drawn from the line is $\mathbf{I}_1$ in Fig. 19.9. Its rms value is

$$I_1 = \frac{P}{VF_p} = \frac{500}{(100)(0.95)} = 5.26 \text{ A}$$

This is almost 1A, or about 16%, less than was drawn from the line before the correction.

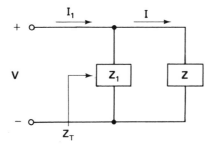

FIGURE 19.9 *Circuit for correcting the power factor.*

One Procedure: A direct method for finding the necessary $Z_1 = jX$ to correct the power factor in Fig. 19.9 is to find the equivalent impedance Z_T as shown. This value is given by

$$Z_T = \frac{(jX)Z}{jX + Z} \qquad\qquad (19.42)$$

where $X = -X_C$ if the power factor is lagging and $X = X_L$ if it is leading. Putting Z_T in the polar form

$$Z_T = |Z_T|\underline{\big/\theta_T} \qquad\qquad (19.43)$$

yields its angle θ_T, which is the corrected power factor angle. If the corrected power factor is to be F_p (given), we have

$$\cos \theta_T = F_P \qquad\qquad (19.44)$$

This equation contains the unknown X, which may thus be found.

Example 19.8: A load with a 0.8-lagging power factor absorbs an average power of 60 W from a 60 Hz-line with an rms voltage of 100 V. Find the parallel capacitor necessary in Fig. 19.9 to correct the power factor to 0.9 lagging.

Solution: The rms value of the load current is

$$I = \frac{P}{VF_p} = \frac{60}{(100)(0.8)} = 0.75 \text{ A}$$

so that the magnitude of the load impedance is

$$|Z| = \frac{V}{I} = \frac{100}{0.75} = \frac{400}{3}\ \Omega$$

The angle of Z is the power factor angle given by

$$\theta = \tan^{-1} 0.8 = 36.9°$$

Therefore, the load impedance is

$$Z = \frac{400}{3}\underline{\big/36.9°}\ \Omega$$

which in rectangular form is

$$Z = \frac{400}{3}(\cos 36.9° + j\sin 36.9°)$$

$$= \frac{80}{3}(4 + j3)\ \Omega$$

Since the load is inductive, the parallel correcting element must have impedance $-jX_C$, so that by (19.42) the equivalent impedance is

$$Z_T = \frac{-jX_C\left(\dfrac{80}{3}\right)(4 + j3)}{-jX_C + \dfrac{80}{3}(4 + j3)}$$

$$= \frac{-j80X_C(4 + j3)}{320 + j(240 - 3X_C)} \cdot \frac{320 - j(240 - 3X_C)}{320 - j(240 - 3X_C)}$$

which may be simplified to

$$Z_T = \frac{80X_C}{320^2 + (240 - 3X_C)^2}[12X_C + j(9X_C - 2000)]$$

The angle θ_T of Z_T is the angle of the number in brackets, because its coefficient is a positive real number with a zero angle. Therefore, we have

$$\theta_T = \tan^{-1}\left(\frac{9X_C - 2000}{12X_C}\right) \tag{19.45}$$

The corrected power factor of 0.9 satisfies the relation

$$0.9 = \cos\theta_T$$

from which

$$\theta_T = \cos^{-1}0.9 = 25.84°$$

From (19.45) we have

$$\tan\theta_T = \tan 25.84° = \frac{9X_C - 2000}{12X_C}$$

or

$$\frac{9X_C - 2000}{12X_C} = 0.4843$$

Solving this equation for X_C results in

$$X_C = 627 \ \Omega$$

Finally, to find the capacitance C needed in parallel across the load, we note that

$$X_C = \frac{1}{\omega C}$$

or

$$
\begin{aligned}
C &= \frac{1}{\omega X_C} = \frac{1}{2\pi f X_C} \\
&= \frac{1}{2\pi (60)(627)} \\
&= 4.23 \times 10^{-6} \ \text{F} \\
&= 4.23 \ \mu\text{F}
\end{aligned}
$$

A Second Procedure: A second, and much easier, method of finding the correcting parallel element $\mathbf{Z}_1$ in Fig. 19.9 is to use reactive power. The basis for the method is that the total complex power of a circuit is the sum of the complex powers of each part, as is true of real power.

The reactive element added in parallel draws no average power, but it does draw reactive power. Thus if P and Q are the real and reactive power before correction and P_T and Q_T are those after correction, then

$$P_T = P \tag{19.46}$$

and

$$Q_T = Q + Q_1 \tag{19.47}$$

where Q_1 is the reactive power absorbed by $\mathbf{Z}_1 = jX_1$. Since Q and Q_T may be found from the power triangles shown in Fig. 19.10, with θ and θ_T as the uncorrected

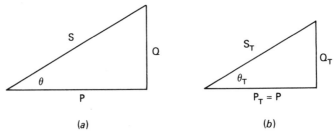

FIGURE 19.10 (a) Uncorrected and (b) corrected power triangles.

and corrected power factor angles, we may find Q_1 from (19.47). Then by (19.37) for $\mathbf{Z}_1 = jX_1$ we know that

$$Q_1 = \frac{V^2}{X_1}$$

so that

$$X_1 = \frac{V^2}{Q_1} \qquad (19.48)$$

Example 19.9: Solve Example 19.8 by using the method of reactive power.

Solution: Since the uncorrected power-factor angle is $\theta = 36.9°$, we have from Fig. 19.10(a),

$$Q = P \tan 36.9°$$
$$= 60 \tan 36.9°$$
$$= 45 \text{ vars}$$

The corrected power-factor angle is $\theta_T = 25.84°$, so that by Fig. 19.10(b) we have

$$Q_T = P \tan 25.84°$$
$$= 60 \tan 25.84°$$
$$= 29.06 \text{ vars}$$

From (19.47) the reactive power to be contributed by $\mathbf{Z}_1$ is

$$Q_1 = Q_T - Q$$
$$= 29.06 - 45$$
$$= -15.94 \text{ vars}$$

This is negative, as expected, since the correcting element must be a capacitor.
    From (19.48) we have

$$X_1 = \frac{V^2}{Q_1} = \frac{(100)^2}{-15.94} = -627 \ \Omega$$

Since $X_1 = -X_C$, this becomes

$$X_C = 627 \ \Omega$$

which is the result obtained in Example 19.8. The value $C = 4.23$ μF follows from

$$X_C = \frac{1}{2\pi f C} = 627$$

as before.

PRACTICE EXERCISES

19-4.1 A load with a power factor of 0.8 lagging has an rms voltage and current of 100 V and 5 A. Find the average power it absorbs.   *Ans.* 400 W

19-4.2 The power factor in Exercise 19-4.1 is to be corrected to 0.9 lagging without changing the power absorbed. Find the rms value of the current drawn from the line after correction.   *Ans.* 4.44 A

19-4.3 Find the parallel capacitance required to make the correction of Exercise 19.4.2 if the frequency is 60 Hz.   *Ans.* 28.2 μF

19.5 MAXIMUM POWER TRANSFER

A *practical* ac voltage source is one consisting of an ideal ac voltage source in series with an internal impedance. This is illustrated in the phasor case by the source of Fig. 19.11, where $\mathbf{V}_g$ is the ideal source voltage and $\mathbf{Z}_g$ is the internal impedance. This case is, of course, very similar to the resistive practical sources considered earlier in Chapter 8.

If the practical source is loaded with an impedance $\mathbf{Z}$, as in Fig. 19.12, the power P delivered to $\mathbf{Z}$ may be found from a knowledge of $\mathbf{V}_g$, $\mathbf{Z}_g$, and $\mathbf{Z}$. It may be shown that if $\mathbf{V}_g$ and $\mathbf{Z}_g$ are fixed, the maximum power delivered to the load $\mathbf{Z}$ occurs when $\mathbf{Z}$ is the conjugate $\mathbf{Z}_g^*$ of the internal impedance $\mathbf{Z}_g$.

That is, if we have

$$\mathbf{Z}_g = R_g + jX_g \tag{19.49}$$

then for maximum power transfer to the load we must have

$$\mathbf{Z} = \mathbf{Z}_g^* = R_g - jX_g \tag{19.50}$$

This result is known as the *maximum power transfer theorem.*

A very similar result was considered for resistive circuits in Practice Exercise 8.4.3, in which case the internal impedance was a resistance R_g and the load was

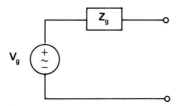

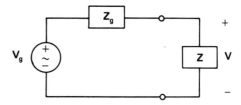

FIGURE 19.11 Practical ac volt-   FIGURE 19.12 Practical source connected
age source.   to a load Z.

a resistance R. The maximum transfer then occurs when $R = R_g$, the special case $X_g = 0$ of (19.50).

Example 19.10: Given that the source in Fig. 19.12 has $\mathbf{V}_g = 40 \,\underline{|0°}$ V and $\mathbf{Z}_g = 10 + j10$ Ω, find the maximum power that the load $\mathbf{Z}$ can draw.

Solution: For maximum power we must have

$$\mathbf{Z} = \mathbf{Z}_g^* = 10 - j10 \text{ } \Omega$$

in which case the current $\mathbf{I}$ is

$$\mathbf{I} = \frac{\mathbf{V}_g}{\mathbf{Z}_g + \mathbf{Z}} = \frac{40\,\underline{|0°}}{10 + j10 + 10 - j10} = 2\,\underline{|0°} \text{ A}$$

Therefore, the power delivered to $\mathbf{Z}$, which is the maximum power, is

$$P = |\mathbf{I}|^2 \, \text{Re } \mathbf{Z}$$
$$= (2)^2(10)$$
$$= 40 \text{ W}$$

Example 19.11: To show that the maximum power transfer theorem is plausible, let the load be (a) $\mathbf{Z} = 10 - j9$ Ω and (b) $\mathbf{Z} = 9 - j10$ Ω in Example 19.10 and find the power it absorbs.

Solution: In (a) the current is

$$\mathbf{I} = \frac{\mathbf{V}_g}{\mathbf{Z}_g + \mathbf{Z}} = \frac{40\,\underline{|0°}}{10 + j10 + 10 - j9} = \frac{40}{20 + j1}$$

which may be simplified to

$$\mathbf{I} = 1.9975 \underline{|-2.86°} \text{ A}$$

so that the average power is

$$P = |\mathbf{I}|^2 \operatorname{Re} \mathbf{Z}$$
$$= (1.9975)^2(10)$$
$$= 39.9 \text{ W}$$

In (b) the current is

$$\mathbf{I} = \frac{40 \underline{|0°}}{10 + j10 + 9 - j10} = 2.105 \underline{|0°} \text{ A}$$

and the power is

$$P = (2.105)^2(9) = 39.88 \text{ W}$$

In both cases P is less than the maximum value of Example 19.10.

PRACTICE EXERCISES

19-5.1 Let $\mathbf{V}_g = 24 \underline{|0°}$ V and $\mathbf{Z}_g = 4 + j3$ Ω in Fig. 19.12. Find $\mathbf{Z}$ for maximum power transfer and find also the maximum power delivered to $\mathbf{Z}$.   *Ans.* $4 - j3$ Ω, 36 W

19-5.2 Let $\mathbf{V}_g = 24 \underline{|0°}$ V and $\mathbf{Z}_g = 4 + j3$ Ω in Fig. 19.12, and find the power delivered to $\mathbf{Z}$ for (a) $\mathbf{Z} = 4 - j4$ Ω, (b) $\mathbf{Z} = 5 - j3$ Ω, and (c) $\mathbf{Z} = 5 - j2$ Ω.
Ans. (a) 35.45 W, (b) 35.56 W, (c) 35.12 W

19.6 POWER MEASUREMENT

A wattmeter, as discussed in Chapter 9, measures power. In the case of ac steady-state circuits it measures the average power P delivered to a load.

Wattmeter Connections: A wattmeter contains two coils—a rotating high-resistance *voltage,* or *potential,* coil and a fixed low-resistance *current* coil. It has four terminals, a pair for each coil, as shown in the sketch of Fig. 19.13. The voltage coil is connected across the load whose power is being measured and the current coil is connected in series with the load. The circuit symbol for a wattmeter is shown in Fig. 19.14 and a typical connection is shown in Fig. 19.15.

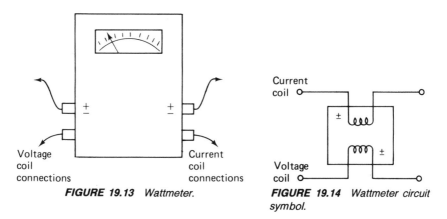

FIGURE 19.13 *Wattmeter.*

FIGURE 19.14 *Wattmeter circuit symbol.*

One terminal of each coil is marked $\pm$, as shown in Figs. 19.13, 19.14, and 19.15. The meter reads

$$P = VI \cos \theta \tag{19.51}$$

where I is the rms value of the current $\mathbf{I}$ entering the $\pm$ terminal of the current coil, V is the rms value of the voltage $\mathbf{V}$ with positive polarity at the $\pm$ terminal of the voltage coil, and θ is the angle between the phasors $\mathbf{V}$ and $\mathbf{I}$. In the case of the meter of Fig. 19.15, it is correctly connected to read the power given by (19.51), which is the power delivered to the load.

Ideally, the voltage across the current coil and the current through the voltage coil are both zero. Thus the presence of the wattmeter does not influence the power it is measuring.

If the terminal connections of either the current coil or voltage coil (but not both) are reversed, a negative, or downscale, reading is indicated. Most meters cannot read downscale—the pointer simply rests on the downscale stop. Such a reading requires reversing the connections of one of the coils, usually the voltage coil.

Other types of meters are available for measuring other types of power. For

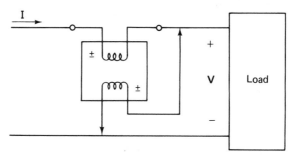

FIGURE 19.15 *Typical connection of a wattmeter.*

example, an *apparent power* or VA meter measures the product of the rms voltage and current, and a *varmeter* measures reactive power.

Example 19.12: Find the wattmeter reading in Fig. 19.15 if the current is $\mathbf{I} = 2\,\underline{|0°}$ A and the impedance of the load is $\mathbf{Z} = 10\,\underline{|60°}$ Ω.

Solution: The voltage is

$$\mathbf{V} = \mathbf{ZI} = (10\,\underline{|60°}\,)(2\,\underline{|0°}\,) = 20\,\underline{|60°}\ \text{V}$$

Therefore, by (19.51) the reading is

$$P = VI \cos\theta$$

$$= (20)(2)\cos 60°$$

$$= 20\ \text{W}$$

PRACTICE EXERCISES

19-6.1 Find the reading of wattmeter *A*.   *Ans.* 87.5 W

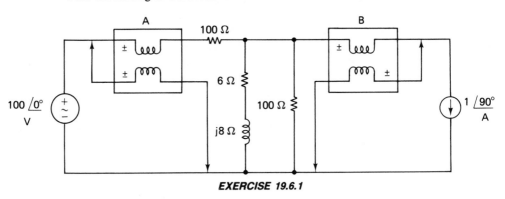

EXERCISE 19.6.1

19-6.2 Find the reading of wattmeter *B* in Exercise 19-6.1   *Ans.* 0

19.7 SUMMARY

In the case of steady-state ac circuits the power read by most wattmeters is average power, which is the average over a cycle of the instantaneous power. However, unlike instantaneous power, average power is not a function of time but depends on the rms values of the load current and voltage phasors and the angle θ between them. The angle θ is called the power-factor angle and cos θ is defined as the power factor of the load.

The power triangle is a right triangle whose side along the real axis is the average power P. Its hypotenuse makes the angle θ with the real axis and its length is the apparent power. The other leg of the triangle, in the imaginary-axis direction, is the reactive power Q. The complex power $\mathbf{S}$ is the complex number $P + jQ$.

A low power factor requires a larger current for a given power P and voltage than a high power factor. Thus it is desirable to correct the power factor (that is, make it higher). This may be done by connecting a reactive element in parallel with the load. If the power factor to be corrected is lagging (an inductive load), the reactive element should be a capacitor, and if the power factor is leading (a capacitive load), the reactive element should be an inductor.

A wattmeter contains a current coil, which is connected in series with the load whose power it measures, and a voltage coil that is connected across the load. With the coils properly connected, the wattmeter reads the average power delivered to the load.

PROBLEMS

19.1 Find the average power delivered to an impedance $\mathbf{Z} = 30 + j40 \ \Omega$ if the voltage across it is $\mathbf{V} = 200 \,\underline{|0°}$ V.

19.2 Find the average power delivered to a load with voltage and current phasors given by

$$\mathbf{V} = 100 \,\underline{|105°} \ \text{V}$$

and

$$\mathbf{I} = \sqrt{2}\,(1 + j1) \ \text{A}$$

19.3 Find the average power delivered by the source.

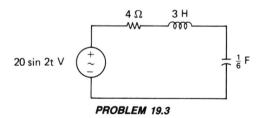

PROBLEM 19.3

19.4 The average power delivered to a load is 400 W. If the rms values of its voltage and current are 100 V and 8 A, find the impedance of the load if its power factor is lagging.

19.5 Find the average power delivered by the source.

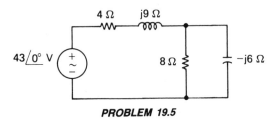

PROBLEM 19.5

19.6 Find the average power absorbed by the 4-Ω resistor and the 8-Ω resistor of Problem 19.5.

19.7 Find the average power delivered to a load whose voltage and current are

$$v = 40 \sin (6t - 20°) \text{ V}$$

and

$$i = 6 \sin (6t + 40°) \text{ A}$$

19.8 Find the average power delivered to a load consisting of a 2-kΩ resistor in series with a 0.5-H inductor, if the current is

$$i = 12 \sin (200t + 10°) \text{ mA}$$

19.9 Find the power factor of the load shown.

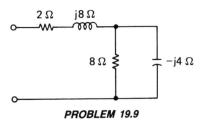

PROBLEM 19.9

19.10 Find the power delivered to the 2-Ω resistor of Problem 19.9 if a source of $\mathbf{V} = 12 \underline{|0°}$ V is connected to the terminals.

19.11 Find the power factor of the load in Problem 19.9 if an impedance of $-j4.8 \ \Omega$ is connected across its terminals.

19.12 Find the power factor of the load shown and the power delivered to it by a source $\mathbf{V} = 42 \underline{|0°}$ V connected across its terminals.

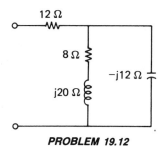

PROBLEM 19.12

19.13 A load has an impedance of $\mathbf{Z} = 6 \underline{|60°}$ Ω and a voltage $\mathbf{V} = 24 \underline{|0°}$ V. Find its average power, its reactive power, and its apparent power.

19.14 A load has a voltage and current given by

$$\mathbf{V} = 50 \underline{|0°} \text{ V}$$

and

$$\mathbf{I} = 10 \underline{|-30°} \text{ A}$$

Find the complex power, the apparent power, the average power, the reactive power, and the power factor.

19.15 A load absorbs an average power of 1200 W. Find the power factor if the reactive power is (a) $Q = 0$, (b) $Q = 900$ vars, (c) $Q = -600$ vars, and (d) $Q = 1600$ vars.

19.16 Find the complex power delivered to a load with a 0.8 lagging power factor and which absorbs an average power of 800 W.

19.17 The complex power of a load is

$$\mathbf{S} = 12 + j16 \text{ VA}$$

If the rms value of its current is 2 A, find the impedance of the load.

19.18 A load has a power factor of 0.8 lagging and an impedance with magnitude

$$|\mathbf{Z}| = 10 \text{ Ω}$$

Find the capacitance needed to connect across the load terminals to correct the power factor to 0.9 lagging. The frequency is 10,000 rad/s.

19.19 A load with a power factor of 0.9 lagging has an rms voltage and current of 60 V and 4 A. Find the parallel capacitance required to connect across the terminals of the load to correct the power factor to 0.95 lagging. The frequency is 60 Hz.

19.20 The practical ac source of Fig. 19.11 has

$$\mathbf{V}_g = 80 \underline{|0°} \text{ V}$$

and

$$\mathbf{Z}_g = 10 + j5 \ \Omega$$

Find the maximum power that can be drawn from the source by a load.

19.21 Let $\mathbf{V}_g = 60 \ \underline{|0°}$ V and $\mathbf{Z}_g = 10 + j6 \ \Omega$ in Fig. 19.12. Find **Z** for maximum power transfer and find also the maximum power delivered to **Z**.

19.22 Find the wattmeter reading.

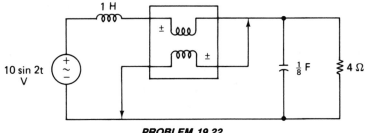

PROBLEM 19.22

19.23 Find the wattmeter reading.

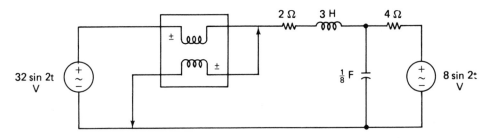

PROBLEM 19.23

20

THREE-PHASE CIRCUITS

One of the most important uses of ac steady-state circuit analysis is its application to power systems, most of which are ac systems. The principal reason for this is that it is much cheaper to transmit power over long distances if the voltages are very high, and it is easier to raise and lower voltages in ac systems than it is in dc systems. As we will see in Chapter 21, alternating voltage may be stepped up for transmission across the country and stepped down for use in homes and industries by *transformers*, which have no moving parts and are relatively easy to construct. On the other hand, with the present technology, rotating machines are generally needed to raise and lower dc voltages.

Also, for reasons of economics and performance, almost all electric power is produced by *polyphase* sources, which are generators that provide voltages with more than one phase. In a single-phase circuit, the instantaneous power delivered to a load is pulsating, even if the current and voltage are in phase. A polyphase system, on the other hand, is somewhat like an automobile engine with many cylinders in that the power delivered is steadier. The result is that there is less vibration in the rotating machinery, which in turn, performs more efficiently. Also, there is an economic advantage in the transmission lines required for a polyphase system. The weight of the conductors and the associated components is much less than that required in a single-phase system that delivers the same power. For these reasons, almost all the power produced in the world is polyphase power. The frequency is usually 50 or 60 Hz, but in the United States 60 Hz is standard.

In this chapter we will consider three-phase systems, which are by far the most common of the polyphase systems. The sources are three-phase ac generators that produce a *balanced* set of voltages, by which we mean three sinusoidal voltages having the same amplitude and frequency but displaced in phase by 120°. Thus the three-phase source is equivalent to three interconnected single-phase sources, each generating a voltage with a different phase. If the three currents drawn from the source also constitute a balanced set, the system is a *balanced* three-phase system. This is the principal case we will consider.

20.1 THREE-PHASE GENERATOR

The three-phase generator has three coils placed 120° apart on the rotor, or armature, as shown in the simplified cross-sectional view of Fig. 20.1. (In an actual machine the voltages are generated in stationary coils and the field coils are on the rotor. The principle, however, is the same. Also an actual machine may have more than one pair of north–south poles, so that "360 electrical degrees" are traversed many times in one rotation of the rotor. The angle of rotation from one north pole through the next south pole to the next north pole is 360 electrical degrees.) Each coil has the same number of turns and rotates at the same speed, so that each generates an identically shaped sinusoid, but at different times, or phases, corresponding to its different position on the rotor.

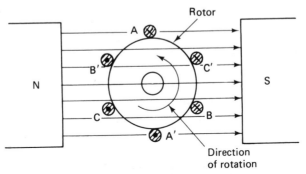

FIGURE 20.1 *Three-phase generator.*

Double-Subscript Notation: Before analyzing the action of the generator of Fig. 20.1, we will reconsider the double-subscript notation, considered earlier, for identifying polarities of voltages and currents. The notation v_{ab} is the voltage of point a with respect to point b, as indicated in Fig. 20.2(a). That is, if v_{ab} is positive, point a is at a potential of v_{ab} higher than point b. Similarly, the current i_{ab}, shown also in Fig. 20.2(a), is the current from a toward b.

The use of double subscripts also makes it easier to add and subtract phasors. This is illustrated in Fig. 20.2(b), where we see that

$$\mathbf{V}_{ab} = \mathbf{V}_{an} + \mathbf{V}_{nb} \qquad (20.1)$$

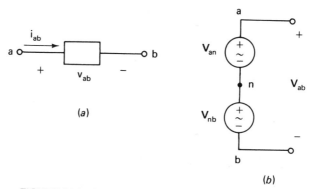

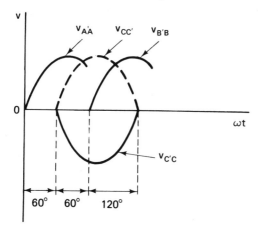

FIGURE 20.2 *Illustrations of the double-subscript notation.*

This equation follows without looking at a circuit, since by Kirchhoff's voltage law the voltage between two points *a* and *b* is the same regardless of the path between them. The path in this case is *a* to *n* and then *n* to *b*. We also note that

$$\mathbf{V}_{nb} = -\mathbf{V}_{bn}$$

so that (20.1) is equivalent to

$$\mathbf{V}_{ab} = \mathbf{V}_{an} - \mathbf{V}_{bn}$$

Three-Phase Voltages: Let us now return to the three-phase generator of Fig. 20.1. Coil *AA'* at the instant shown (which we take as $\omega t = 0$) is linking the maximum flux at a rate which has just reached zero. Thus its voltage is going through zero. If its current is into *A* and out of *A'* as shown by ×, the tail of the arrow, and •, the head of the arrow, the voltage $v_{A'A}$ will increase from zero toward its peak as the rotor turns, as shown in Fig. 20.3. Coil *CC'* will shortly occupy the position of

FIGURE 20.3 *Sequence of the gener-ated voltages.*

AA' at $\omega t = 0$. This will occur at $\omega t = 60°$, but C', rather than C, will be in the position of A, and C will be in the position of A'. Therefore, the voltage $v_{CC'}$ will rise from zero toward its positive peak at $\omega t = 60°$, as is indicated by the dashed curve of Fig. 20.3. Thus the voltage $v_{C'C}$, the negative of $v_{CC'}$, is as indicated by the solid line. Finally, at $\omega t = 120°$, point B is where A was at $\omega t = 0$ and B' is where A' was. Therefore, $v_{B'B}$ will be like $v_{A'A}$, but shifted to the right by 120°, again, as shown in Fig. 20.3.

The completed voltages, $v_{A'A}$, $v_{B'B}$, and $v_{C'C}$, are shown from $\omega t = 0$ through $\omega t = 3\pi$ in Fig. 20.4. If these are the voltages whose terminals are brought out for use, we see that the result is a three-phase generator producing three identical sinusoids 120° apart in phase. That is, we have

$$v_{A'A} = V_m \sin \omega t$$

$$v_{B'B} = V_m \sin (\omega t - 120°) \tag{20.2}$$

$$v_{C'C} = V_m \sin (\omega t - 240°)$$

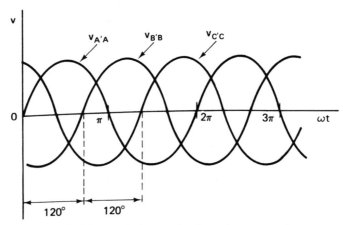

FIGURE 20.4   *Voltages of a three-phase generator.*

Phasor Representation: The phasors representing the sinusoidal voltages of (20.2) are given by

$$\mathbf{V}_{A'A} = V_p \underline{|0°} \text{ V}$$

$$\mathbf{V}_{B'B} = V_p \underline{|-120°} \text{ V} \tag{20.3}$$

$$\mathbf{V}_{C'C} = V_p \underline{|-240°} \text{ V} = V_p \underline{|120°} \text{ V}$$

where $V_p = V_m/\sqrt{2}$ is the rms value of the voltage of each "phase." As discussed earlier, this set of voltages is said to be a *balanced* set, since they are identical except that the phases are 120° apart.

Writing the phasor voltages in rectangular form, we have

$$\mathbf{V}_{A'A} = V_p$$

$$\mathbf{V}_{B'B} = V_p \left[\cos\left(-120°\right) + j\sin\left(-120°\right)\right]$$

$$= V_p \left(-\frac{1}{2} - j\frac{\sqrt{3}}{2}\right)$$

$$\mathbf{V}_{C'C} = V_p \left[\cos\left(-240°\right) + j\sin\left(-240°\right)\right]$$

$$= V_p \left(-\frac{1}{2} + j\frac{\sqrt{3}}{2}\right)$$

Therefore, their sum is

$$\mathbf{V}_{A'A} + \mathbf{V}_{B'B} + \mathbf{V}_{C'C} = V_p \left(1 - \frac{1}{2} - j\frac{\sqrt{3}}{2} - \frac{1}{2} + j\frac{\sqrt{3}}{2}\right) \tag{20.4}$$

$$= 0$$

This is true in general of balanced three-phase voltages.

PRACTICE EXERCISES

20-1.1 Given the voltages

$$\mathbf{V}_{an} = 10 + j6 \text{ V}$$

$$\mathbf{V}_{bn} = 20 - j8 \text{ V}$$

use the double-subscript notation to find (a) $\mathbf{V}_{ab}$ and (b) $\mathbf{V}_{na} + \mathbf{V}_{nb}$.
Ans. (a) $-10 + j14$ V, (b) $-30 + j2$ V

20-1.2 If the voltage across terminals a and b of an element is $\mathbf{V}_{ab} = 20 \,\underline{|0°}$ V, find the currents $\mathbf{I}_{ab}$ and $\mathbf{I}_{ba}$ if the element is (a) a 10-Ω resistor and (b) a $j5$-Ω inductor.
Ans. (a) 2 A, -2 A, (b) $-j4$ A, $j4$ A

20-1.3 If $V_p = 100$ V in (20.3), find (a) $\mathbf{V}_{A'A} + \mathbf{V}_{B'B}$ and (b) $\mathbf{V}_{A'A} + \mathbf{V}_{BB'}$.
Ans. (a) $100 \,\underline{|-60°}$ V, (b) $100 \sqrt{3} \,\underline{|30°}$ V

20.2 Y-CONNECTED GENERATOR

If the coils AA', BB', and CC' of the generator of Fig. 20.1 are connected as shown in Fig. 20.5, the result is a source with *line* terminals A', B', and C' and a common point (A, B, and C tied together) called the *neutral* terminal. In this case, the source is said to be *Y-connected* (connected in a Y, as shown).

The resulting generator is equivalent to three single-phase generators, as shown in Fig. 20.6(a), which we will consider as the representation of a Y-connected source. For convenience we have replaced the letters A', B', and C' by a, b, and c, and

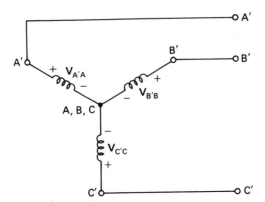

FIGURE 20.5 *Y-connected generator.*

labeled the neutral point *n*. Using this notation the so-called *phase voltages*, by (20.3), are

$$\mathbf{V}_{an} = V_p \underline{|0°} \text{ V}$$

$$\mathbf{V}_{bn} = V_p \underline{|-120°} \text{ V} \qquad (20.5)$$

$$\mathbf{V}_{cn} = V_p \underline{|120°} \text{ V}$$

An equivalent representation of a Y-connected source that is somewhat easier to draw is that of Fig. 20.6(b). Each representation has the generators connected to a common point with accessible line terminals *a*, *b*, and *c*, and a neutral terminal *n*. Such a source is called a *Y-connected three-phase, four-wire generator.*

Positive and Negative Phase Sequences: The sequence of voltages in (20.5) is called the *positive sequence*, or *abc sequence*. The phasor diagram is shown in

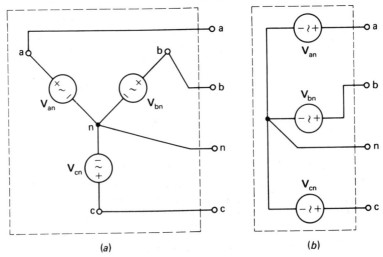

(a)   (b)

FIGURE 20.6 *Two representations of a Y-connected source.*

Fig. 20.7(a), where we may see that if the phasors rotate counterclockwise, the voltages labeled *an*, *bn*, and *cn* cross the real axis in that order.

The set of voltages in the phasor diagram of Fig. 20.7(b) is

$$\mathbf{V}_{an} = V_p \lfloor 0° \text{ V}$$
$$\mathbf{V}_{bn} = V_p \lfloor 120° \text{ V} \qquad (20.6)$$
$$\mathbf{V}_{cn} = V_p \lfloor -120° \text{ V}$$

In Fig. 20.7(b), counterclockwise rotation of the voltages yields the subscript order *an*, *cn*, and *bn*. The rotation must be in the negative direction (clockwise) to have the sequence *an*, *bn*, and *cn*. Thus this sequence is called the *negative sequence*, or the *acb sequence*.

The two sequences are alike except for the way the coils are connected. Therefore, we will generally use the positive sequence.

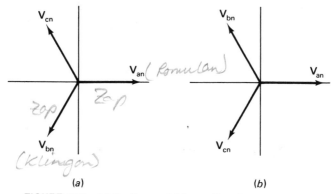

(a)   (b)

FIGURE 20.7   (a) Positive and (b) negative phase sequence.

Line Voltages: The *line-to-line* voltages, or simply *line* voltages, in Fig. 20.6 are $\mathbf{V}_{ab}$, $\mathbf{V}_{bc}$, $\mathbf{V}_{ca}$, which may be found from the phase voltages. For example, using the double-subscript notation, we have

$$\mathbf{V}_{ab} = \mathbf{V}_{an} + \mathbf{V}_{nb}$$
$$= \mathbf{V}_{an} - \mathbf{V}_{bn}$$
$$= V_p \lfloor 0° - V_p \lfloor -120°$$
$$= V_p - V_p \left[\cos(-120°) + j \sin(-120°) \right]$$
$$= V_p - V_p \left(-\frac{1}{2} - j\frac{\sqrt{3}}{2} \right)$$
$$= V_p \left(1 + \frac{1}{2} + j\frac{\sqrt{3}}{2} \right)$$

or finally,

$$\mathbf{V}_{ab} = V_p \left(\frac{3}{2} + j\frac{\sqrt{3}}{2} \right)$$

In polar form this is

$$\mathbf{V}_{ab} = V_p \sqrt{\left(\frac{3}{2}\right)^2 + \left(\frac{\sqrt{3}}{2}\right)^2} \left| \tan^{-1}\frac{\sqrt{3}/2}{3/2} \right.$$

or

$$\mathbf{V}_{ab} = \sqrt{3}\, V_p \underline{|30°} \tag{20.7}$$

We may also find $\mathbf{V}_{ab}$ graphically by completing the parallelogram with sides $\mathbf{V}_{an}$ and $-\mathbf{V}_{bn}$ ($\mathbf{V}_{nb}$ reversed), as shown in Fig. 20.8. The parallelogram contains two identical triangles and since by the construction shown, the angle of the parallelogram at 0 is 60°, that of each triangle is 30°. Thus the angle of $\mathbf{V}_{ab}$ is 30°. Also, half the length of $\mathbf{V}_{ab}$ (from 0 to d) is a leg of a right triangle with an angle of 30°. Therefore, we have

$$\frac{1}{2}|\mathbf{V}_{ab}| = |\mathbf{V}_{an}| \cos 30°$$

$$= V_p \frac{\sqrt{3}}{2}$$

Thus the magnitude of $\mathbf{V}_{ab}$ is $\sqrt{3}\, V_p$, so that (20.7) holds.

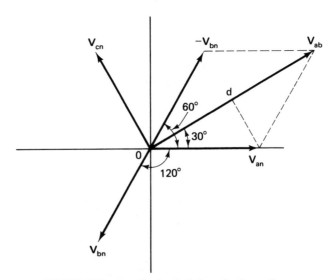

FIGURE 20.8 *Graphical calculation of a line voltage.*

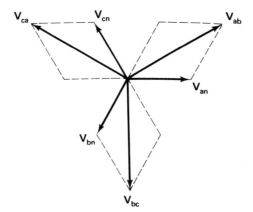

FIGURE 20.9 *Phasor diagram of phase and line voltages.*

The other two line voltages $\mathbf{V}_{bc}$ and $\mathbf{V}_{ca}$ may be found algebraically as in (20.7), but they are easier to see graphically. The construction is done as in Fig. 20.8, with the results shown in Fig. 20.9. From this figure and the result for $\mathbf{V}_{ab}$ already obtained, we may write

$$\mathbf{V}_{ab} = \sqrt{3}\ V_p \underline{|\,30°}$$
$$\mathbf{V}_{bc} = \sqrt{3}\ V_p \underline{|-90°} \qquad\qquad (20.8)$$
$$\mathbf{V}_{ca} = \sqrt{3}\ V_p \underline{|\,150°}$$

The line voltages, therefore, are also a balanced set. They have the same amplitude and their phases are spaced 120° apart. The line voltages lead the phase voltages by 30°. That is, $\mathbf{V}_{ab}$ leads $\mathbf{V}_{an}$ by 30°, $\mathbf{V}_{bc}$ leads $\mathbf{V}_{bn}$ by 30°, and $\mathbf{V}_{ca}$ leads $\mathbf{V}_{cn}$ by 30°. Therefore, if all the phase voltages are rotated by some angle, we may still obtain the line voltages by adding 30° to the angles of the phase voltages and multiplying their magnitudes by $\sqrt{3}$.

If we denote the magnitude of the line voltages by V_L, then we have

$$V_L = \sqrt{3}\ V_p \qquad\qquad (20.9)$$

and the line voltage phasors are

$$\mathbf{V}_{ab} = V_L \underline{|\,30°}$$
$$\mathbf{V}_{bc} = V_L \underline{|-90°} \qquad\qquad (20.10)$$
$$\mathbf{V}_{ca} = V_L \underline{|\,150°}$$

Example 20.1: Find the line voltages if the phase voltages are given by

$$\mathbf{V}_{an} = 100 \underline{|\,10°}\ \text{V}$$
$$\mathbf{V}_{bn} = 100 \underline{|-110°}\ \text{V}$$
$$\mathbf{V}_{cn} = 100 \underline{|\,130°}\ \text{V}$$

Solution: We add 30° to the phase voltage angles and multiply their magnitudes by $\sqrt{3}$ to obtain

$$\mathbf{V}_{ab} = 100 \sqrt{3} \,\underline{|40°}\ \text{V}$$

$$\mathbf{V}_{bc} = 100 \sqrt{3} \,\underline{|-80°}\ \text{V}$$

$$\mathbf{V}_{ca} = 100 \sqrt{3} \,\underline{|160°}\ \text{V}$$

PRACTICE EXERCISES

20-2.1 Find the phase voltages $\mathbf{V}_{bn}$ and $\mathbf{V}_{cn}$ in a balanced, positive-sequence Y-connected source if

$$\mathbf{V}_{an} = 200 \,\underline{|0°}\ \text{V}$$

Ans. $200 \,\underline{|-120°}$ V, $200 \,\underline{|120°}$ V

20-2.2 Find the line voltages $\mathbf{V}_{ab}$, V_{bc}, and $\mathbf{V}_{ca}$ of the source of Exercise 20.2.1.
Ans. $200 \sqrt{3} \,\underline{|30°}$ V, $200 \sqrt{3} \,\underline{|-90°}$ V, $200 \sqrt{3} \,\underline{|150°}$ V

20-2.3 Find the phase voltages $\mathbf{V}_{an}$, V_{bn}, and $\mathbf{V}_{cn}$ if $\mathbf{V}_{ab} = 50 \sqrt{3} \,\underline{|70°}$ V in a positive sequence Y-connected source.   *Ans.* $50 \,\underline{|40°}$ V, $50 \,\underline{|-80°}$ V, $50 \,\underline{|160°}$ V

20.3 Y–Y SYSTEMS

Three-phase loads may also be in the form of a Y and connected to the three-phase Y-connected source, as shown in Fig. 20.10. Such an arrangement is called a *Y–Y three-phase system*. The three load impedances are called *phase impedances* and may be equal, as in Fig. 20.10, or they may be unequal.

Balanced System: If the phase impedances are equal, say to $\mathbf{Z}_p$, as in Fig. 20.10, and the source voltages form a balanced set, then the system is called a *balanced* system. The circuit of Fig. 20.10 is thus a balanced Y–Y, three-phase, four-wire system. The fourth wire is the neutral line *n-N*, which may be omitted to form a three-wire system.

Line Currents: Because of the presence of the neutral line in Fig. 20.10, we see that the voltages of the sources are also across the loads. This is more easily seen in Fig. 20.11, which shows only *phase a* (the phase containing $\mathbf{V}_{an}$ and its load between A and N). The *line current* from a to A in line aA is denoted by $\mathbf{I}_{aA}$, and by Fig. 20.11 is

$$\mathbf{I}_{aA} = \frac{\mathbf{V}_{an}}{\mathbf{Z}_p} \qquad\qquad (20.11)$$

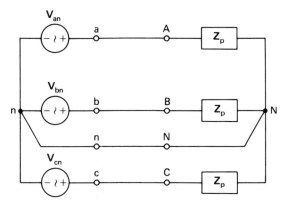

FIGURE 20.10 *Balanced Y-Y system.*

The other line currents, obtained in the same way, are

$$\mathbf{I}_{bB} = \frac{\mathbf{V}_{bn}}{\mathbf{Z}_p}$$

$$\mathbf{I}_{cC} = \frac{\mathbf{V}_{cn}}{\mathbf{Z}_p}$$

(20.12)

If the phase impedances are given by

$$\mathbf{Z}_p = |\mathbf{Z}_p| \underline{\,\theta\,}$$

(20.13)

and we replace the phase voltages by their values in (20.5), then (20.11) and (20.12) become

$$\mathbf{I}_{aA} = \frac{V_p}{|\mathbf{Z}_p|} \underline{\,-\theta\,}$$

$$\mathbf{I}_{bB} = \frac{V_p}{|\mathbf{Z}_p|} \underline{\,-120° - \theta\,}$$

(20.14)

$$\mathbf{I}_{cC} = \frac{V_p}{|\mathbf{Z}_p|} \underline{\,120° - \theta\,}$$

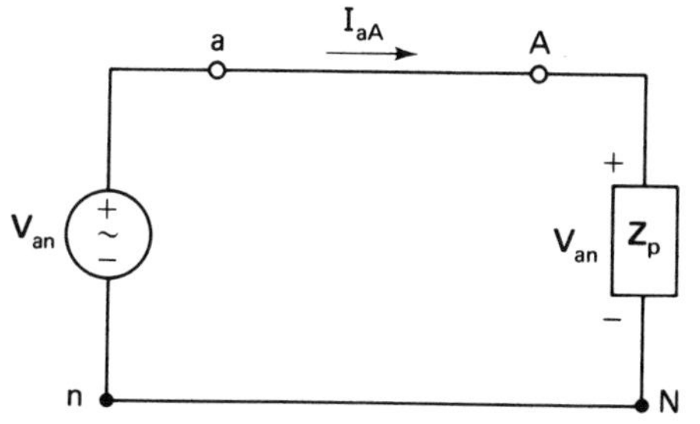

FIGURE 20.11 *Phase a of Fig. 20.10.*

Finally, if we take the line current magnitude to be I_L, then

$$I_L = \frac{V_p}{|\mathbf{Z}_p|} \tag{20.15}$$

and we have

$$\mathbf{I}_{aA} = I_L \underline{|-\theta}$$
$$\mathbf{I}_{bB} = I_L \underline{|-120° - \theta} \tag{20.16}$$
$$\mathbf{I}_{cC} = I_L \underline{|120° - \theta}$$

The line currents thus have equal magnitudes and phase angles that are spaced 120° apart. Therefore, they also form a balanced set.

Neutral Line Current: As we saw in (20.4), the sum of the elements of a balanced set is zero. Therefore, since the line currents form a balanced set, their sum is zero. That is, we have

$$\mathbf{I}_{aA} + \mathbf{I}_{bB} + \mathbf{I}_{cC} = 0 \tag{20.17}$$

Referring back to Fig. 20.10, we see that Kirchhoff's current law at node n yields

$$\mathbf{I}_{aA} + \mathbf{I}_{bB} + \mathbf{I}_{cC} + \mathbf{I}_{nN} = 0$$

Therefore, the neutral line current $\mathbf{I}_{nN}$ is given by

$$\mathbf{I}_{nN} = -(\mathbf{I}_{aA} + \mathbf{I}_{bB} + \mathbf{I}_{cC})$$

However, by (20.17) the right member of the equation is zero, so that we have

$$\mathbf{I}_{nN} = 0$$

Therefore, in a balanced, four-wire Y–Y system the neutral line carries no current. The neutral line thus could contain resistance, be a short circuit, or be removed entirely, without changing anything else in the circuit.

Phase Currents: From Fig. 20.10 we note that the currents in the lines *aA, bB,* and *cC* are also the *phase currents* (the currents carried by the phase impedances). If the magnitude of the phase currents is I_p, then for the Y-connected load we have

$$I_L = I_p \tag{20.18}$$

Example 20.2: Find the line currents in the Y–Y system of Fig. 20.12.

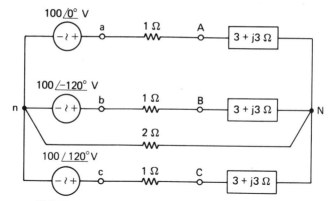

FIGURE 20.12 *Balanced system with line impedances.*

Solution: This system has more realistic lines, with line impedances, which in this case are 1-Ω resistors. However, since the line impedance is in series with the phase impedance $3 + j3$ Ω, as is always the case in a Y-connected load, we may combine the two into an equivalent impedance of

$$\mathbf{Z}_p = 1 + (3 + j3) = 4 + j3$$
$$= 5\underline{|36.9°}\ \Omega \tag{20.19}$$

which is the effective load impedance with ideal conducting lines as before.

Since in this equivalent case the lines have no impedance, we have the same balanced system as before, except that the neutral line has resistance. However, the neutral current is zero because the system is balanced. Therefore, the neutral may be replaced by a perfect conductor, so that the system is equivalent to that of Fig. 20.10, with $\mathbf{Z}_p$ given by (20.19) and the source voltages as in Fig. 20.12.

By (20.14), therefore, the line currents are

$$\mathbf{I}_{aA} = \frac{\mathbf{V}_{an}}{\mathbf{Z}_p} = \frac{100\underline{|0°}}{5\underline{|36.9°}} = 20\underline{|-36.9°}\ \text{A}$$

$$\mathbf{I}_{bB} = \frac{\mathbf{V}_{bn}}{\mathbf{Z}_p} = \frac{100\underline{|-120°}}{5\underline{|36.9°}} = 20\underline{|-156.9°}\ \text{A}$$

$$\mathbf{I}_{cC} = \frac{\mathbf{V}_{cn}}{\mathbf{Z}_p} = \frac{100\underline{|120°}}{5\underline{|36.9°}} = 20\underline{|83.1°}\ \text{A}$$

which is a balanced set.

Per-Phase Basis: This example, like Example 20.1, was solved on a "per-phase" basis. Since the neutral current is zero in a balanced Y–Y system, the impedance of the neutral has no effect. Therefore, we may replace the neutral by a short circuit. We may do this even if the neutral is not present (a three-wire system). We may

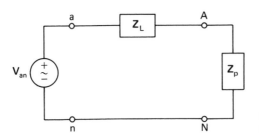

FIGURE 20.13 *Single phase for a per-phase analysis.*

then look at only one phase, say phase *a*, as in Fig. 20.13, consisting of the source V_{an} in series with the line impedance denoted by Z_L, and the phase impedance Z_p. The line current I_{aA}, the phase voltage $I_{aA}Z_p$, and the voltage drop $I_{aA}Z_L$ in the line may all be found from this single-phase analysis. The other voltages and currents may be found in a similar way, or from the previous results, since the system is balanced.

Three-Phase Power: The average power delivered to each phase of the Y-connected load of Fig. 20.10 is denoted by P_p, and is given by

$$P_p = V_p I_p \cos \theta$$
$$= I_p^2 \operatorname{Re} Z_p \tag{20.20}$$

where θ is the angle of Z_p. The total power delivered to the load (to all three phases) thus is

$$P = 3P_p = 3 V_p I_p \cos \theta \tag{20.21}$$

The angle θ of the phase impedance is therefore the power-factor angle of the total three-phase load, as well as that of a single phase.

Example 20.3: A balanced Y–Y system has line voltage $V_L = 200$ V (rms value) and a three-phase power $P = 600$ W at a power factor of 0.9 lagging. Find the line current I_L and the phase impedance Z_p.

Solution: From (20.21) the phase power is

$$P_p = \frac{P}{3} = \frac{600}{3} = 200 \text{ W}$$

The phase voltage V_p, from (20.9), is

$$V_p = \frac{V_L}{\sqrt{3}} = \frac{200}{\sqrt{3}} \text{ V}$$

Therefore, by (20.20), the phase current is

$$I_p = \frac{P_p}{V_p \cos\theta} = \frac{200}{(200/\sqrt{3})(0.9)} = 1.925 \text{ A}$$

Since in a Y–Y system the line current is the phase current, we have

$$I_L = 1.925 \text{ A}$$

The magnitude of $\mathbf{Z}_p$ is given by

$$|\mathbf{Z}_p| = \frac{V_p}{I_p} = \frac{200/\sqrt{3}}{1.925} = 60 \ \Omega$$

and its angle is the power-factor angle given by

$$\theta = \cos^{-1} 0.9 = 25.84°$$

Therefore, the phase impedance is

$$\mathbf{Z}_p = 60 \lfloor 25.84° \ \Omega$$

PRACTICE EXERCISES

20-3.1 A balanced Y–Y system has $\mathbf{V}_{an} = 100 \lfloor 0°$ V, phase impedance $\mathbf{Z}_p = 7 + j6 \ \Omega$, and line impedance $\mathbf{Z}_L = 1 \lfloor 0° \ \Omega$. Find the line current I_L and the total power delivered by the source.   *Ans.* 10 A, 2.4 kW

20-3.2 Find the power absorbed by the three-phase load in Exercise 20-3.1   *Ans.* 2.1 kW

20-3.3 In Fig. 20.10, the line currents form a balanced set with $\mathbf{I}_{aA} = 5 \lfloor 0°$ A. If $\mathbf{Z}_p = 2 \lfloor 30° \ \Omega$, find the line voltages $\mathbf{V}_{ab}$, $\mathbf{V}_{bc}$, $\mathbf{V}_{ca}$, and the power delivered to the three-phase load. (The phase sequence is *abc*.)
Ans. $10 \sqrt{3} \lfloor 60°$ V, $10 \sqrt{3} \lfloor -60°$ V, $10 \sqrt{3} \lfloor 180°$ V, 129.9 W

20.4 DELTA-CONNECTED LOAD

Another method of connecting a three-phase load to a line is the *delta*, or Δ, connection. A *balanced* Δ-connected load is one for which the three phase impedances are equal. An example, with phase impedances of $\mathbf{Z}_p$, is drawn in Fig. 20.14(a) in a way that

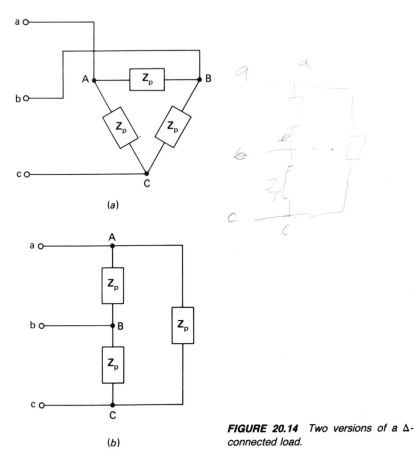

(a)

(b)

FIGURE 20.14 *Two versions of a Δ-connected load.*

resembles a Δ. An equivalent sketch, which is somewhat easier to draw, is shown in Fig. 20.14(b).

Advantages and Disadvantages: An advantage of a Δ-connected load is that elements may be added or removed more readily on a single phase of a Δ, since the loads are connected directly across the lines. This is not true of a Y-connected load, which has a neutral point that may or may not be accessible. Also, as we will see, for a given power delivered to the load the phase currents in a Δ are smaller than those in a Y. On the other hand, the Δ phase voltages are higher than those of the Y connection, which is a disadvantage.

Generators are rarely Δ-connected, because if the voltages are not perfectly balanced, there will be a net voltage, and thus a circulating current, around the Δ. This, of course, causes undesirable heating effects in the generating machinery. Also, the phase voltages are lower in the Y-connected generator, as in the Y-connected load, and thus less insulation is required.

The Δ connection has no neutral point, so that Δ-connected loads are always a part of a three-wire system.

Phase Voltages and Currents: From Fig. 20.14 we see that for a Δ-connected load the line voltages are the same as the phase load voltages. Therefore, we have

$$V_p = V_L \tag{20.22}$$

where V_p and V_L are the load phase voltage and line-voltage magnitudes.

In the Y–Δ system (Y-connected source and Δ-connected load) of Fig. 20.15, if the generator phase voltages are given by

$$\mathbf{V}_{an} = V_g \underline{|0°}$$
$$\mathbf{V}_{bn} = V_g \underline{|-120°} \tag{20.23}$$
$$\mathbf{V}_{cn} = V_g \underline{|120°}$$

then the line voltages, and thus the phase load voltages, by (20.8) are

$$\mathbf{V}_{ab} = \sqrt{3}\, V_g \underline{|30°} = V_L \underline{|30°}$$
$$\mathbf{V}_{bc} = \sqrt{3}\, V_g \underline{|-90°} = V_L \underline{|-90°} \tag{20.24}$$
$$\mathbf{V}_{ca} = \sqrt{3}\, V_g \underline{|150°} = V_L \underline{|150°}$$

where

$$V_L = V_p = \sqrt{3}\, V_g \tag{20.25}$$

If $\mathbf{Z}_p = |\mathbf{Z}_p| \underline{|\theta}$ in Fig. 20.14, the phase currents are

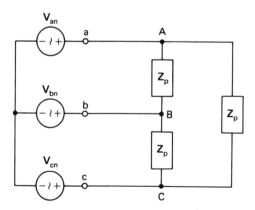

FIGURE 20.15 Y-Δ system.

$$\mathbf{I}_{AB} = \frac{\mathbf{V}_{ab}}{\mathbf{Z}_p} = I_p \underline{|30° - \theta}$$

$$\mathbf{I}_{BC} = \frac{\mathbf{V}_{bc}}{\mathbf{Z}_p} = I_p \underline{|-90° - \theta} \qquad (20.26)$$

$$\mathbf{I}_{CA} = \frac{\mathbf{V}_{ca}}{\mathbf{Z}_p} = I_p \underline{|150° - \theta}$$

where I_p is the rms value of the phase currents given by

$$I_p = \frac{V_L}{|\mathbf{Z}_p|} = \frac{V_p}{|\mathbf{Z}_p|} \qquad (20.27)$$

We see from (20.26) that the phase currents have the same magnitudes and are 120° apart in phase. Therefore, they form a balanced set.

Line Currents: The current in line aA of Fig. 20.14 is given by

$$\mathbf{I}_{aA} = \mathbf{I}_{AB} + \mathbf{I}_{AC} = \mathbf{I}_{AB} - \mathbf{I}_{CA}$$

The subtraction may be done graphically, as in the case of the line voltages for the Y connection in Figs. 20.8 and 20.9. The work is shown in Fig. 20.16, where we see that $\mathbf{I}_{aA}$ lags $\mathbf{I}_{AB}$ by 30° and its magnitude is $\sqrt{3}$ times that of $\mathbf{I}_{aA}$. Applying the same procedure to $\mathbf{I}_{bB}$ and $\mathbf{I}_{cC}$ (also shown in Fig. 20.16) yields the line currents

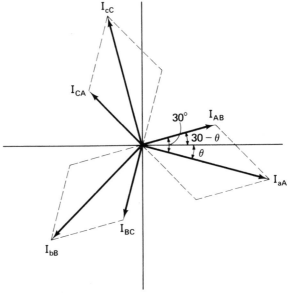

FIGURE 20.16   *Phase and line currents for a Δ load.*

$$\mathbf{I}_{aA} = I_L \underline{|-\theta}$$

$$\mathbf{I}_{bB} = I_L \underline{|-120° - \theta} \qquad (20.28)$$

$$\mathbf{I}_{cC} = I_L \underline{|120° - \theta}$$

where I_L is the line-current magnitude given by

$$I_L = \sqrt{3} \, I_p$$

Thus the line currents also form a balanced set.

The phase angles in (20.24), (20.26), and (20.28) are based on those of the generator voltages of (20.23). If we add an amount to the phase of the generator voltages, we must add the same amount to the phase of all the other quantities.

Example 20.4: In the Y–Δ system of Fig. 20.15 the source voltages are $\mathbf{V}_{an} = 100 \underline{|0°}$ V, $\mathbf{V}_{bn} = 100 \underline{|-120°}$ V, and $\mathbf{V}_{cn} = 100 \underline{|120°}$ V, and the phase impedance is $\mathbf{Z}_p = 10 \underline{|60°}$ Ω. Find the magnitudes V_L, I_L, and I_p of the line voltages, the line currents, and the load currents, and the power delivered to the load.

Solution: The line voltage is given by

$$V_L = \sqrt{3} \, (100) = 173.2 \text{ V}$$

and the phase current is

$$I_p = \frac{V_p}{|\mathbf{Z}_p|} = \frac{V_L}{|\mathbf{Z}_p|} = \frac{173.2}{10} = 17.32 \text{ A}$$

From this we have the line current given by

$$I_L = \sqrt{3} \, I_p = \sqrt{3} \, (17.32) = 30 \text{ A}$$

Finally, the power P_p delivered to each phase of the load is

$$P_p = V_p I_p \cos \theta$$

Since $V_p = V_L = 173.2$ V and $\theta = 60°$ (the angle of $\mathbf{Z}_p$), we have

$$P_p = (173.2)(17.32) \cos 60° = 1500 \text{ W}$$

The power delivered to the load is therefore

$$P = 3P_p = 3(1500) = 4500 \text{ W} = 4.5 \text{ kW}$$

TABLE 20.1 *Three-phase voltage and current relationships*

Load	Relation of Line Voltage V_L and Phase Voltage V_p	Relation of line Current I_L and Phase Current I_p
Y	$V_L = \sqrt{3}\ V_p$	$I_L = I_p$
Δ	$V_L = V_p$	$I_L = \sqrt{3}\ I_p$

Summary: A summary of the characteristics of the line and phase voltages and currents is given in Table 20.1. As we have seen, the line and phase currents are the same in the Y-connected case, as are the line and phase voltages in the Δ-connected case. The line voltages in the Y-connected case and the line currents in the Δ-connected case, however, are $\sqrt{3}$ times the phase quantities.

Y–Δ Transformations: Some three-phase circuit problems may be easier to solve if the load is Y-connected, and some may be easier if the load is Δ-connected. In these cases we may want to use the Y–Δ transformations discussed in Chapter 18, and given in (18.5) and (18.7). For the special case of a balanced load, if the phase impedance is $\mathbf{Z}_p = \mathbf{Z}_y$ in the Y-connected case and $\mathbf{Z}_p = \mathbf{Z}_\Delta$ in the Δ-connected case, we have by (18.8) and (18.9),

$$\mathbf{Z}_\Delta = 3\mathbf{Z}_y \tag{20.29}$$

Example 20.5: Find the magnitude I_L of the line currents in the Y–Δ system of Fig. 20.17.

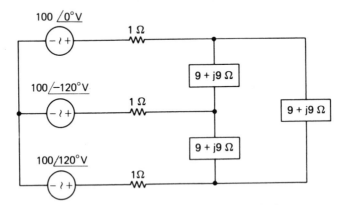

FIGURE 20.17 *Y-Δ system with losses in the lines.*

Solution: Since the lines have resistances of 1 Ω, we must account for a voltage drop in the lines. This can be done by finding the load voltages in terms of the source voltages and the line currents. An easier way, however, is to convert the Δ load to an equivalent Y and then combine the phase impedances of the Y with the impedances of the lines.

To see this, we note from (20.29) that the phase impedance $\mathbf{Z}_y$ of the equivalent Y is $\mathbf{Z}_\Delta/3$, where $\mathbf{Z}_\Delta = 9 + j9$ Ω is the phase impedance of the Δ. Therefore we have

$$\mathbf{Z}_y = \frac{9+j9}{3} = 3 + j3 \ \Omega$$

with the resulting equivalent Y–Y system of Fig. 20.18.

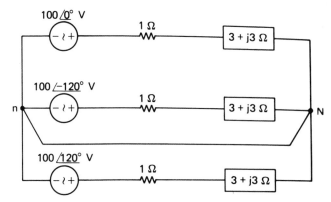

FIGURE 20.18 *Y-Y equivalent of Fig. 20.17.*

Combining the series impedances of 1 Ω and $3 + j3$ Ω, we have the equivalent phase impedance

$$\mathbf{Z}_p = 4 + j3$$
$$= 5 \underline{|36.9°} \ \Omega$$

We have also added a neutral line nN, which we may do (since it carries no current) without changing any of the currents and voltages. Therefore, on a "per-phase" basis, the line current is

$$I_L = \frac{100}{|\mathbf{Z}_p|} = \frac{100}{5} = 20 \ \text{A}$$

General Method of Finding Power: Whether the load is Y-connected or Δ-connected, we have the total power given by

$$P = 3P_p = 3 V_p I_p \cos \theta \qquad (20.30)$$

In the Y-connected case, $V_p = V_L/\sqrt{3}$ and $I_p = I_L$, and in the Δ-connected case we have $V_p = V_L$ and $I_p = I_L/\sqrt{3}$. In either case (20.30) becomes

$$P = 3 \frac{V_L I_L}{\sqrt{3}} \cos \theta$$

or

$$P = \sqrt{3}\ V_L I_L \cos\theta \tag{20.31}$$

Example 20.6: A Δ-connected load has $V_L = 250$ V and $\mathbf{Z}_p = 100\lfloor 36.9°\ \Omega$. Find, in two ways, the power delivered to the load.

Solution: The phase current is

$$I_p = \frac{V_L}{|\mathbf{Z}_p|} = \frac{250}{100} = 2.5\ \text{A}$$

and the phase voltage is

$$V_p = V_L = 250\ \text{V}$$

Since $\cos\theta = \cos 36.9° = 0.8$, we have by (20.30)

$$P = 3V_p I_p \cos\theta$$
$$= 3(250)(2.5)(0.8)$$
$$= 1500\ \text{W}$$

The line current is

$$I_L = \sqrt{3}\ I_p = 2.5\ \sqrt{3}\ \text{A}$$

Therefore, using (20.31) the power is

$$P = \sqrt{3}\ V_L I_L \cos\theta$$
$$= \sqrt{3}\ (250)(2.5\ \sqrt{3})(0.8)$$
$$= 1500\ \text{W}$$

PRACTICE EXERCISES

20-4.1 In Fig. 20.15 we have $\mathbf{V}_{an} = 200\lfloor 0°$ V and $\mathbf{Z}_p = 30 + j40\ \Omega$. Find the phase current I_p, the line current I_L, and the power delivered to the three-phase load.
Ans. $4\sqrt{3}$ A, 12 A, 4.32 kW

20-4.2 A balanced Δ-connected load has $\mathbf{Z}_p = 24 + j18\ \Omega$ and a line voltage of 200 V (rms value). If the lines are perfect conductors, find the power delivered to the three-phase load.
Ans. 3.2 kW

20-4.3 Solve Exercise 20-4.2 if the lines have resistances of 4 Ω. (Assume that the line voltages
are at the generator end.)   *Ans.* $\frac{16}{9}$ kW

20.5 POWER MEASUREMENT

To measure the power delivered to a three-phase load, we may connect a wattmeter
to measure the power delivered to each phase, as shown in Fig. 20.19, for a Y connection. Each wattmeter has its current coil in series with one phase of the load and
its voltage coil across the phase of the load. It would be better if we could make
the measurements using only the lines *a, b,* and *c,* because it may be difficult to
have access to the neutral point *N.* (In the case of a Δ connection there is, of course,
no neutral point.)

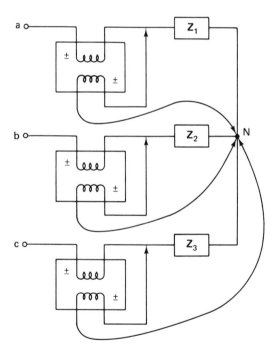

FIGURE 20.19 *Power measurement
using three wattmeters.*

Two-Wattmeter Method: It is possible to measure the total power delivered to
the three phases by considering only line voltages. In fact, such a method is particularly
attractive because it requires only two wattmeters instead of the three used in Fig.
20.19. The method is general and applies to both Y and Δ loads, which may be
balanced or unbalanced.

To illustrate the method, let us consider the three-phase load of Fig. 20.20,

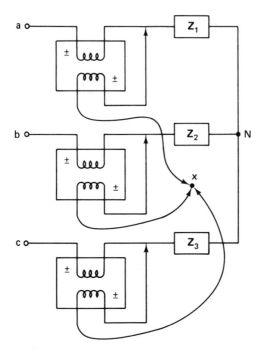

FIGURE 20.20 *Three wattmeters connected to a common point.*

which has three wattmeters, each with its current coil in one line and its voltage coil between that line and a common point x. The sum of the wattmeter readings is P, which is the average value of the instantaneous power p given by

$$p = i_{aN}v_{ax} + i_{bN}v_{bx} + i_{cN}v_{cx} \qquad (20.32)$$

Using the double-subscript notation we may write

$$v_{ax} = v_{aN} + v_{Nx}$$

$$v_{bx} = v_{bN} + v_{Nx}$$

$$v_{cx} = v_{cN} + v_{Nx}$$

which may be substituted into (20.32) to give

$$p = i_{aN}(v_{aN} + v_{Nx}) + i_{bN}(v_{bN} + v_{Nx}) + i_{cN}(v_{cN} + v_{Nx})$$

Rearranging the right member we have

$$\begin{aligned} p = {} & i_{aN}v_{aN} + i_{bN}v_{bN} + i_{cN}v_{cN} \\ & + v_{Nx}(i_{aN} + i_{bN} + i_{cN}) \end{aligned} \qquad (20.33)$$

However, by Kirchhoff's current law we have

$$i_{aN} + i_{bN} + i_{cN} = 0$$

so that (20.33) becomes

$$p = i_{aN}v_{aN} + i_{bN}v_{bN} + i_{cN}v_{cN} \qquad (20.34)$$

The three terms in the right member of (20.34) are the instantaneous powers delivered to the three loads. Thus p is the total instantaneous power and its average value P is the total three-phase load power. Thus no matter where the point x is, the algebraic sum of the three wattmeter readings is the total average power delivered to the load.

Since the point x in Fig. 20.20 is arbitrary, we may place it on one of the lines. Then the meter whose current coil is in that line will read zero because the voltage across its voltage coil is zero. Therefore, the total power delivered to the load is measured by the other two wattmeters and the wattmeter reading zero may be removed. As an example, the point x is placed on line b in Fig. 20.21, and the total load power is

$$P = P_A + P_C$$

where P_A and P_C are the readings of wattmeters A and C. There are two other ways of reading P with two wattmeters. In one of these the point x is placed on line a and in the other it is placed on line c.

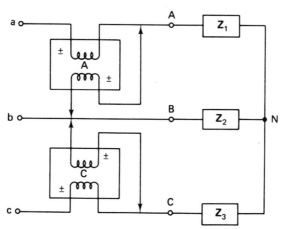

FIGURE 20.21 Two wattmeters reading the total load power.

Example 20.7: In Fig. 20.21 the line voltages are

$$\mathbf{V}_{ab} = 100\sqrt{3}\lfloor 0° \text{ V}$$

$$\mathbf{V}_{bc} = 100\sqrt{3}\lfloor -120° \text{ V}$$

$$\mathbf{V}_{ca} = 100\sqrt{3}\lfloor 120° \text{ V}$$

and the phase impedances are

$$\mathbf{Z}_1 = \mathbf{Z}_2 = \mathbf{Z}_3 = 10 + j10 = 10\sqrt{2}\lfloor 45° \text{ }\Omega$$

Find the wattmeter readings P_A and P_C and the power delivered to the load.

Solution: The phase voltage $\mathbf{V}_{AN}$ may be found from the line voltage $\mathbf{V}_{ab}$. Since $\mathbf{V}_{AN}$ lags $\mathbf{V}_{ab}$ by 30° and its magnitude is $V_p = V_L/\sqrt{3} = 100$ V, we have

$$\mathbf{V}_{AN} = 100\lfloor -30° \text{ V}$$

Therefore, the other phase voltages are

$$\mathbf{V}_{BN} = 100\lfloor -150° \text{ V}$$

$$\mathbf{V}_{CN} = 100\lfloor -270° \text{ V}$$

The line current $\mathbf{I}_{aA}$ is given by

$$\mathbf{I}_{aA} = \frac{\mathbf{V}_{AN}}{\mathbf{Z}_1} = \frac{100\lfloor -30°}{10\sqrt{2}\lfloor 45°} = 5\sqrt{2}\lfloor -75° \text{ A}$$

Thus the other line currents are

$$\mathbf{I}_{bB} = 5\sqrt{2}\lfloor -195° \text{ A}$$

$$\mathbf{I}_{cC} = 5\sqrt{2}\lfloor -315° \text{ A} = 5\sqrt{2}\lfloor 45° \text{ A}$$

The meter readings are

$$P_A = |\mathbf{V}_{ab}| \cdot |\mathbf{I}_{aA}| \cos (\text{ang } \mathbf{V}_{ab} - \text{ang } \mathbf{I}_{aA}) \tag{20.35}$$

and

$$P_C = |\mathbf{V}_{cb}| \cdot |\mathbf{I}_{cC}| \cos (\text{ang } \mathbf{V}_{cb} - \text{ang } \mathbf{I}_{cC}) \tag{20.36}$$

The voltage $\mathbf{V}_{cb}$ is given by

$$\mathbf{V}_{cb} = -\mathbf{V}_{bc} = -100\sqrt{3}\,\underline{|-120°}$$

$$= 100\sqrt{3}\,\underline{|-120° + 180°}$$

$$= 100\sqrt{3}\,\underline{|60°}\ \ \mathbf{V}$$

Therefore, by (20.35) and (20.36) we have

$$P_A = (100\sqrt{3})(5\sqrt{2})\cos(0 + 75°)$$

$$= 317\ \mathrm{W}$$

and

$$P_C = (100\sqrt{3})(5\sqrt{2})\cos(60° - 45°)$$

$$= 1183\ \mathrm{W}$$

The total power is therefore

$$P = P_A + P_C = 317 + 1183 = 1500\ \mathrm{W}$$

As a check, the power delivered to phase a is

$$P_P = |\mathbf{V}_{AN}| \cdot |\mathbf{I}_{aA}| \cos(\mathrm{ang}\ \mathbf{V}_{AN} - \mathrm{ang}\ \mathbf{I}_{aA})$$

$$= (100)(5\sqrt{2})\cos(-30° + 75°)$$

$$= 500\ \mathrm{W}$$

Since the system is balanced, the total power is

$$P = 3P_p = 1500\ \mathrm{W}$$

PRACTICE EXERCISES

20-5.1   In Fig. 20.19, let $\mathbf{Z}_1 = \mathbf{Z}_2 = \mathbf{Z}_3 = 10\,\underline{|30°}\ \Omega$ and let the line voltages be

$$\mathbf{V}_{ab} = 200\,\underline{|0°}\ \ \mathrm{V}$$

$$\mathbf{V}_{bc} = 200\,\underline{|-120°}\ \ \mathrm{V}$$

$$\mathbf{V}_{ca} = 200\,\underline{|120°}\ \ \mathrm{V}$$

Find each wattmeter reading.   *Ans.* $2/\sqrt{3} = 1.155$ kW

20-5.2 If the power in Exercise 20-5.1 is measured by the two wattmeters A and C in Fig. 20.21, find the readings P_A and P_C and the total power P.

Ans. $2/\sqrt{3} = 1.155$ kW, $4/\sqrt{3} = 2.31$ kW, $6/\sqrt{3} = 3.465$ kW

20.6 UNBALANCED LOADS

If a three-phase load is unbalanced (unequal phase impedances), then the shortcut procedures we have considered thus far do not apply. There is a shortcut method that may be used for unbalanced loads, which is called the *symmetrical components* method, and which is usually considered in more advanced treatments of three-phase circuits. However, we will be content to use ordinary analysis methods and analyze unbalanced circuits exactly as we would any other circuit. After all, a three-phase circuit is still a circuit and all the analysis methods we have used are applicable.

Example 20.8: Find the line currents $\mathbf{I}_{aA}$, $\mathbf{I}_{bB}$, and $\mathbf{I}_{cC}$ in the unbalanced three-phase circuit of Fig. 20.22.

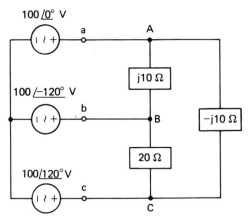

FIGURE 20.22 *Unbalanced three-phase network.*

Solution: Since the load is Δ-connected, we may find the phase currents as the line voltages divided by the phase impedances. From the phase currents we may find the line currents using Kirchhoff's current law.

The line voltages, by (20.8), are

$$\mathbf{V}_{ab} = 100\sqrt{3}\,\underline{|30°}\ \text{V}$$

$$\mathbf{V}_{bc} = 100\sqrt{3}\,\underline{|-90°}\ \text{V} \tag{20.37}$$

$$\mathbf{V}_{ca} = 100\sqrt{3}\,\underline{|150°}\ \text{V}$$

Therefore, by Ohm's law the phase currents are

$$I_{AB} = \frac{V_{ab}}{j10} = \frac{100\sqrt{3}\,\lfloor 30° }{10\lfloor 90° } = 10\sqrt{3}\,\lfloor -60°\ \text{A}$$

$$I_{BC} = \frac{V_{bc}}{20} = \frac{100\sqrt{3}\,\lfloor -90° }{20} = 5\sqrt{3}\,\lfloor -90°\ \text{A}$$

$$I_{CA} = \frac{V_{ca}}{-j10} = \frac{100\sqrt{3}\,\lfloor 150° }{10\lfloor -90° } = 10\sqrt{3}\,\lfloor 240°\ \text{A}$$

In rectangular form these are

$$I_{AB} = 10\sqrt{3}\,[\cos(-60°) + j\sin(-60°)] = 8.66 - j15\ \text{A}$$

$$I_{BC} = 5\sqrt{3}\,[\cos(-90°) + j\sin(-90°)] = -j8.66\ \text{A}$$

$$I_{CA} = 10\sqrt{3}\,(\cos 240° + j\sin 240°) = -8.66 - j15\ \text{A}$$

Finally, by Kirchhoff's current law the line currents are

$$I_{aA} = I_{AB} + I_{AC} = I_{AB} - I_{CA}$$
$$= 8.66 - j15 - (-8.66 - j15)$$
$$= 17.32\ \text{A}$$

$$I_{bB} = I_{BA} + I_{BC} = -I_{AB} + I_{BC}$$
$$= -(8.66 - j15) - j8.66$$
$$= -8.66 + j6.34\ \text{A}$$

$$I_{cC} = I_{CB} + I_{CA} = -I_{BC} + I_{CA}$$
$$= -(-j8.66) - 8.66 - j15$$
$$= -8.66 - j6.34\ \text{A}$$

As a check on the work we see that the sum of the line currents is

$$I_{aA} + I_{bB} + I_{cC} = 17.32 - 8.66 + j6.34 - 8.66 - j6.34$$
$$= 0$$

as it should be.

Use of Y–Δ Transformations:

If the load is Y-connected we may convert it to an equivalent Δ by a Y–Δ transformation, and find the line currents as in Example 20.8. Also, if the lines have impedances they may be combined with the phase impedances of the Y-connected load before the conversion is made to an equivalent Δ.

PRACTICE EXERCISE

20-6.1 Find the line currents $\mathbf{I}_{aA}$, $\mathbf{I}_{bB}$, and $\mathbf{I}_{cC}$.

Ans: 17.32$\lfloor 0°$ A, 10.73$\lfloor 143.8°$ A, 10.73$\lfloor -143.8°$ A

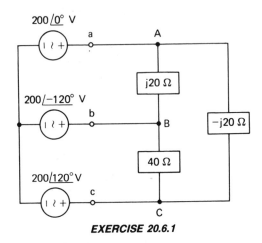

EXERCISE 20.6.1

20.7 SUMMARY

A balanced three-phase generator is equivalent to three single-phase generators, each of which generates a sinusoidal voltage. The voltages are identical except that they are 120° apart in phase. The most common source is the Y-connected source, in which the three single-phase generators are connected in a Y. The generated voltages in this case form a balanced set.

The three-phase load consists of three loads connected either in a Y or in a Δ. If the three loads are identical, the currents generated by the source will also be a balanced set, in which case the system is a balanced Y–Y system, or a balanced Y–Δ system. A Y–Y system may also have a neutral line connecting the neutral points of the source and load, in addition to the three lines from the source to the three phases of the load.

In a balanced system the line voltages are $\sqrt{3}$ times the phase voltages of the source. The line currents for a Y load are the same as the phase currents (the currents in the three phases), and the line currents for a Δ load are $\sqrt{3}$ times the phase currents. The Δ phase voltages are the same as the line voltages, but in the Y-connected load case the line voltages are $\sqrt{3}$ times the phase voltages.

In the case of unbalanced loads (unequal load impedances) the circuits may be analyzed by ordinary circuit analysis methods, or Y–Δ transformations may be used to obtain equivalent Δ loads with perfect conducting lines. In this case the line and phase currents may be found by Ohm's and Kirchhoff's laws.

Three wattmeters may always be used to measure the power delivered to the three-phase load. However, properly connected, only two wattmeters are needed. The algebraic sum of their readings is the total three-phase power.

PROBLEMS

20.1 In Fig. 20.10, the generator voltages are

$$\mathbf{V}_{an} = 200\underline{|0°}\ \text{V}$$

$$\mathbf{V}_{bn} = 200\underline{|-120°}\ \text{V}$$

$$\cdot\ \mathbf{V}_{cn} = 200\underline{|120°}\ \text{V}$$

and the phase impedance is $\mathbf{Z}_p = 10\underline{|60°}\ \Omega$. Find the line current I_L and the power delivered to the three-phase load.

20.2 In Fig. 20.10 the line currents form a balanced, positive sequence set with $\mathbf{I}_{aA} = 10\underline{|0°}$ A. If $\mathbf{Z}_p = \sqrt{3}\underline{|30°}\ \Omega$, find the line voltages and the power delivered to the three-phase load.

20.3 A balanced Y-connected load has a 240-V line voltage and a phase impedance of $4\underline{|60°}\ \Omega$. Find the total power delivered to the load.

20.4 Repeat Problem 20.3 if the load is Δ-connected and the phase impedance is $12\underline{|60°}\ \Omega$.

20.5 In Fig. 20.10 the source is a positive sequence source with $\mathbf{V}_{an} = 100\underline{|0°}$ V and $\mathbf{Z}_p = 10\underline{|30°}\ \Omega$. Find the line voltage V_L, the line current I_L, and the power delivered to the load.

20.6 Repeat Problem 20.5 if the load impedance is $\mathbf{Z}_p = 6 + j8\ \Omega$.

20.7 A balanced Y–Y three-wire, positive sequence system has $\mathbf{V}_{an} = 200\underline{|0°}$ V and $\mathbf{Z}_p = 3 - j4\ \Omega$. The lines each have a resistance of $1\ \Omega$. Find the line current I_L and the power delivered to the load.

20.8 A balanced three-phase Y-connected load draws 6 kW at a power factor of 0.8 lagging. If the line voltages are a balanced 200-V set, find the line current I_L.

20.9 In Fig. 20.10, the source is balanced, with positive phase sequence, and $\mathbf{V}_{an} = 100\underline{|0°}$ V. Find $\mathbf{Z}_p$ if the source delivers 2.4 kW at a power factor of 0.8 lagging.

20.10 Repeat Problem 20.9 if the load is a balanced Δ.

20.11 Repeat Problem 20.10 if each line contains a resistance of $1\ \Omega$.

20.12 In the Y–Δ system of Fig. 20.15, the source is positive sequence with $\mathbf{V}_{an} = 200\underline{|0°}$ V, and the phase impedance is $\mathbf{Z}_p = 4 + j3\ \Omega$. Find the line voltage V_L, the line current I_L, and the power delivered to the load.

20.13 The balanced Y–Δ system of Fig. 20.15 has $\mathbf{V}_{an} = 120\underline{|0°}$ V, positive phase sequence, $\mathbf{Z}_p = 6 - j9\ \Omega$, and a resistance of $1\ \Omega$ in the lines. Find the power delivered to the load.

20.14 In Fig. 20.15 the source is positive sequence, $V_{ab} = 200 \underline{|0°}$ V, and the power delivered to the load is 4800 W at a power factor of 0.8 lagging. Find the phase currents.

20.15 A balanced Δ-connected load has $Z_p = 12 + j9$ Ω and a line voltage $V_L = 225$ V. Find the power delivered to the load.

20.16 Solve Problem 20.15 if the voltage V_L is at the generator end and the lines have resistances of 2 Ω.

20.17 In Fig. 20.21 the line voltages form a balanced positive-sequence set with $V_{ab} = 100 \underline{|0°}$ V, and

$$Z_1 = Z_2 = Z_3 = 6 + j8 \text{ Ω}$$

Find the wattmeter readings P_A and P_C and the power delivered to the load.

20.18 Repeat Problem 20.17 if the impedances are

$$Z_1 = Z_2 = Z_3 = 10 \underline{|30°} \text{ Ω}$$

20.19 Repeat Problem 20.17 if the impedances are

$$Z_1 = Z_2 = Z_3 = 10 \underline{|60°} \text{ Ω}$$

Note that in this case wattmeter *C* reads the total power.

20.20 Find the wattmeter readings P_A and P_B and the total power delivered to the load if the line voltages form a balanced positive-sequence set with $V_{ab} = 100 \underline{|0°}$ V and $Z_p = 6 + j8$ Ω. (Note that this is the case of Fig. 20.20 with the point *x* placed on line *c*.)

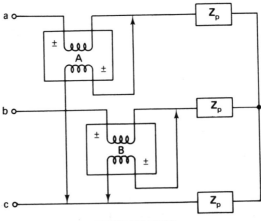

PROBLEM 20.20

20.21 A balanced three-phase, positive-sequence source with $V_{ab} = 200 \underline{|0°}$ V is supplying a Δ-connected load, $Z_{AB} = 50$ Ω, $Z_{BC} = 20 + j20$ Ω, and $Z_{CA} = 30 - j40$ Ω. Find the line currents.

21

TRANSFORMERS

In studying inductance in Chapter 14, we noted that a changing current in a coil produces a changing magnetic field, which in turn produces a voltage in the coil. If two or more coils are near enough to share a common magnetic flux, they are said to be *mutually coupled*. In this case a changing current in one coil will produce a changing flux which causes a voltage in all the coils.

As we saw in Chapter 14, the inductance L is a measure of the capacity of a coil to produce a voltage induced by a changing current in the coil. In the same manner, the capacity of a coil to produce a voltage induced by the current in another coil is measured by the so-called *mutual inductance* that exists between the coils. To distinguish the two types of inductance, we will call L the *self inductance* of its coil. Thus self inductance depends on the number of turns of the coil, the permeability of its core, and its shape (the length and cross-sectional area of the winding). Mutual inductance depends on these properties of the mutually coupled coils, as well as on how close the coils are to each other and how they are oriented toward each other.

A set of two or more mutually coupled coils that are wound on a single form, or core, is commonly called a *transformer*. The most common type is one with two coils, which is used to raise or lower voltages by producing a voltage across one coil that is higher or lower than that across the other coil. Transformers are available in a wide variety of sizes and shapes that are designed for countless uses. Devices as small as an aspirin tablet, for example, are common in radios, television sets, and stereos, and transformers designed for use in 60-Hz power applications may be

FIGURE 21.1 *Variable transformer (Courtesy, Ohmite Manufacturing Company).*

larger than an automobile. An example of one such large transformer was shown in Fig. 1.2. Examples of smaller transformers are shown in Figs. 21.1 and 21.2.

In this chapter we will define mutual inductance and see how mutually coupled coils, or transformers, may be used to raise and lower voltages. Finally, we will consider equivalent circuits that are useful in representing transformers for easier circuit analysis.

21.1 MUTUAL INDUCTANCE

In Chapter 14 we saw that the flux linkages $N\phi$ of a coil with N turns was related to the current i producing the flux ϕ by

$$N\phi = Li \tag{21.1}$$

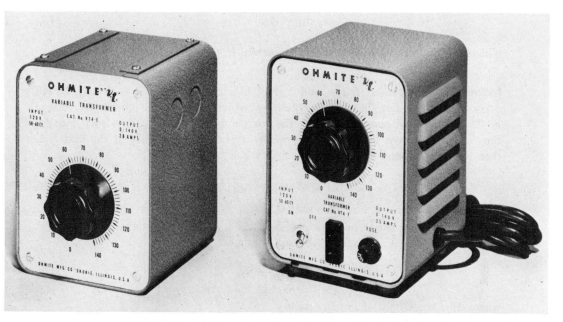

FIGURE 21.2 *Portable transformers (Courtesy, Ohmite Manufacturing Company).*

where L is the inductance in henrys of the coil. Let us consider now what happens when a coil with N_1 turns, carrying a current i_1, is placed near a second coil of N_2 turns, as shown in Fig. 21.3.

Mutual and Leakage Flux: The current i_1 produces a flux that we will denote by ϕ_{11}. Some of this flux, shown as ϕ_{21}, links the other coil, and is called the *mutual flux*. The remainder, shown as ϕ_{L1}, is called the *leakage flux,* since it does not stay in the path, or core, between the two coils. Thus ϕ_{21} is the flux linking the second coil and produced by the current in the first coil, and ϕ_{L1} is leakage flux produced by the current in the first coil. Their sum is

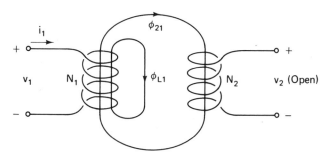

FIGURE 21.3 *Mutually coupled coils.*

$$\phi_{11} = \phi_{21} + \phi_{L1} \tag{21.2}$$

which is the total flux produced by the current in the first coil.

Induced Voltages: If the second coil of Fig. 21.3 is open, as indicated, it will carry no current, but the changing flux ϕ_{21} which links it will induce the voltage v_2. This voltage, of course, is due to the current i_1 in the first coil. Also i_1 induces the voltage v_1 across the first coil. If we let L_1 be the inductance of the first coil, then by (21.1) its flux linkage is

$$N_1\phi_{11} = L_1 i_1 \tag{21.3}$$

The flux linkage in the second coil is $N_2\phi_{21}$, which is also proportional to the current i_1 which produces it. Therefore, we may say that

$$N_2\phi_{21} = M i_1 \tag{21.4}$$

where M is a constant of proportionality.

Since by Faraday's law the induced voltage is the rate of change of $N\phi$, we see from Fig. 21.3 and (21.3) and (21.4) that the induced coil voltages are

$$v_1 = L_1 \frac{di_1}{dt} \tag{21.5}$$

and

$$v_2 = M \frac{di_1}{dt} \tag{21.6}$$

Of course, if the current in a coil is constant, such as dc, its rate of change is zero. Therefore, it induces no voltage either in its own coil or any other coil in its vicinity.

Self and Mutual Inductance: From (21.5) and (21.6) we see that L_1 and M have the same units, and therefore M, like L_1, must be measured in henrys (H). Thus both L_1 and M are types of inductance. To distinguish the two, L_1 is called the *self inductance* of the first coil and M is called the *mutual inductance* between the coils.

Transformer Action: Let us now consider the case where both coils are carrying a varying current. An example is the transformer of Fig. 21.4, consisting of two mutually coupled coils wound on a common core. We will call the coil on the left the *primary winding*, or coil 1, and that on the right the *secondary winding*, or coil

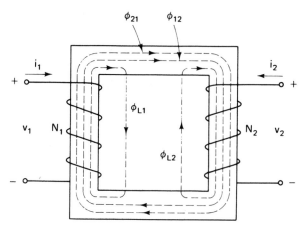

FIGURE 21.4 *Transformer.*

2. Thus the primary has N_1 turns, a voltage v_1, and a current i_1. The secondary has N_2 turns, a voltage v_2, and a current i_2.

The current i_1 produces a flux ϕ_{11} given by

$$\phi_{11} = \phi_{21} + \phi_{L1} \tag{21.7}$$

and i_2 produces a flux ϕ_{22} given by

$$\phi_{22} = \phi_{12} + \phi_{L2} \tag{21.8}$$

where ϕ_{21} and ϕ_{12} link both coils and ϕ_{L1} and ϕ_{L2} are leakage fluxes. The total flux in coil 1 is seen from Fig. 21.4 to be ϕ_1 given by

$$\phi_1 = \phi_{21} + \phi_{L1} + \phi_{12}$$

which by (21.7) is

$$\phi_1 = \phi_{11} + \phi_{12}$$

Therefore, its flux linkage is

$$N_1\phi_1 = N_1\phi_{11} + N_1\phi_{12} \tag{21.9}$$

The first term $N_1\phi_{11}$ is due to the current i_1 and is given by

$$N_1\phi_{11} = L_1 i_1 \tag{21.10}$$

where L_1 is the self inductance of coil 1. The second term $N_1\phi_{12}$ is due to i_2 and is given by

$$N_1\phi_{12} = Mi_2 \tag{21.11}$$

where M is the mutual inductance between the coils. Therefore, (21.9) may be written

$$N_1\phi_1 = L_1 i_1 + Mi_2 \tag{21.12}$$

The total flux in coil 2 is seen from Fig. 21.4 to be ϕ_2 given by

$$\phi_2 = \phi_{21} + \phi_{12} + \phi_{L2}$$

Using (21.8) we may write its flux linkage as

$$\begin{aligned} N_2\phi_2 &= N_2(\phi_{21} + \phi_{12} + \phi_{L2}) \\ &= N_2(\phi_{21} + \phi_{22}) \\ &= N_2\phi_{21} + N_2\phi_{22} \end{aligned} \tag{21.13}$$

The second term $N_2\phi_{22}$ is due to i_2 and is given by

$$N_2\phi_{22} = L_2 i_2 \tag{21.14}$$

where L_2 is the self inductance of coil 2. The other term $N_2\phi_{21}$ is due to i_1 and is given by

$$N_2\phi_{21} = Mi_1 \tag{21.15}$$

where, again, M is the mutual inductance. Therefore, (21.13) may be written

$$N_2\phi_2 = Mi_1 + L_2 i_2 \tag{21.16}$$

Transformer Voltages: By Faraday's law the voltage v_1 across coil 1 in Fig. 21.4 is the rate of change of the flux linkages $N_1\phi_1$ given in (21.12). Similarly, the voltage v_2 is the rate of change of $N_2\phi_2$, the flux linkages of coil 2 given by (21.16). From these results we have the primary and secondary voltages

$$\begin{aligned} v_1 &= L_1\frac{di_1}{dt} + M\frac{di_2}{dt} \\ v_2 &= M\frac{di_1}{dt} + L_2\frac{di_2}{dt} \end{aligned} \tag{21.17}$$

The quantities di_1/dt and di_2/dt, as in Chapter 14, are the rates of change of the currents.

Coefficient of Coupling: If two coils are not coupled (such as if they are shielded from one another or very widely separated), then the mutual inductance M is zero. If, on the other hand, the coils are very close together and there is almost no leakage flux, then M is relatively large. To have a measurement indicating when M is high or low, let us consider the ratio

$$\frac{M^2}{L_1 L_2} = \frac{M}{L_1} \cdot \frac{M}{L_2} \qquad (21.18)$$

From (21.11) and (21.15) we have

$$M = \frac{N_1 \phi_{12}}{i_2}$$

and

$$M = \frac{N_2 \phi_{21}}{i_1}$$

and from (21.10) and (21.14) we have

$$L_1 = \frac{N_1 \phi_{11}}{i_1}$$

and

$$L_2 = \frac{N_2 \phi_{22}}{i_2}$$

Substituting these values into the right member of (21.18) results in

$$\frac{M^2}{L_1 L_2} = \frac{N_1 \phi_{12}/i_2}{N_1 \phi_{11}/i_1} \cdot \frac{N_2 \phi_{21}/i_1}{N_2 \phi_{22}/i_2}$$

which simplifies to

$$\frac{M^2}{L_1 L_2} = \frac{\phi_{12} \phi_{21}}{\phi_{11} \phi_{22}} \qquad (21.19)$$

The *coefficient of coupling* k is defined by

$$k = \frac{M}{\sqrt{L_1 L_2}} \qquad (21.20)$$

which by (21.19) is

$$k = \sqrt{\frac{\phi_{12}\phi_{21}}{\phi_{11}\phi_{22}}}$$

By (21.7) and (21.8) this last result is

$$k = \sqrt{\frac{\phi_{21}}{(\phi_{21} + \phi_{L1})} \cdot \frac{\phi_{12}}{(\phi_{12} + \phi_{L2})}} \qquad (21.21)$$

If there is no leakage flux ($\phi_{L1} = \phi_{L2} = 0$), all the flux links both coils, and the coupling is said to be *perfect*. In this case we see from (21.21) that $k = 1$. If there is no mutual flux ($\phi_{12} = \phi_{21} = 0$), then all the flux is leakage flux, as far as the two coils are concerned. In this case $k = 0$. Since these two cases are the extremes, we must have

$$0 \le k \le 1 \qquad (21.22)$$

By (21.20) we have

$$M = k\sqrt{L_1 L_2} \qquad (21.23)$$

so that M may vary from 0 ($k = 0$) to $\sqrt{L_1 L_2}$ ($k = 1$). The coefficient of coupling k is therefore a measure of how closely the coils are coupled. If $k = 1$, the coils are perfectly coupled, and if k is near 1, they are *tightly* coupled. If k is near 0, the coils are *loosely* coupled, and if $k = 0$, the coils are not coupled, or *uncoupled*.

Example 21.1: Find v_1 and v_2 in Fig. 21.4 if $L_1 = 2$ H, $L_2 = 8$ H, $k = 0.75$, and the rates of change of the currents are

$$\frac{di_1}{dt} = 20 \text{ A/s}$$

and

$$\frac{di_2}{dt} = -6 \text{ A/s}$$

Solution: By (21.23) we have

$$M = k\sqrt{L_1 L_2} = 0.75\sqrt{2(8)} = 3 \text{ H}$$

Therefore, the voltages are by (21.17)

$$v_1 = L_1 \frac{di_1}{dt} + M \frac{di_2}{dt}$$

$$= 2(20) + 3(-6) = 22 \text{ V}$$

and

$$v_2 = M \frac{di_1}{dt} + L_2 \frac{di_2}{dt}$$

$$= 3(20) + 8(-6) = 12 \text{ V}$$

PRACTICE EXERCISES

21-1.1 Find v_1 and v_2 in Fig. 21.4 if $L_1 = 2$ H, $L_2 = 5$ H, $M = 3$ H, and the currents i_1 and i_2 are changing at the rates

$$\frac{di_1}{dt} = 10 \text{ A/s} \quad \text{and} \quad \frac{di_2}{dt} = -2 \text{ A/s}$$

Ans. 14 V, 20 V

21-1.2 Find the coefficient of coupling if $L_1 = 0.02$ H, $L_2 = 0.125$ H, and $M = 0.04$ H.

Ans. 0.8

21-1.3 Find M if $L_1 = 0.4$ H, $L_2 = 0.9$ H, and (a) $k = 1$, (b) $k = 0.5$, and (c) $k = 0.01$.

Ans. (a) 0.6 H, (b) 0.3 H, (c) 6 mH

21.2 TRANSFORMER PROPERTIES

In the transformer of Fig. 21.4, the two coils are wound in such a way that the currents produce fluxes in the same direction. Thus ϕ_{12} and ϕ_{21} add to produce the voltages v_1 and v_2, and consequently the mutual terms $M \, di_1/dt$ and $M \, di_2/dt$ in (21.17) have positive signs. If one of the coils (but not both) had been wound in the opposite way, the fluxes ϕ_{12} and ϕ_{21} would have opposed each other and the voltage terms due to each current would not have added, but one term would have been subtracted from the other. Since it is usually not convenient to draw the mutually coupled coils in a circuit so that the directions of the windings can be seen, the transformer circuit symbols have dots placed on the terminals, as shown in Fig. 21.5. The dots may then be used to write the circuit equations, as we will see.

The Dot Convention: Figure 21.5(a) and (b) show two circuit symbols for the transformer of Fig. 21.4, where the dots on the terminals are used in place of the directions of the windings to write the circuit equations. The rule, or *dot convention,* for use in writing the equations, is as follows.

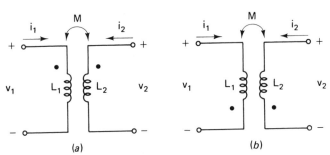

FIGURE 21.5 *Circuit symbols for the transformer of Fig. 21.4.*

A current i entering a dotted (or undotted) terminal in one winding induces a voltage M di/dt with positive polarity at the dotted (or undotted) terminal of the other winding.

To illustrate the use of the dot convention, let us write the expressions for v_1 and v_2 in Fig. 21.5(a). The voltage v_1 has two parts, the part $L_1\, di_1/dt$ due to i_1 because of the self inductance L_1 and the part $M\, di_2/dt$ due to i_2 because of the mutual inductance M. The only problem is to get the signs right for the two terms. The self-inductance term is obtained as in Chapter 14. Since the current enters the positive terminal, the sign of $L_1\, di_1/dt$ is plus. In the case of the mutual-inductance term, we use the dot convention. The current i_2 producing the mutual-inductance term enters the dotted terminal. Therefore, the positive polarity of the voltage $M\, di_2/dt$ is at the dotted terminal of the primary. Since this agrees with the polarity of v_1, we have a plus sign on the mutual term. Thus we have

$$v_1 = L_1 \frac{di_1}{dt} + M \frac{di_2}{dt} \qquad (21.24)$$

In a similar way, i_1 enters the dotted terminal, so that its effect $M\, di_1/dt$ has positive polarity at the dotted terminal of the secondary. The other part of v_2 is $L_2\, di_2/dt$, which also has its positive polarity at the upper terminal. Thus we have

$$v_2 = M \frac{di_1}{dt} + L_2 \frac{di_2}{dt} \qquad (21.25)$$

These two results agree with those of (21.17).

The circuit of Fig. 21.5(b) is equivalent to that of Fig. 21.5(a). The self-inductance terms $L_1\, di_1/dt$ and $L_2\, di_2/dt$ have the polarities of v_1 and v_2, as in (21.24) and (21.25). The current i_2 enters the undotted terminal and thus $M\, di_2/dt$ has its positive polarity at the other undotted terminal. Since this agrees with the polarity of v_1, Equation (21.24) is valid. Similarly, i_1 enters the undotted terminal so that $M\, di_1/dt$ has its positive polarity at the other undotted terminal. This agrees with that of v_2, so that (21.25) holds.

In Fig. 21.6(a) and (b), one of the dots has been moved from its position in

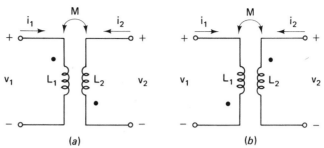

FIGURE 21.6 *Circuit symbols for a transformer with one winding reversed from that of Fig. 21.5.*

Fig. 21.5(a) and (b) to the other terminal. This indicates that the winding of one coil has been reversed. The self-inductance terms $L_1\, di_1/dt$ and $L_2\, di_2/dt$ in (21.24) and (21.25) remain unchanged in sign because the current and voltage polarities are the same in Figs. 21.5 and 21.6. However, the signs of the mutual terms are changed. To see this we note that in Fig. 21.6(a) the current i_2 enters the undotted side. Thus its induced voltage $M\, di_2/dt$ has its positive polarity at the undotted terminal of v_1, and its sign therefore is the opposite of v_1. In a similar way, i_1 enters the undotted terminal, so that $M\, di_1/dt$ has the opposite polarity of v_2. We therefore have for Fig. 21.6(a),

$$v_1 = L_1\frac{di_1}{dt} - M\frac{di_2}{dt}$$

$$v_2 = -M\frac{di_1}{dt} + L_2\frac{di_2}{dt}$$

(21.26)

By applying the dot convention of Fig. 21.6(b), we may see that (21.26) also holds for it. Therefore Fig. 21.6(a) and (b) are equivalent representations of the same transformer. In other words, the first dot may be placed on either terminal, but its position determines that of the second dot.

Phasor Circuits: If the currents and voltages are sinusoids with frequency ω, we know that the voltage $L\, di/dt$ in the time domain carries over to the phasor domain as $j\omega L\mathbf{I}$, where $\mathbf{I}$ is the phasor representation of i. Since $M\, di/dt$ has the same form as $L\, di/dt$, its phasor representation will be $j\omega M\mathbf{I}$. Therefore, we may represent the phasor circuit of Fig. 21.5(a) by that of Fig. 21.7(a), and that of Fig. 21.6(a) by that of Fig. 21.7(b). From (21.24) to (21.26), the phasor equations for Fig. 21.7(a) are

$$\mathbf{V}_1 = j\omega L_1\mathbf{I}_1 + j\omega M\mathbf{I}_2$$

$$\mathbf{V}_2 = j\omega M\mathbf{I}_1 + j\omega L_2\mathbf{I}_2$$

(21.27)

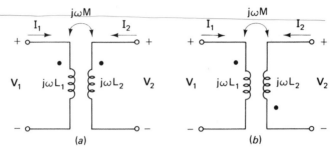

FIGURE 21.7 *Transformer phasor circuits.*

and those for Fig. 21.7(b) are

$$V_1 = j\omega L_1 I_1 - j\omega M I_2$$
$$V_2 = -j\omega M I_1 + j\omega L_2 I_2$$

(21.28)

The phasors V_1, V_2, I_1, and I_2 are those of v_1, v_2, i_1, and i_2. Equations (21.27) and (21.28) may also be obtained directly from the phasor circuits using the dot convention. The terms $\pm j\omega M I_1$ and $\pm j\omega M I_2$ are the mutual terms induced in one coil by the current in the other coil.

Example 21.2: Find the steady-state voltage v in the circuit of Fig. 21.8.

Solution: The frequency is $\omega = 2$ rad/s and the phasor circuit is shown in Fig. 21.9 with loop currents I_1 and I_2. Since $V = 1 I_2 = I_2$, we need only find the loop current I_2. The currents both enter the dots, so that the transformer voltages V_{ab} and V_{dc} are given by

$$V_{ab} = j8 I_1 + j2 I_2$$

and

$$V_{dc} = -V = j2 I_1 + j4 I_2$$

The loop equations therefore are

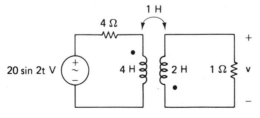

FIGURE 21.8 *Circuit containing a transformer.*

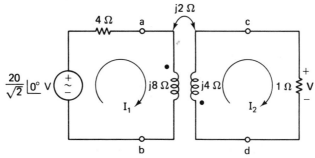

FIGURE 21.9 *Phasor circuit of Fig. 21.8.*

$$4\mathbf{I}_1 + \mathbf{V}_{ab} = 4\mathbf{I}_1 + j8\mathbf{I}_1 + j2\mathbf{I}_2 = \frac{20}{\sqrt{2}}\underline{|0°}$$

$$1\mathbf{I}_2 + \mathbf{V}_{dc} = \mathbf{I}_2 + j2\mathbf{I}_1 + j4\mathbf{I}_2 = 0$$

or

$$(4 + j8)\mathbf{I}_1 + j2\mathbf{I}_2 = \frac{20}{\sqrt{2}}$$

$$j2\mathbf{I}_1 + (1 + j4)\mathbf{I}_2 = 0$$

From the second of these we have

$$\mathbf{I}_1 = -\frac{1 + j4}{j2}\,\mathbf{I}_2 = \left(-2 + j\frac{1}{2}\right)\mathbf{I}_2 \qquad (21.29)$$

which substituted into the first yields

$$(4 + j8)\left(-2 + j\frac{1}{2}\right)\mathbf{I}_2 + j2\mathbf{I}_2 = \frac{20}{\sqrt{2}}$$

Carrying out the multiplication and collecting terms, we have

$$(-12 - j12)\,\mathbf{I}_2 = \frac{20}{\sqrt{2}}$$

or

$$\mathbf{I}_2 = \frac{20/\sqrt{2}}{-12 - j12} = \frac{20/\sqrt{2}}{12\sqrt{2}\underline{|-135°}} = \frac{5}{6}\underline{|135°}\ \text{A} \qquad (21.30)$$

Therefore, we have

$$\mathbf{V} = 1\mathbf{I}_2 = \frac{5}{6} \lfloor 135° \quad \mathbf{V}$$

and in the time domain the voltage is

$$v = \frac{5\sqrt{2}}{6} \sin(2t + 135°) \, \mathbf{V}$$

Stored Energy: The energy stored in two coupled coils depends on the current, the self and mutual inductances, and the position of the dots. In the case of Fig. 21.5, where both currents enter, or both leave, the dotted terminals, the stored energy w at any time t is

$$w = \frac{1}{2} L_1 i_1^2 + \frac{1}{2} L_2 i_2^2 + M i_1 i_2 \qquad (21.31)$$

where i_1 and i_2 are the currents at time t. In the case of Fig. 21.6, where one current enters and the other leaves a dotted terminal, the energy is

$$w = \frac{1}{2} L_1 i_1^2 + \frac{1}{2} L_2 i_2^2 - M i_1 i_2 \qquad (21.32)$$

Example 21.3: Find the energy stored in the transformer of Fig. 21.8 at $t = 0$.

Solution: From the phasor $\mathbf{I}_2$ of (21.30) we may find i_2 given by

$$i_2 = \frac{5\sqrt{2}}{6} \sin(2t + 135°) \, \mathbf{A}$$

Therefore at $t = 0$, we have

$$i_2 = \frac{5\sqrt{2}}{6} \sin 135° = \frac{5}{6} \, \mathbf{A}$$

From (21.29) and (21.30) we may find $\mathbf{I}_1$ given by

$$\mathbf{I}_1 = \left(-2 + j\frac{1}{2}\right)\left(\frac{5}{6} \lfloor 135°\right)$$

$$= (2.062 \lfloor 165.96°)\left(\frac{5}{6} \lfloor 135°\right)$$

$$= 1.72 \lfloor 300.96° \quad \mathbf{A}$$

Therefore, the time-domain current is

$$i_1 = 1.72\sqrt{2}\sin(2t + 300.96°)$$

which at $t = 0$ is

$$i_1 = 1.72\sqrt{2}\sin 300.96° = -2.09 \text{ A}$$

Since both currents enter the dotted terminals, the energy stored is given by (21.31). At $t = 0$ the result is

$$w = \frac{1}{2}L_1 i_1^2 + \frac{1}{2}L_2 i_2^2 + M i_1 i_2$$

$$= \frac{1}{2}(4)(-2.09)^2 + \frac{1}{2}(2)\left(\frac{5}{6}\right)^2 + 1(-2.09)\left(\frac{5}{6}\right)$$

$$= 7.69 \text{ J}$$

Constant Currents: As noted earlier, constant currents, such as dc, induce no voltages in coils. Therefore, we cannot use transformers to step up or step down dc voltages. It is primarily for this reason that power is transmitted across long distances in the form of ac voltages and currents. The voltages are easily raised to the high values necessary for economical transmission and lowered for use at their destinations.

PRACTICE EXERCISES

21-2.1 Find the phasor currents I_1 and I_2.   *Ans.* $4 - j4$ A, $j2$ A

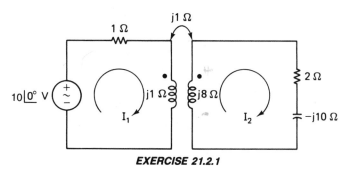

EXERCISE 21.2.1

21-2.2 Find the energy stored in the transformer of Exercise 21-2.1 at $t = 0$ if the frequency is $\omega = 2$ rad/s. (*Suggestion:* $j\omega L_1 = j1$ so that $L_1 = 1\omega = \frac{1}{2}$ H, etc.)   *Ans.* 32 J

21-2.3 Find the energy stored in the transformer of Fig. 21.5(a) at a time when $i_1 = 2$ A and $i_2 = 4$ A if $L_1 = 2$ H, $L_2 = 10$ H, and $M = 4$ H.   *Ans.* 116 J

21-2.4   Repeat Exercise 21-2.3 if one of the dots is moved to the other terminal.

Ans. 52 J

21-2.5   Given the two coils as shown. (a) Connect points *b* and *c* and find the equivalent inductance L_{ad} seen at terminals *a* and *d*. (b) Connect terminals *b* and *d* and find the equivalent inductance L_{ac} seen at terminals *a* and *c*. (c) Find *M* from L_{ad} and L_{ac}.   *Ans.* (a) $L_1 + L_2 + 2M$, (b) $L_1 + L_2 - 2M$, (c) $(L_{ad} - L_{ac})/4$

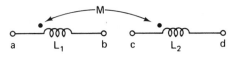

EXERCISE 21.2.5

21.3   IDEAL TRANSFORMERS

The transformer phasor circuit of Fig. 21.7(a) was described earlier by (21.27). If we reverse the direction of $\mathbf{I}_2$, as in Fig. 21.10, then the equations relating the phasor voltages and currents will be (21.27) with $\mathbf{I}_2$ replaced by $-\mathbf{I}_2$. Thus for the transformer of Fig. 21.10 we have

$$\mathbf{V}_1 = j\omega L_1 \mathbf{I}_1 - j\omega M \mathbf{I}_2$$
$$\mathbf{V}_2 = j\omega M \mathbf{I}_1 - j\omega L_2 \mathbf{I}_2 \tag{21.33}$$

Perfect Coupling: In the case of perfect coupling, the coefficient of coupling is $k = 1$, so that by (21.23) we have

$$M = \sqrt{L_1 L_2} \tag{21.34}$$

Substituting this value for *M* into (21.33) and finding the ratio $\mathbf{V}_2/\mathbf{V}_1$, we have

$$\frac{\mathbf{V}_2}{\mathbf{V}_1} = \frac{j\omega \sqrt{L_1 L_2}\mathbf{I}_1 - j\omega L_2 \mathbf{I}_2}{j\omega L_1 \mathbf{I}_1 - j\omega\sqrt{L_1 L_2}\,\mathbf{I}_2}$$

Canceling $j\omega$ from the numerator and denominator and factoring out $\sqrt{L_2}$ in the numerator and $\sqrt{L_1}$ in the denominator yields

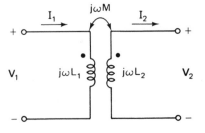

FIGURE 21.10 *Transformer circuit for Fig. 21.7(a) with* I_2 *reversed.*

$$\frac{\mathbf{V}_2}{\mathbf{V}_1} = \frac{\sqrt{L_2}\,(\sqrt{L_1}\,\mathbf{I}_1 - \sqrt{L_2}\mathbf{I}_2)}{\sqrt{L_1}\,(\sqrt{L_1}\,\mathbf{I}_1 - \sqrt{L_2}\,\mathbf{I}_2)}$$

or

$$\frac{\mathbf{V}_2}{\mathbf{V}_1} = \frac{\sqrt{L_2}}{\sqrt{L_1}} \tag{21.35}$$

In Chapter 14 we saw that the inductance of a coil was given by

$$L = \frac{N^2 \mu A}{l}$$

where N is the number of turns, μ is the permeability of the core, and A and l are the area and length of the flux path. In the case of the transformer, the flux path is the same for both coils if the coupling is perfect, so that the primary and secondary inductances are

$$L_1 = \frac{N_1^2 \mu A}{l}$$

and

$$L_2 = \frac{N_2^2 \mu A}{l}$$

(The permeability, area and length of the flux path are the same for both coils and are denoted by μ, l, and A.)

Substituting these inductance values into (21.35), we have

$$\frac{\mathbf{V}_2}{\mathbf{V}_1} = \frac{\sqrt{N_2^2 \mu A/l}}{\sqrt{N_1^2 \mu A/l}}$$

which after simplification is

$$\frac{\mathbf{V}_2}{\mathbf{V}_1} = \frac{N_2}{N_1}$$

Turns Ratio: The ratio N_2/N_1 is called the *turns ratio* and is denoted by a. That is, we have

$$\frac{\mathbf{V}_2}{\mathbf{V}_1} = \frac{N_2}{N_1} = a \tag{21.36}$$

in the case of a perfectly coupled transformer. In the time domain (21.36) still holds, so that

$$\frac{v_2}{v_1} = \frac{N_2}{N_1} = a \qquad (21.37)$$

Iron-Core Transformer: Perfect coupling is an ideal that cannot be achieved, but if the core is of a very high permeability material, such as iron, there is almost no leakage flux. This is because the flux follows the path of least reluctance, and it is much easier to establish a flux in iron than in other materials, such as air. Iron-core transformers may have coefficients of coupling in excess of 0.98, which is very close to perfect coupling and may be considered perfect in most applications. The symbol for an iron-core transformer is shown in Fig. 21.11, where the dots are still to be added.

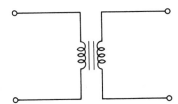

FIGURE 21.11 *Iron-core transformer.*

Ideal Transformer Conditions: For the perfectly coupled transformer of Fig. 21.10, we have by the first of (21.33)

$$\frac{\mathbf{V}_1}{j\omega L_1} = \mathbf{I}_1 - \frac{j\omega M \mathbf{I}_2}{j\omega L_1}$$

$$= \mathbf{I}_1 - \frac{j\omega \sqrt{L_1 L_2}\, \mathbf{I}_2}{j\omega L_1} \qquad (21.38)$$

$$= \mathbf{I}_1 - \sqrt{\frac{L_2}{L_1}}\, \mathbf{I}_2$$

From (21.35) and (21.36) we have

$$\frac{\sqrt{L_2}}{\sqrt{L_1}} = \frac{N_2}{N_1} = a \qquad (21.39)$$

so that (21.38) becomes

$$\frac{\mathbf{V}_1}{j\omega L_1} = \mathbf{I}_1 - a\mathbf{I}_2 \qquad (21.40)$$

If L_1 and L_2 are extremely large quantities (ideally infinite) with ratio $L_2/L_1 = a^2$, as in (21.39), then the left member of (21.40) is zero, or very nearly so. In this case we have

$$\mathbf{I}_1 - a\mathbf{I}_2 = 0$$

or

$$\mathbf{I}_1 = a\mathbf{I}_2 \qquad\qquad (21.41)$$

Substituting $a = N_2/N_1$ into this result, we have

$$\mathbf{I}_1 = \frac{N_2}{N_1}\mathbf{I}_2$$

or

$$N_1\mathbf{I}_1 = N_2\mathbf{I}_2 \qquad\qquad (21.42)$$

Thus the ampere turns are the same for both the primary and the secondary.

An *ideal transformer* is one for which the coupling is perfect and the inductances L_1 and L_2 of the primary and secondary are extremely large, with ratio $L_2/L_1 = a^2$. Thus the transformer of Fig. 21.10 is an example, provided that (21.36) and (21.41) or (21.42) hold. That is,

$$\frac{\mathbf{V}_2}{\mathbf{V}_1} = \frac{N_2}{N_1} = a \qquad \frac{\mathbf{I}_1}{\mathbf{I}_2} = \frac{N_2}{N_1} = a \qquad\qquad (21.43)$$

The symbol for such an ideal transformer is shown in Fig. 21.12(a). The vertical lines symbolize the iron core and $1:a$ denotes the turns ratio.

Equations (21.43) are also valid in the time domain. That is, we have

$$\frac{v_2}{v_1} = \frac{N_2}{N_1} = a \qquad \frac{i_1}{i_2} = \frac{N_2}{N_1} = a$$

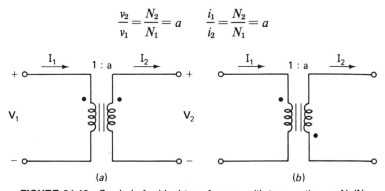

(a)   (b)

FIGURE 21.12   Symbols for ideal transformers with turns ratio $a = N_2/N_1$.

In other words, for an ideal transformer the voltages are in the same ratio as the turns, and the ampere-turns N_1I_1 (or N_1i_1) of the primary are the same as those, N_2I_2 (or N_2i_2), of the secondary.

If one of the polarity dots is moved, as in the ideal transformer of Fig. 21.12(b), Equations (21.43) change to

$$\frac{V_2}{V_1} = -a \qquad \frac{I_1}{I_2} = -a \tag{21.44}$$

This may be shown by going through the same steps that led to (21.43). The polarity dots may be omitted, in which case it is assumed that we are considering Fig. 21.12(a) and Equations (21.43).

Example 21.4: An ideal transformer has 100 turns on the primary side, and 600 turns on the secondary side. If the primary voltage is $V_1 = 100 \underline{|0°}$ V and the primary current is $I_1 = 2 \underline{|10°}$ A, find the turns ratio, the secondary voltage, and the secondary current. [Since the location of the dots is not mentioned, it is assumed that they are placed as in Fig. 21.12(a).]

Solution: The turns ratio is

$$a = \frac{N_2}{N_1} = \frac{600}{100} = 6$$

so that by (21.43) the secondary voltage and current are

$$V_2 = aV_1 = 6(100 \underline{|0°}) = 600 \underline{|0°} \text{ V}$$

and

$$I_2 = \frac{1}{a}I_1 = \frac{1}{6}(2 \underline{|10°}) = \frac{1}{3} \underline{|10°} \text{ A}$$

Thus this transformer "steps up" the voltage and "steps down" the current.

Stepping Up and Stepping Down the Voltage: In the general case the secondary voltage and current are given in the case of Fig. 21.12(a) by

$$V_2 = aV_1 \qquad I_2 = \frac{I_1}{a} \tag{21.45}$$

Therefore, if as in Example 21.4, the turns ratio a is greater than 1 (in which case $N_2 > N_1$), the primary voltage V_1 is stepped up to a higher secondary voltage V_2.

On the other hand, if a is between 0 and 1 ($N_1 > N_2$), the voltage $\mathbf{V}_1$ is stepped down to a lower secondary voltage $\mathbf{V}_2$. We see also from (21.45) that if the voltage is stepped up, the current is stepped down ($\mathbf{I}_2 < \mathbf{I}_1$), and if the voltage is stepped down, the current is stepped up ($\mathbf{I}_2 > \mathbf{I}_1$).

Power: From (21.45) we have

$$\mathbf{V}_2\mathbf{I}_2 = (a\mathbf{V}_1)\left(\frac{\mathbf{I}_1}{a}\right) = \mathbf{V}_1\mathbf{I}_1 \tag{21.46}$$

The power P_1 delivered to the primary is

$$P_1 = |\mathbf{V}_1| \cdot |\mathbf{I}_1| \cos \theta \tag{21.47}$$

where θ is the angle between $\mathbf{V}_1$ and $\mathbf{I}_1$. Since by (21.45), the angle between $\mathbf{V}_2$ and $\mathbf{I}_2$ is the same as that between $\mathbf{V}_1$ and $\mathbf{I}_1$, it is also θ, and the power P_2 delivered to the secondary is

$$P_2 = |\mathbf{V}_2| \cdot |\mathbf{I}_2| \cos \theta$$

By (21.46) we see that this result is the same as (21.47), so that

$$P_1 = P_2$$

That is, in the ideal case the power delivered to the primary is transferred exactly to the secondary.

Example 21.5: Find the power delivered to the primary and to the secondary in the transformer of Example 21.4.

Solution: The primary power is given by

$$
\begin{aligned}
P_1 &= |\mathbf{V}_1| \cdot |\mathbf{I}_1| \cos \theta \\
&= (100)(2) \cos 10° \\
&= 197 \text{ W}
\end{aligned}
$$

The secondary power is equal to the primary power and therefore is

$$P_2 = 197 \text{ W}$$

As a check, we have

$$\mathbf{V}_2 = 600 \underline{|0°} \text{ V} \qquad \mathbf{I}_2 = \frac{1}{3}\underline{|10°} \text{ A}$$

so that the secondary power is

$$P_2 = |\mathbf{V}_2| \cdot |\mathbf{I}_2| \cos \theta$$

$$= (600)\left(\frac{1}{3}\right) \cos 10°$$

$$= 197 \text{ W}$$

PRACTICE EXERCISES

21-3.1 An ideal transformer has $N_1 = 100$ turns, $N_2 = 1000$ turns, $\mathbf{V}_1 = 50 \underline{|0°}$ V, $\mathbf{I}_2 = 0.5 \underline{|30°}$ A. If the dots are placed as in Fig. 21.12(a), find a, $\mathbf{V}_2$, and $\mathbf{I}_1$.

Ans. 10, 500 $\underline{|0°}$ V, 5 $\underline{|30°}$ A

21-3.2 Solve Exercise 21-3.1 if the dots are placed as in Fig. 21.12(b).

Ans. 10, $-500 \underline{|0°}$ V, $-5 \underline{|30°}$ A

21-3.3 Find the primary and secondary power in Exercises 21-3.1 and 21-3.2.

Ans. 216.5 W

21.4 EQUIVALENT CIRCUITS

In many cases it is possible to replace transformer circuits by equivalent circuits without transformers. To see how this may be done, we will first consider the circuit of Fig. 21.13, which contains an ideal transformer.

Reflected Impedance: Let us define the impedance seen at the primary terminals x–y of the transformer by $\mathbf{Z}_1$, and note from Fig. 21.13 that it is given by

$$\mathbf{Z}_1 = \frac{\mathbf{V}_1}{\mathbf{I}_1} \tag{21.48}$$

Also, the load impedance $\mathbf{Z}_L$ is given by

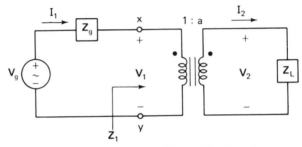

FIGURE 21.13 *Circuit containing an ideal transformer.*

$$\mathbf{Z}_L = \frac{\mathbf{V}_2}{\mathbf{I}_2} \tag{21.49}$$

Using (21.43), we may write (21.48) in the form

$$\mathbf{Z}_1 = \frac{\mathbf{V}_1}{\mathbf{I}_1} = \frac{\mathbf{V}_2/a}{a\mathbf{I}_2} = \frac{1}{a^2} \cdot \frac{\mathbf{V}_2}{\mathbf{I}_2}$$

which by (21.49) is equivalent to

$$\mathbf{Z}_1 = \frac{\mathbf{Z}_L}{a^2} \tag{21.50}$$

We may obtain an equivalent circuit of Fig. 21.13, therefore, by replacing everything to the right of the primary terminals of the transformer by $\mathbf{Z}_1$, as shown in Fig. 21.14. The impedance $\mathbf{Z}_1 = \mathbf{Z}_L/a^2$ is called the *reflected impedance* because it may be thought of as the impedance inserted into, or *reflected* into, the primary by the secondary.

If we are given $\mathbf{V}_g$, $\mathbf{Z}_g$, $\mathbf{Z}_L$, and a in Fig. 21.13, we may easily find the primary and secondary voltages from the equivalent circuit of Fig. 21.14. In the latter case we have by Kirchoff's voltage law,

$$-\mathbf{V}_g + \mathbf{Z}_g\mathbf{I}_1 + \frac{\mathbf{Z}_L}{a^2}\mathbf{I}_1 = 0$$

or by (21.45)

$$\mathbf{I}_1 = a\mathbf{I}_2 = \frac{\mathbf{V}_g}{\mathbf{Z}_g + \mathbf{Z}_L/a^2} \tag{21.51}$$

Also, from the equivalent circuit and (21.45), we have

$$\mathbf{V}_1 = \frac{\mathbf{V}_2}{a} = \frac{\mathbf{Z}_L}{a^2}\mathbf{I}_1 = \frac{\mathbf{Z}_L/a^2}{\mathbf{Z}_g + \mathbf{Z}_L/a^2}\mathbf{V}_g \tag{21.52}$$

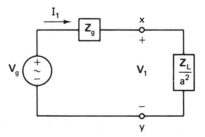

FIGURE 21.14 *Equivalent circuit of Fig. 21.13.*

Example 21.6: In Fig. 21.13 we have $V_g = 120 \underline{|0°}$ V, $Z_g = 10 \underline{|0°}$ Ω, $Z_L = 500 \underline{|0°}$ Ω, and $a = 10$. Find V_1, V_2, I_1, and I_2.

Solution: By (21.52) we have

$$V_1 = \frac{500 \underline{|0°} / 100}{10 \underline{|0°} + 500 \underline{|0°} / 100} (120 \underline{|0°}) = 40 \underline{|0°} \text{ V}$$

and

$$V_2 = aV_1 = 10(40 \underline{|0°}) = 400 \underline{|0°} \text{ V}$$

The currents are by (21.51)

$$I_1 = \frac{120 \underline{|0°}}{10 \underline{|0°} + 500 \underline{|0°} / 100} = 8 \underline{|0°} \text{ A}$$

and

$$I_2 = \frac{I_1}{a} = \frac{8 \underline{|0°}}{10} = 0.8 \underline{|0°} \text{ A}$$

Different Dot Assignment:

If one of the coils in Fig. 21.13 is wound in the opposite way, one of the dots will be assigned to the opposite terminal. In this case, as we have seen, the effect is to replace a by $-a$. Thus the reflected impedance of (21.50) is unchanged, but (21.51) and (21.52) for the currents and voltages will change. To illustrate this case, we will give an example using the circuit of Fig. 21.15.

Example 21.7: Find V_1, V_2, I_1, and I_2 in Fig: 21.15.

Solution: The load impedance is

$$Z_L = 100 - j75 \text{ Ω}$$

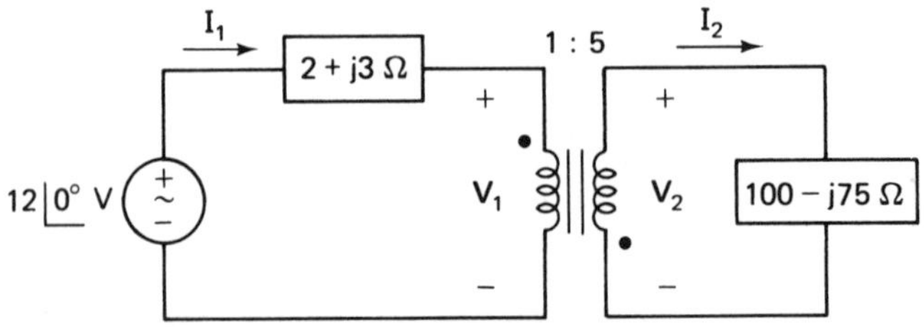

FIGURE 21.15 *Transformer circuit with a different dot arrangement.*

so that the reflected impedance is

$$\mathbf{Z}_1 = \frac{\mathbf{Z}_L}{a^2} = \frac{100 - j75}{(5)^2} = 4 - j3 \ \Omega$$

Using this result, we may draw the equivalent circuit of Fig. 21.16.

From Fig. 21.16 we have

$$\mathbf{I}_1 = \frac{12\underline{|0^\circ}}{(2 + j3) + (4 - j3)} = \frac{12\underline{|0^\circ}}{6} = 2\underline{|0^\circ} \ \text{A}$$

and by voltage division

$$\mathbf{V}_1 = \frac{4 - j3}{(2 + j3) + (4 - j3)} \cdot 12\underline{|0^\circ} = \frac{(5\underline{|-36.9^\circ})(12)}{6}$$

or

$$\mathbf{V}_1 = 10\underline{|-36.9^\circ} \ \text{V}$$

Because of the location of the dots in Fig. 21.15, the secondary current and voltage are

$$\mathbf{I}_2 = -\frac{\mathbf{I}_1}{a} = -\frac{2\underline{|0^\circ}}{5} = 0.4\underline{|180^\circ} \ \text{A}$$

and

$$\mathbf{V}_2 = -a\mathbf{V}_1 = -5(10\underline{|-36.9^\circ}) = 50\underline{|143.1^\circ} \ \text{V}$$

Impedance Matching: In Example 21.7 we saw that the reflected impedance $\mathbf{Z}_1 = 4 - j3 \ \Omega$ due to the load impedance and the turns ratio was the impedance seen at the primary terminals of the transformer. Thus $\mathbf{Z}_1$ may be thought of as the load seen by the generator consisting of the 12-V source and the source impedance $2 + j3$ of Fig. 21.15. Therefore, the turns ratio, which in this case is $a = 5$, may be adjusted to vary the load $\mathbf{Z}_1$ by any real factor we choose. In this way we may

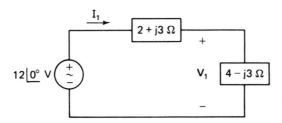

FIGURE 21.16 Equivalent circuit of Fig. 21.15.

match the load impedance of the secondary to the primary to draw various amounts of power from the source. It may even be possible to use this type of *impedance matching* to draw the maximum power from the source, as we will see.

As we know from Chapter 18, the maximum power is drawn from a source, such as $\mathbf{V}_g$ in series with $\mathbf{Z}_g$ in Fig. 21.13, when the load $\mathbf{Z}_1$ is given by

$$\mathbf{Z}_1 = \mathbf{Z}_g^*$$

where $\mathbf{Z}_g^*$ is the complex conjugate of $\mathbf{Z}_g$. If $\mathbf{Z}_g$ is a resistance, say $\mathbf{Z}_g = R_g$, then the maximum power is drawn when

$$\mathbf{Z}_1 = \mathbf{Z}_g = R_g$$

In this case, by (21.50), we need only make the load $\mathbf{Z}_L$ a resistance, say R_L, and adjust the turns ratio a so that

$$\mathbf{Z}_1 = R_g = \frac{\mathbf{Z}_L}{a^2} = \frac{R_L}{a^2}$$

That is, solving for a we have

$$a = \sqrt{\frac{R_L}{R_g}} \qquad (21.53)$$

Example 21.8: Find the turns ratio a in Fig. 21.17 so that the maximum power is drawn from the source. Also find the maximum power.

Solution: The generator has $\mathbf{V}_g = 12\,\underline{/0°}$ V and $\mathbf{Z}_g = R_g = 4\,\underline{/0°}$ Ω, and the load is $R_L = 10$ kΩ. Therefore, by (21.53) the turns ratio is

$$a = \sqrt{\frac{R_L}{R_g}} = \sqrt{\frac{10{,}000}{4}} = 50$$

The reflected impedance thus is

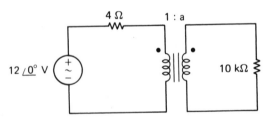

FIGURE 21.17 *Circuit for impedance matching.*

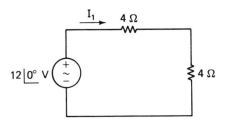

FIGURE 21.18   Equivalent circuit of Fig. 21.17.

$$\mathbf{Z}_1 = \frac{R_L}{a^2} = \frac{10,000}{(50)^2} = 4 \ \Omega$$

so that the equivalent circuit is that of Fig. 21.18.

The load impedance is now matched to the generator impedance of 4 Ω, so that the maximum power is drawn from the source. From Fig. 21.18 the current is

$$\mathbf{I}_1 = \frac{12 \underline{|0°}}{4 + 4} = 1.5 \underline{|0°} \ \text{A}$$

and thus the power, which is maximum, is

$$P = 12|\mathbf{I}_1| \cos 0° = 12(1.5) = 18 \ \text{W}$$

PRACTICE EXERCISES

21-4.1   In Fig. 21.13, if $\mathbf{V}_g = 100 \ \underline{|0°}$ V, $\mathbf{Z}_g = 6 + j3$ Ω, $\mathbf{Z}_L = 400 - j300$ Ω, and $a = 10$, find $\mathbf{I}_1$, $\mathbf{V}_1$, $\mathbf{I}_2$, and $\mathbf{V}_2$.   *Ans.* $10 \ \underline{|0°}$ A, $50 \ \underline{|-36.9°}$ V, $1 \ \underline{|0°}$ A, $500 \ \underline{|-36.9°}$ V

21-4.2   If in Exercise 21-4.1 the load is changed to $\mathbf{Z}_L = j300$ Ω, find $\mathbf{I}_1$, $\mathbf{V}_1$, $\mathbf{I}_2$, and $\mathbf{V}_2$.   *Ans.* $50/3 \ \sqrt{2} \ \underline{|-45°}$ A, $50/\sqrt{2} \ \underline{|45°}$ V, $5/3 \ \sqrt{2} \ \underline{|-45°}$ A, $500/\sqrt{2} \ \underline{|45°}$ V

21-4.3   Find the power delivered to the load $\mathbf{Z}_L$ in Exercise 21-4.1 and in Exercise 21-4.2.   *Ans.* 400 W, 0

21-4.4   Find the turns ratio so that the maximum power is drawn from the source. Also find the maximum power.   *Ans.* 100, 20 W

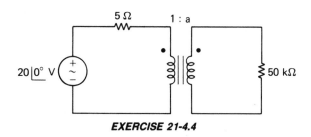

EXERCISE 21-4.4

21.5 TYPES OF TRANSFORMERS

As stated earlier, transformers are available in a variety of shapes and sizes, and are designed for many different uses. Some common types are power transformers, audio transformers, intermediate-frequency (IF) transformers and radio-frequency (RF) transformers. Power transformers, used in transmitting and distributing power, are relatively large, whereas IF and RF transformers, used in radio and television receivers and transmitters, are relatively small.

Isolation: One great advantage of the transformers we have considered thus far is their action as an *isolation* device. That is, the primary and secondary circuits are isolated from each other in that there is no physical connection between them. They are connected only by their mutual flux. Thus a step-down transformer could be included as part of a transmission line to step the line voltage down to a safe value in order to measure it. It is obviously safer to apply a voltmeter across a 100-V secondary, which is isolated from the primary, than it would be to apply it across a 50-kV primary.

Autotransformer: A power transformer which does not have the isolation feature, but which instead uses a common coil for both primary and secondary windings, is called an *autotransformer*. An example is the step-down autotransformer of Fig. 21.19(a), where the secondary terminal 2 is *tapped* to the primary winding at node

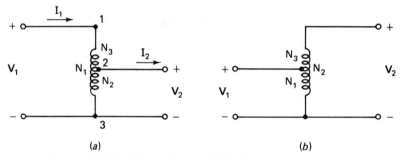

(a)   (b)

FIGURE 21.19 (a) Step-down and (b) step-up autotransformers.

2. The secondary has N_2 turns, as shown, and the number of primary turns is N_1 given by

$$N_1 = N_2 + N_3$$

Thus we have

$$\frac{V_2}{V_1} = \frac{N_2}{N_1} = \frac{N_2}{N_2 + N_3} \tag{21.54}$$

and

562

$$N_1\mathbf{I}_1 = (N_2 + N_3)\mathbf{I}_1 = N_2\mathbf{I}_2 \qquad (21.55)$$

Fig. 21.19(b) is a step-up autotransformer since the primary turns are fewer in number than the secondary turns (N_1 as opposed to $N_2 = N_1 + N_3$).

Autotransformers have the advantages of being compact and efficient, and are more economical because one winding is used rather than two. Their big disadvantage, of course, is the lack of isolation between the primary and secondary that is provided by the other transformers.

Multiple-Load Transformers: More than one load may be connected to the secondary by tapping into the secondary winding, or, as in Fig. 21.20, by connecting separate

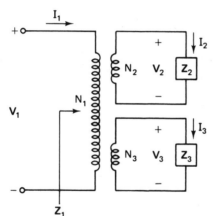

FIGURE 21.20 *Transformer with multiple loads.*

secondary windings. In the case of the *multiple-load* transformer of Fig. 21.20, the voltage ratios are

$$\frac{\mathbf{V}_1}{\mathbf{V}_2} = \frac{N_1}{N_2} \qquad \frac{\mathbf{V}_1}{\mathbf{V}_3} = \frac{N_1}{N_3} \qquad (21.56)$$

where N_1, N_2, and N_3 are the number of turns of the coils. We may also find $\mathbf{V}_2/\mathbf{V}_3$ from (21.56) as

$$\frac{\mathbf{V}_2}{\mathbf{V}_3} = \frac{(N_2/N_1)\mathbf{V}_1}{(N_3/N_1)\mathbf{V}_1}$$

which simplifies to

$$\frac{\mathbf{V}_2}{\mathbf{V}_3} = \frac{N_2}{N_3} \tag{21.57}$$

Thus in every case the voltages are in the same ratio as the turns.

Since the number of ampere turns of the primary is the same as those of the secondary, we have in Fig. 21.20,

$$N_1\mathbf{I}_1 = N_2\mathbf{I}_2 + N_3\mathbf{I}_3 \tag{21.58}$$

From this result and the voltage ratios we may find the input impedance $\mathbf{Z}_1$, given by

$$\mathbf{Z}_1 = \frac{\mathbf{V}_1}{\mathbf{I}_1}$$

Substituting for $\mathbf{V}_1$ from (21.56) and for $\mathbf{I}_1$ from (21.58), we have

$$\mathbf{Z}_1 = \frac{(N_1/N_2)\mathbf{V}_2}{(N_2/N_1)\mathbf{I}_2 + (N_3/N_1)\mathbf{I}_3} = \frac{N_1/N_2}{(N_2/N_1)(\mathbf{I}_2/\mathbf{V}_2) + (N_3/N_1)(\mathbf{I}_3/\mathbf{V}_2)} \tag{21.59}$$

From Fig. 21.20 we have

$$\mathbf{I}_2 = \frac{\mathbf{V}_2}{\mathbf{Z}_2} \qquad \mathbf{I}_3 = \frac{\mathbf{V}_3}{\mathbf{Z}_3}$$

which substituted into (21.59) results in

$$\mathbf{Z}_1 = \frac{N_1/N_2}{N_2/N_1\mathbf{Z}_2 + (N_3/N_1\mathbf{Z}_3)(\mathbf{V}_3/\mathbf{V}_2)}$$

Finally, substituting for $\mathbf{V}_3/\mathbf{V}_2$ from (21.57), we have

$$\mathbf{Z}_1 = \frac{N_1/N_2}{N_2/N_1\mathbf{Z}_2 + N_3^2/N_1N_2\mathbf{Z}_3} = \frac{N_1^2}{N_2^2/\mathbf{Z}_2 + N_3^2/\mathbf{Z}_3} \tag{21.60}$$

If we define the turns ratios a_2 and a_3 by

$$a_2 = \frac{N_2}{N_1} \qquad a_3 = \frac{N_3}{N_1} \tag{21.61}$$

then (21.60) may be written

$$\mathbf{Z}_1 = \frac{1}{(N_2/N_1)^2/\mathbf{Z}_2 + (N_3/N_1)^2/\mathbf{Z}_3} = \frac{1}{a_2^2/\mathbf{Z}_2 + a_3^2/\mathbf{Z}_3} \qquad (21.62)$$

or

$$\frac{1}{\mathbf{Z}_1} = \frac{a_2^2}{\mathbf{Z}_2} + \frac{a_3^2}{\mathbf{Z}_3} = \frac{1}{\mathbf{Z}_2/a_2^2} + \frac{1}{\mathbf{Z}_3/a_3^2} \qquad (21.63)$$

Equation (21.63) is the relation for the equivalent impedance $\mathbf{Z}_1$ of two parallel impedances $\mathbf{Z}_2/a_2^2$ and $\mathbf{Z}_3/a_3^2$. Therefore, an equivalent circuit of Fig. 21.20 is that of Fig. 21.21, where it may be seen that the reflected impedances $\mathbf{Z}_2/a_2^2$ and $\mathbf{Z}_3/a_3^2$ of Fig. 21.20 "reflect" into the primary as parallel impedances.

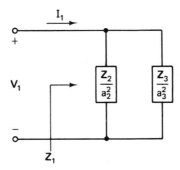

FIGURE 21.21   *Equivalent circuit of Fig. 21.20.*

Example 21.9:   Find the input impedance $\mathbf{Z}_1$ and the currents $\mathbf{I}_1$, $\mathbf{I}_2$, and $\mathbf{I}_3$ in Fig. 21.20, if $\mathbf{V}_1 = 144\,\underline{|0°}$ V, $\mathbf{Z}_2 = 6\,\underline{|0°}$ Ω, $\mathbf{Z}_3 = j2$ Ω, $N_1 = 1200$, $N_2 = 600$, and $N_3 = 400$.

Solution: By (21.60) the input impedance is

$$\mathbf{Z}_1 = \frac{(1200)^2}{(600)^2/6 + (400)^2/j2} = \frac{144}{36/6 + 16/j2} = \frac{144}{6 - j8}$$

or

$$\mathbf{Z}_1 = \frac{144}{10\,\underline{|-53.1°}} = \frac{72}{5}\,\underline{|53.1°}\ \Omega$$

The current $\mathbf{I}_1$ is given by

$$\mathbf{I}_1 = \frac{\mathbf{V}_1}{\mathbf{Z}_1} = \frac{144\,\underline{|0°}}{72/5\,\underline{|53.1°}} = 10\,\underline{|-53.1°}\ \text{A}$$

By (21.56) the secondary voltages are

$$\mathbf{V}_2 = \frac{N_2\mathbf{V}_1}{N_1} = \frac{600(144)}{1200} = 72\underline{|0°}\ \text{V}$$

and

$$\mathbf{V}_3 = \frac{N_3\mathbf{V}_1}{N_1} = \frac{400(144)}{1200} = 48\underline{|0°}\ \text{V}$$

Therefore, the secondary currents are

$$\mathbf{I}_2 = \frac{\mathbf{V}_2}{\mathbf{Z}_2} = \frac{72\underline{|0°}}{6} = 12\underline{|0°}\ \text{A}$$

and

$$\mathbf{I}_3 = \frac{\mathbf{V}_3}{\mathbf{Z}_3} = \frac{48\underline{|0°}}{j2} = -j24 = 24\underline{|-90°}\ \text{A}$$

As a check the secondary ampere turns are

$$N_2\mathbf{I}_2 + N_3\mathbf{I}_3 = (600)(12) - j(400)(24)$$

$$= 7200 - j9600$$

$$= 12,000\underline{|-53.1°}$$

and the primary ampere turns are

$$N_1\mathbf{I}_1 = 1200(10\underline{|-53.1°}) = 12,000\underline{|-53.1°}$$

These are the same, as they should be.

PRACTICE EXERCISES

21-5.1 In Fig. 21.19(a) the autotransformer has $N_1 = 1000$, $N_2 = 400$, and $N_3 = 600$ turns. Find $\mathbf{I}_2$ and $\mathbf{V}_2$ if $\mathbf{V}_1 = 100\underline{|0°}$ V and $\mathbf{I}_1 = 4\underline{|30°}$ A.   *Ans.* $10\underline{|30°}$ A, $40\underline{|0°}$ V

21-5.2 In Fig. 21-20 we have $\mathbf{V}_1 = 100\underline{|0°}$ V, $\mathbf{Z}_2 = 8 + j6\ \Omega$, $\mathbf{Z}_3 = 5\ \Omega$, $N_1 = 800$, $N_2 = 400$, and $N_3 = 160$. Find $\mathbf{I}_1$, $\mathbf{I}_2$, and $\mathbf{I}_3$.   *Ans.* $2.8 - j1.5$ A, $4 - j3$ A, 4 A

21.6 SUMMARY

The changing current in a coil produces a changing magnetic flux, which in turn produces a voltage across the terminals of the coil in accordance with Faraday's law. If one or more other coils are in the vicinity and thus are linked by the flux, they too will have voltages produced across their terminals. In this case the coils are said to be *mutually coupled* and to have a *mutual inductance M* among them. The mutual inductance plays the same role and has the same unit as the inductance *L*, which is called the *self inductance,* of a single coil. That is, the voltage v_1 produced in one coil by a current i_2 in another coil is

$$v_1 = M \frac{di_2}{dt}$$

If most, or ideally all, of the flux produced in one coil links another coil, the coils are *closely coupled,* or ideally *perfectly coupled.* Coupling is measured by a *coefficient of coupling k* which varies from 0 (no coupling) to 1 (perfect coupling).

If the directions in which the coils are wound can be determined, the polarities of the induced voltages can be ascertained. However, in the circuit symbols for the coils, the directions of the windings cannot be determined. Therefore, a dot convention is used to determine the polarities. Dots are also needed in many cases for the actual devices when the coils are enclosed in a container and the windings are not visible.

Two or more mutually coupled coils that are wound on a single core constitute a *transformer,* which is used to raise or lower voltages by producing a voltage across one coil that is higher or lower than that across another coil. If the coupling is nearly perfect and the self inductances of the coils are extremely high, the transformer approaches a so-called *ideal* transformer. In this case the voltages are in the same ratio as the number of turns in their coils. That is, if v_1 and N_1 are the voltage and turns of one coil in a two-coil transformer, and v_2 and N_2 are the voltage and turns of the other coil, then

$$\frac{v_2}{v_1} = \frac{N_2}{N_1} = a$$

The number a is the turns *ratio.* Also, in this case the ampere-turns are the same for both coils. That is, if i_1 and i_2 are the currents, then

$$N_1 i_1 = N_2 i_2$$

or

$$i_1 = \frac{N_2}{N_1} i_2 = a i_2$$

Also, the coils may be wound to reverse the polarities in these last two results. In this case, a is replaced by $-a$.

PROBLEMS

21.1 Find v_1 and v_2 in Fig. 21.5(a) if $L_1 = 3$ H, $L_2 = 2$ H, $M = 1.5$ H, and the rates of change of the currents are

$$\frac{di_1}{dt} = 40 \text{ A/s} \qquad \frac{di_2}{dt} = -10 \text{ A/s}$$

21.2 Find v_1 and v_2 in Fig. 21.5(a) if $L_1 = 40$ mH, $L_2 = 90$ mH, the coefficient of coupling is $k = 0.5$, and the rates of change of the currents are as in Problem 21.1.

21.3 Repeat Problem 21.1 if the circuit is that of Fig. 21.6(a).

21.4 Repeat Problem 21.2 if the circuit is that of Fig. 21.6(a).

21.5 Find the coefficient of coupling for a transformer with $L_1 = 20$ mH and $L_2 = 80$ mH, if (a) $M = 5$ mH, (b) $M = 30$ mH, and (c) $M = 40$ mH.

21.6 Find the mutual inductance M for a transformer with $L_1 = 10$ mH and $L_2 = 40$ mH, if the coefficient of coupling is (a) $k = 0.1$, (b) $k = 0.8$, and (c) $k = 1$.

21.7 Find $\mathbf{V}_1$ and $\mathbf{V}_2$ in Fig. 21.7(a) if $L_1 = 3$ H, $L_2 = 2$ H, $M = 2$ H, $\mathbf{I}_1 = 2\underline{|0°}$ A, $\mathbf{I}_2 = 4 + j3$ A, and the frequency is $\omega = 4$ rad/s.

21.8 Repeat Problem 21.7 if the circuit is that of Fig. 21.7(b).

21.9 Find the steady-state currents i_1 and i_2 if $M = 1$ H and $R = 2$ Ω.

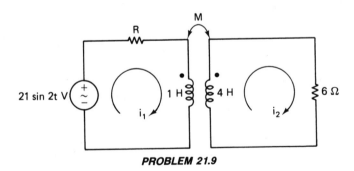

PROBLEM 21.9

21.10 Repeat Problem 21.9 if $M = 2$ H and $R = \frac{15}{8}$ Ω.

21.11 Find the energy stored in the transformer at an instant when $i_1 = 2$ A and $i_2 = 3$ A.

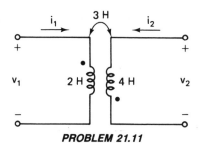

PROBLEM 21.11

21.12 Repeat Problem 21.11 if the dot on the secondary side is moved to the upper terminal.

21.13 Find the phasor currents I_1 and I_2 if $R = 1\ \Omega$ and $X = 6\ \Omega$.

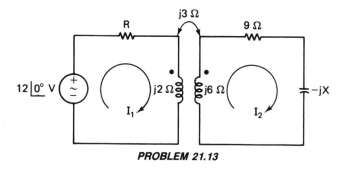

PROBLEM 21.13

21.14 Solve Problem 21.13 if $R = 1.5\ \Omega$ and $X = 18\ \Omega$.

21.15 An ideal transformer has 600 turns on the primary side and 1200 turns on the secondary side. If the primary voltage and current are $V_1 = 50\underline{|0°}$ V and $I_1 = 4\underline{|60°}$ A, find the turns ratio, the secondary voltage, and the secondary current. [Assume that the dots are as in Fig. 21.12(a).]

21.16 Solve Problem 21.15 if the dots are as in Fig. 21.12(b).

21.17 Find the power delivered to the primary terminals and to the load of the transformer of Problem 21.15.

21.18 Find the power delivered to the $300 - j500 -\Omega$ load using reflected impedance.

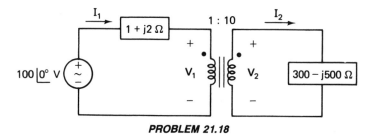

PROBLEM 21.18

21.19 Find $\mathbf{I}_1$ and $\mathbf{I}_2$ in Problem 21.18 using two loop equations. (*Suggestion:* Note that $\mathbf{V}_2 = 10\ \mathbf{V}_1$, etc.)

21.20 Find $\mathbf{V}_1$ and $\mathbf{V}_2$ in Problem 21.18.

21.21 Find the turns ratio a so that the power delivered to the 20-$k\Omega$ load is a maximum if (a) $R_g = 2\ \Omega$ and (b) $R_g = 8\ \Omega$. Find the maximum power in each case.

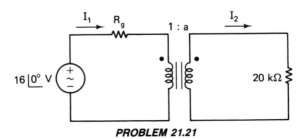

PROBLEM 21.21

21.22 Find the currents $\mathbf{I}_1$ and $\mathbf{I}_2$ in Problem 21.21 if $a = 20$ and $R_g = 14\ \Omega$.

21.23 The autotransformer of Fig. 21.19(a) has $N_1 = 800$, $N_2 = 200$, and $N_3 = 600$ turns. Find $\mathbf{V}_2$ and $\mathbf{I}_2$ if $\mathbf{V}_1 = 60\,\underline{|0°}$ V and $\mathbf{I}_1 = 3\,\underline{|10°}$ A.

21.24 Find $\mathbf{V}_2$ and $\mathbf{I}_2$ in the autotransformer of Fig. 21.19(b) if $N_1 = 400$ turns, $N_3 = 600$ turns, $\mathbf{V}_1 = 20\,\underline{|15°}$ V, and $\mathbf{I}_1 = 5\,\underline{|6°}$ A.

21.25 Find $\mathbf{Z}_1$ in Fig. 21.20 if $N_1 = 1000$ turns, $N_2 = 200$ turns, $N_3 = 400$ turns, $\mathbf{Z}_2 = 2\ \Omega$, and $\mathbf{Z}_3 = j8\ \Omega$.

21.26 Find $\mathbf{I}_1$, $\mathbf{I}_2$, and $\mathbf{I}_3$ in Problem 21.25 if $\mathbf{V}_1 = 50\,\underline{|0°}$ V.

22

FILTERS

In this last chapter we will consider a very important class of circuits known as *electric filters*. These are circuits that *pass* signals of certain frequencies and *block* or *attenuate* signals of other frequencies. For example, when we turn a radio selection knob or a television channel selector we are tuning an electric filter to the frequency of a certain radio station or television channel. The desired signal comes through and the other signals are filtered out.

Whether a filter passes or blocks a frequency such as ω_1 rad/s (or $f_1 = \omega_1/2\pi$ Hz) may be determined by the output signal of the filter at the frequency ω_1. For example, if the output signal is a sinusoidal voltage *v(t)*, then its phasor **V**, which we denote by $\mathbf{V}(j\omega)$, will depend on the filter circuit and the frequency ω. If the amplitude at ω_1, given by $|\mathbf{V}(j\omega_1)|$, is relatively large, then ω_1 passes, and if $|\mathbf{V}(j\omega_1)|$ is relatively small, then ω_1 is blocked. Thus in filter theory, we will need to consider the amplitude $|\mathbf{V}(j\omega)|$ as a function of the frequency ω. We will do this in the first section before we consider the different types of filters.

Closely associated with filter theory is the concept of *resonance*, which we will also consider in this chapter. The frequency which is of primary interest in a filter is the one that the filter concentrates on passing, or in some types, concentrates on blocking. This is the resonant frequency, and when the circuit is driven at this frequency it is said to be in resonance.

22.1 AMPLITUDE AND PHASE RESPONSES

If the frequency ω rad/s, or $f = \omega/2\pi$ Hz, is not fixed, but is allowed to change in a given ac circuit, then the amplitudes and phases of the current and voltage phasors will change. For example, let us consider the time-domain circuit of Fig. 22.1(a), where the input is

$$v_g = \sqrt{2}\,\sin \omega t\ \text{V}$$

and the output is the current i. The frequency ω is unspecified and may be any value.

In the corresponding phasor circuit of Fig. 22.1(b), the input is

$$\mathbf{V}_g = 1\,\underline{|0°}\ \text{V}$$

and the output phasor is therefore

$$\mathbf{I} = \frac{\mathbf{V}_g}{4 + j2\omega} = \frac{1}{4 + j2\omega}$$

Putting this result in polar form we have

$$\mathbf{I} = \frac{1}{\sqrt{4^2 + (2\omega)^2}\,\underline{|\tan^{-1} 2\omega/4}}$$

or

$$\mathbf{I} = \frac{1}{2\sqrt{4 + \omega^2}}\,\underline{|-\tan^{-1} \omega/2} = |\mathbf{I}|\,\underline{|\phi}\ \text{A} \tag{22.1}$$

The time-domain current, therefore, is

$$i = \sqrt{2}\,|\mathbf{I}|\,\sin (\omega t + \phi)\ \text{A}$$

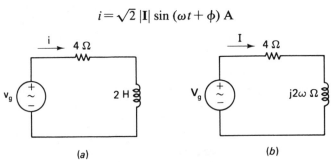

(a)   (b)

FIGURE 22.1 (a) Time-domain circuit and (b) its corresponding phasor circuit.

where by (22.1) the amplitude is

$$|\mathbf{I}| = \frac{1}{2\sqrt{4+\omega^2}} \tag{22.2}$$

and the phase is

$$\phi = -\tan^{-1}\frac{\omega}{2} \tag{22.3}$$

Thus both the amplitude and phase, which are also called the *amplitude response* and the *phase response,* are functions of the frequency ω.

Sketch of the Responses: To see more clearly how the amplitude and phase responses vary with the frequency, we may plot them versus ω, as shown in Fig. 22.2. In the case of the amplitude response of Fig. 22.2(a), we see from (22.2) that $|\mathbf{I}|$ is a maximum when its denominator is a minimum, which occurs at $\omega = 0$. For any other frequency, ω^2 is positive and the denominator is larger. Therefore, the amplitude response starts at its peak value of $1/2\sqrt{4} = \frac{1}{4}$ when $\omega = 0$, and steadily declines as ω increases.

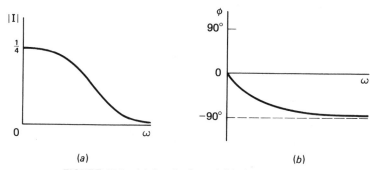

(a)   (b)

FIGURE 22.2   *(a) Amplitude and (b) phase response.*

In the case of the phase response, we see from (22.3) that when $\omega = 0$, the phase is $\phi = -\tan^{-1} 0 = 0$. As ω increases the phase decreases, approaching $-90°$ as ω gets large. This is because

$$-\tan\phi = \frac{\omega}{2}$$

and $\tan 90° = \infty$. The phase response is shown in Fig. 22.2(b).

Example 22.1: Find and sketch the amplitude response for the circuit of Fig. 22.3(a) if the input is

$$v_1 = \sqrt{2}\, \sin \omega t \text{ V} \qquad\qquad (22.4)$$

and the output is v_2.

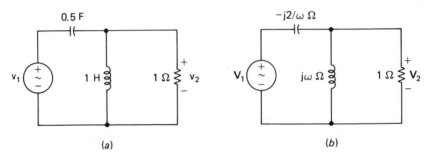

(a)   (b)

FIGURE 22.3 (a) Time-domain circuit with two reactive elements and (b) its corresponding phasor circuit.

Solution: The phasor circuit is shown in Fig. 22.3(b), from which we see that the impedance Z_1 of the parallel combination of the inductor and resistor is

$$Z_1 = \frac{1(j\omega)}{1 + j\omega} = \frac{j\omega}{1 + j\omega}$$

Therefore, by voltage division we have

$$\frac{V_2}{V_1} = \frac{\dfrac{j\omega}{1 + j\omega}}{-j\dfrac{2}{\omega} + \dfrac{j\omega}{1 + j\omega}} = \frac{j\omega}{-j\dfrac{2}{\omega} + 2 + j\omega}$$

which, since $V_1 = 1\,\underline{|0^\circ} = 1$, may be simplified to

$$V_2 = \frac{j\omega^2}{2\omega + j(\omega^2 - 2)}$$

From this result the amplitude may be found to be

$$|V_2| = \frac{\omega^2}{\sqrt{(2\omega)^2 + (\omega^2 - 2)^2}}$$

$$= \frac{\omega^2}{\sqrt{4\omega^2 + \omega^4 - 4\omega^2 + 4}}$$

or

$$|V_2| = \frac{\omega^2}{\sqrt{4 + \omega^4}} \qquad\qquad (22.5)$$

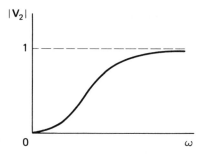

FIGURE 22.4 Amplitude response for Fig. 22.3.

To sketch the amplitude response more easily, we may divide the numerator and denominator by ω^2, resulting in

$$|\mathbf{V_2}| = \frac{1}{\sqrt{4 + \omega^4}/\omega^2} = \frac{1}{\sqrt{(4 + \omega^4)/\omega^4}}$$

or

$$|\mathbf{V_2}| = \frac{1}{\sqrt{(4/\omega^4) + 1}} \tag{22.6}$$

Now we may see from (22.5) that the response starts at 0 when $\omega = 0$, and from (22.6) that as ω increases, the denominator of the response decreases toward 1. Therefore, the response steadily rises from 0 at $\omega = 0$ and approaches $1/1 = 1$ for large ω. The response is sketched in Fig. 22.4.

PRACTICE EXERCISES

22-1.1 Find the amplitude response $|\mathbf{V_1}|$, where $\mathbf{V_1}$ is the phasor of v_1.

$$Ans. \ \frac{0.2|\omega|}{\sqrt{\omega^4 - 1.96\,\omega^2 + 1}}$$

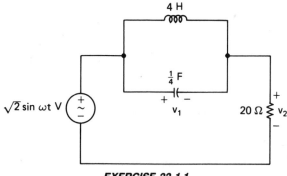

EXERCISE 22-1.1

22-1.2 Find the amplitude response $|V_2|$, where V_2 is the phasor of v_2 in Exercise 22-1.1.

$$Ans. \ \frac{|1 - \omega^2|}{\sqrt{\omega^4 - 1.96\,\omega^2 + 1}}$$

22.2 RESONANCE

A physical system, such as an electric circuit, which has a sinusoidal type of source-free, or *natural,* response, reacts vigorously when an input is applied with a frequency at or near its *natural frequency* (the frequency of its natural response). This type of vigorous behavior is known as *resonance* and a circuit with such an input is a *resonant* circuit.

In this section we will consider resonance in electric circuits, but there are many other examples of resonance. For instance, a singer's voice may break a crystal goblet if a note is properly produced at exactly the right frequency. Also, a bridge may be destroyed if it is subjected to a periodic force with the same frequency as one of its natural frequencies. Such a force may be produced by a high, pulsating wind, as in the case of the destruction of the Tacoma Narrows Bridge over Puget Sound in 1940. It is also possible for marching troops to set resonant forces in motion to destroy a bridge. For this reason no thoughtful commander will march troops in step across a bridge.

Resonance in Electric Circuits: We will define an ac electric circuit to be in resonance when the amplitude of its output function attains a pronounced maximum value, such as in the case of $|V|$ at the frequency f_r Hz, shown in Fig. 22.5(a). The frequency f_r Hz (or $\omega_r = 2\pi f_r$ rad/s) where the peak occurs is called the *resonant frequency.*

Depending on the output function, resonance may also be defined to exist when a pronounced minimum amplitude is attained, as in Fig. 22.5(b). Again, the resonant frequency is f_r.

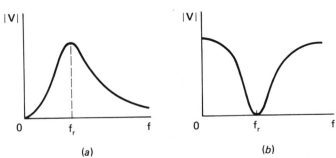

FIGURE 22.5 *Examples of amplitude responses in the case of resonance.*

Series Resonance: To see how the frequency may be varied to achieve resonance, let us consider the RLC series circuit of Fig. 22.6, where the input is the phasor $\mathbf{V}_g$, given by

$$\mathbf{V}_g = 1 \underline{|0°} \ \text{V}$$

and the output is the phasor **I**. The frequency is ω rad/s or $f = \omega/2\pi$ Hz.

The impedance seen by the source is given by

$$\mathbf{Z} = R + j\omega L - j\frac{1}{\omega C}$$

so that the current is

$$\mathbf{I} = \frac{\mathbf{V}}{\mathbf{Z}} = \frac{1}{R + j(\omega L - 1/\omega C)}$$

The amplitude therefore is

$$|\mathbf{I}| = \frac{1}{\sqrt{R^2 + (\omega L - 1/\omega C)^2}} \tag{22.7}$$

To find the resonant frequency $\omega_r = 2\pi f_r$ we note in (22.7) that the maximum amplitude occurs when the denominator in the right member is a minimum. Since ω is the only variable, this occurs when we have

$$\omega L - \frac{1}{\omega C} = 0$$

or

$$\omega L = \frac{1}{\omega C} \tag{22.8}$$

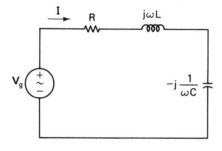

FIGURE 22.6 *RLC series circuit.*

That is, at resonance the inductive reactance $X_L = \omega L$ exactly cancels the capacitive reactance $X_C = 1/\omega C$.

We may write (22.8) in the form

$$\omega^2 = \frac{1}{LC}$$

so that the resonant frequency is

$$\omega = \omega_r = \frac{1}{\sqrt{LC}} \, \text{rad/s} \qquad\qquad (22.9)$$

In hertz, the resonant frequency is

$$f_r = \frac{\omega_r}{2\pi} = \frac{1}{2\pi \sqrt{LC}} \, \text{Hz} \qquad\qquad (22.10)$$

Substituting (22.8) into (22.7) we have the maximum amplitude $|\mathbf{I}|_{max}$ given by

$$|\mathbf{I}|_{max} = \frac{1}{\sqrt{R^2}} = \frac{1}{R}$$

Thus the circuit of Fig. 22.6 is in resonance when the frequency is f_r Hz, with a maximum amplitude of $1/\dot{R}$. If the voltage $\mathbf{V}_g$ has amplitude V rather than 1, the peak amplitude will be V/R. In either case, because the circuit is an RLC series circuit, the resonant condition is called *series resonance*.

Sketching the Amplitude Response: The amplitude response (22.7) may be readily sketched by noting the peak value at resonance, and by considering the circuit of Fig. 22.6. If the frequency is $f = 0$, the capacitor is an open circuit so that $\mathbf{I} = 0$. Also, for very high frequencies the inductor approaches an open circuit, so that again $\mathbf{I} = 0$. Thus the response is as shown in Fig. 22.7. It rises from 0 at $f = 0$ to the peak $1/R$ at $f = f_r$. It then decreases and approaches 0 again as f increases.

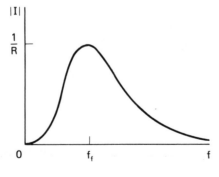

FIGURE 22.7 Amplitude response of the circuit of Fig. 22.6.

Parallel Resonance: Let us now consider the *RLC* parallel circuit of Fig. 22.8, where the input is I_g, the output is **V**, and the frequency is ω. The condition of resonance in this case is called *parallel resonance* and it occurs when the amplitude $|V|$ reaches its peak value.

The voltage **V** is given by

$$V = ZI_g$$

where **Z** is the impedance seen at the terminals of the source. Since *R, L,* and *C* are in parallel, we may write

$$\frac{1}{Z} = \frac{1}{R} + \frac{1}{j\omega L} + \frac{1}{-j(1/\omega C)}$$

$$= \frac{1}{R} + \frac{1}{j\omega L} + j\omega C$$

Therefore, we have

$$Z = \frac{1}{(1/R) + j(\omega C - 1/\omega L)} \qquad (22.11)$$

If we take the source phasor as

$$I_g = 1\underline{|0°}\ A$$

then the voltage **V** is given by

$$V = Z = \frac{1}{(1/R) + j(\omega C - 1/\omega L)}$$

The amplitude is therefore

$$|V| = \frac{1}{\sqrt{(1/R^2) + (\omega C - 1/\omega L)^2}} \qquad (22.12)$$

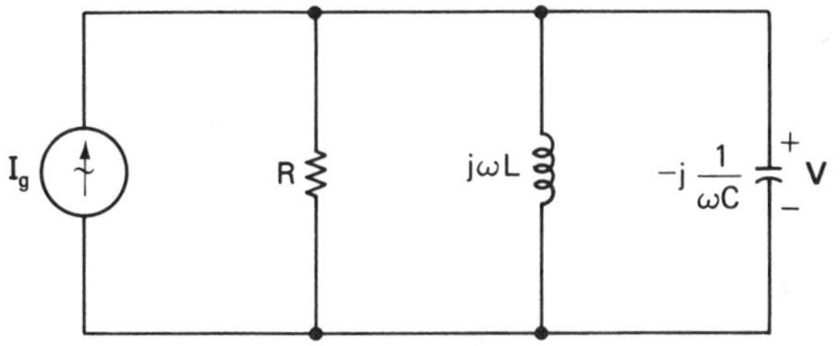

FIGURE 22.8 *RLC parallel circuit.*

This is very similar in form to the amplitude of (22.7) for the *RLC* series circuit. Thus by the same reasoning used in that case we see that the peak amplitude is $|V|_{max}$ given by

$$|V|_{max} = \frac{1}{\sqrt{1/R^2}} = R$$

occurring at the resonant frequency ω_r satisfying

$$\omega C - \frac{1}{\omega L} = 0$$

or, as before,

$$\omega_r = \frac{1}{\sqrt{LC}} \text{ rad/s}$$

In hertz we have

$$f_r = \frac{1}{2\pi \sqrt{LC}} \text{ Hz} \qquad (22.13)$$

Thus, as in the *RLC* series circuit, resonance occurs when the impedance seen at the source terminals is real. That is, its reactive parts exactly cancel, and the current and voltage at the input terminals are in phase. In both cases the resonant frequency has the same value, given by (22.10) and (22.13). The sketch of $|V|$ in the parallel case is identical in shape to that of $|I|$ in the series case, shown in Fig. 22.7. The only difference is the value of the peak points.

Example 22.2: For the *RLC* parallel circuit òf Fig. 22.8, $R = 50$ kΩ, $L = 4$ mH, $C = 100$ nF, and $I_g = 2 \underline{|0°}$ mA. Find the resonant frequency and the amplitude $|V|$ of the output voltage at resonance.

Solution: The resonant frequency is given by

$$f_r = \frac{1}{2\pi \sqrt{LC}} = \frac{1}{2\pi \sqrt{(4 \times 10^{-3})(100 \times 10^{-9})}}$$

$$= 7958 \text{ Hz}$$

At resonance the impedance, by (22.11), is

$$\mathbf{Z} = R = 50 \text{ k}\Omega$$

Therefore, the voltage is

$$\mathbf{V} = \mathbf{ZI}_g = (50 \text{ k}\Omega)(2 \underline{|0°} \text{ mA}) = 100 \underline{|0°} \text{ V}$$

and thus the amplitude at resonance is

$$|\mathbf{V}| = 100 \text{ V}$$

Finding L or C from f_r: If we are given the resonant frequency f_r and the capacitance C in either the RLC series or parallel circuit, we may find the necessary value of L. Similarly, given f_r and L, we may find C. If we square both sides of (22.13), we have

$$f_r^2 = \frac{1}{(2\pi)^2 LC} = \frac{1}{4\pi^2 LC}$$

Solving for L, we have

$$L = \frac{1}{4\pi^2 f_r^2 C} \tag{22.14}$$

and solving for C, we have

$$C = \frac{1}{4\pi^2 f_r^2 L} \tag{22.15}$$

Example 22.3:   Find the inductance needed in the RLC series circuit if $C = 20$ pF and resonance is to occur at 40 kHz.

Solution: By (22.14) we have

$$L = \frac{1}{\omega^2 C}$$

$$= \frac{1}{4\pi^2 (40 \times 10^3)^2 (20 \times 10^{-12})} \text{ H}$$

$$= 792 \text{ mH}$$

PRACTICE EXERCISES

22.2.1   Find the resonant frequency and the peak current amplitude in the RLC series circuit of Fig. 22.6 if $R = 10 \ \Omega$, $L = 8$ mH, $C = 0.2 \ \mu$F, and $\mathbf{V}_g = 20 \underline{|0°}$ V.

Ans. 4 kHz, 2 A

22-2.2 In the *RLC* parallel circuit of Fig. 22.8 we have $L = 10$ mH and $|\mathbf{I}_g| = 20$ mA. Find R and C if the peak voltage amplitude is $|\mathbf{V}| = 100$ V at a resonant frequency of 1500 Hz. *Ans.* 5 kΩ, 1.1 μF

22-2.3 Find the inductance needed in the *RLC* parallel circuit if $C = 0.01$ μF and resonance is to occur at 2 kHz. *Ans.* 0.63 H

22.3 BANDPASS FILTERS

In the case of the amplitude response of the RLC circuit shown in Fig. 22.7, we see that frequencies clustered around f_r correspond to relatively large amplitudes, while those near zero and larger than f_r correspond to relatively small amplitudes. Thus the *RLC* series circuit with output $\mathbf{I}$ is an example of a *bandpass filter*, which passes the *band* of frequencies centered around f_r and rejects the frequencies that are lower and higher than those in the band.

Network Functions: The response of Fig. 22.7 was sketched for an input $\mathbf{V}_g = 1\underline{|0°}$ V. If the input had been the more general value $\mathbf{V}_g = V\underline{|0°}$ V, the response would still have had the same general shape. The values on the vertical axis would simply be multiplied by V. For this reason, we may sketch $|\mathbf{I}/\mathbf{V}_g|$ rather than $|\mathbf{I}|$ and not lose any information in the amplitude response. Dividing out the input amplitude $|\mathbf{V}_g|$ also has the advantage that only the effect of the circuit is being considered, rather than the circuit and the input.

We will denote the ratio of the output phasor to the input phasor by $\mathbf{H}(j\omega)$, and define it as the *network function*. (Another term that is sometimes used is *transfer function*.) The units of $\mathbf{H}$ may be ohms (if the output is a voltage and the input is a current) or mhos (output is a current and the input is a voltage, as in Fig. 22.7), or $\mathbf{H}$ may be dimensionless (ratio of two voltages or two currents).

General Case: In the general bandpass amplitude $|\mathbf{H}(j\omega)|$, shown in Fig. 22.9, we will refer to the resonant frequency ω_r or f_r as the *center frequency* ω_0 rad/s or

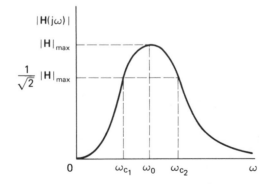

FIGURE 22.9 *General bandpass amplitude response.*

f_0 Hz. The band of frequencies that passes, or the *passband*, is defined to be

$$\omega_{c_1} \leq \omega \leq \omega_{c_2}$$

where ω_{c_1} and ω_{c_2} are the *cutoff points*, or *cutoff frequencies*, defined as the frequencies at which the amplitude is $1/\sqrt{2} = 0.707$ times the maximum amplitude. The width of the passband, given by

$$B = \omega_{c_2} - \omega_{c_1} \text{ rad/s} \qquad (22.16)$$

is called the *bandwidth*. In hertz we have

$$B = f_{c_2} - f_{c_1} \text{ Hz} \qquad (22.17)$$

where $f_{c_1} = \omega_{c_1}/2\pi$ and $f_{c_2} = \omega_{c_2}/2\pi$.

Because power is often associated with the square of the amplitude function, and $(1/\sqrt{2})^2 = 1/2$, the cutoff frequencies are also called the *half-power* points or frequencies.

RLC Series Circuit Example: In the case of the RLC series circuit of Fig. 22.6, if the output is the current $\mathbf{I}$ and the input is the source voltage $\mathbf{V}_g$, the network function is

$$\mathbf{H}(j\omega) = \frac{\mathbf{I}}{\mathbf{V}_g} = \frac{1}{\mathbf{Z}} = \frac{1}{R + j(\omega L - 1/\omega C)} \qquad (22.18)$$

with amplitude response, by (22.7),

$$|\mathbf{H}(j\omega)| = \frac{1}{\sqrt{R^2 + (\omega L - 1/\omega C)^2}} \qquad (22.19)$$

The response, of course, is that of Fig. 22.7 or Fig. 22.9.

We already know that the center frequency or resonant frequency is given by

$$\omega_0 = \frac{1}{\sqrt{LC}} \qquad (22.20)$$

To find the cutoff points we note that

$$|\mathbf{H}|_{max} = \frac{1}{R}$$

and thus at ω_{c_1} and ω_{c_2} we must have an amplitude of

$$|\mathbf{H}| = \frac{1}{\sqrt{2}} |\mathbf{H}|_{\text{max}} = \frac{1}{\sqrt{2}\,R}$$

This happens, as we see in (22.19), when

$$\omega L - \frac{1}{\omega C} = \pm R$$

which is the equation satisfied by the cutoff points. Transposing the term $\pm R$ to the left side and multiplying through by ω/L, we have

$$\omega^2 \mp \frac{R}{L}\omega - \frac{1}{LC} = 0 \qquad\qquad (22.21)$$

Using the negative sign on R, we have by the quadratic formula

$$\omega = \frac{(R/L) \pm \sqrt{(R/L)^2 + 4/LC}}{2}$$

For a positive value of ω we reject the minus sign, resulting in

$$\omega_{c_2} = \frac{(R/L) + \sqrt{(R/L)^2 + 4/LC}}{2} \qquad\qquad (22.22)$$

By similar calculations using the positive sign on R in (22.21), we have

$$\omega_{c_1} = \frac{-(R/L) + \sqrt{(R/L)^2 + 4/LC}}{2} \qquad\qquad (22.23)$$

We have labeled the answers ω_{c_2} and ω_{c_1} because they are the cutoff points and the value in (22.22) is larger than that in (22.23).

Bandwidth: We may find the bandwidth B from (22.16), (22.22), and (22.23). The result is

$$B = \omega_{c_2} - \omega_{c_1}$$
$$= \left(\frac{R}{2L} + \frac{\sqrt{(R/L)^2 + 4/LC}}{2}\right) - \left(-\frac{R}{2L} + \frac{\sqrt{(R/L)^2 + 4/LC}}{2}\right)$$

or

$$B = \frac{R}{L}\ \text{rad/s} \qquad\qquad (22.24)$$

Thus to make the passband wider or narrower, we may adjust R. (We could also adjust L but this will change the center frequency $\omega_0 = 1/\sqrt{LC}$ unless C is also adjusted.)

Quality Factor: A good measure of *selectivity* or *sharpness of peak* in a resonant circuit is the so-called *quality factor Q*, which is defined as the ratio of the resonant frequency to the bandwidth. That is, we have

$$Q = \frac{\omega_0}{B} \tag{22.25}$$

(The letter Q is also our symbol for reactive power, as the reader will recall. However, the two quantities are used in completely different applications, so there should be no confusion.)

From (22.25) we see that a low Q corresponds to a large bandwidth B and a high Q corresponds to a small bandwidth. Thus a high Q (sometimes arbitrarily taken as 10 or more) indicates a very *selective* circuit (a very narrow passband).

In the case of the RLC series circuit, substitution of (22.24) into (22.25) yields

$$Q = \frac{\omega_0}{R/L}$$

or

$$Q = \frac{\omega_0 L}{R} \tag{22.26}$$

At resonance $L = 1/\omega_0^2 C$, so that (22.26) becomes

$$Q = \frac{\omega_0 (1/\omega_0^2 C)}{R}$$

or

$$Q = \frac{1}{\omega_0 RC} \tag{22.27}$$

Since we have at resonance

$$\omega_0 L = X_L$$

and

$$\frac{1}{\omega_0 C} = X_C$$

we may write (22.26) and (22.27) in the forms, for the RLC series circuit,

$$Q = \frac{X_L}{R} \tag{22.28}$$

and

$$Q = \frac{X_C}{R} \tag{22.29}$$

Approximate Cutoff Points for High Q: Using (22.20) and (22.24), we may obtain the cutoff points of (22.22) and (22.23) in terms of the center frequency and the bandwidth. The results are

$$\omega_{c_1} = \frac{-B + \sqrt{B^2 + 4\omega_0^2}}{2} \tag{22.30}$$

and

$$\omega_{c_2} = \frac{B + \sqrt{B^2 + 4\omega_0^2}}{2} \tag{22.31}$$

If Q is high, then B is low, and B^2 may be neglected in the sum $B^2 + 4\omega_0^2$ in (22.30) and (22.31). Good approximations to the cutoff points in this case are

$$\omega_{c_1} = \frac{-B + \sqrt{4\omega_0^2}}{2} = \omega_0 - \frac{B}{2} \tag{22.32}$$

and

$$\omega_{c_2} = \frac{B + \sqrt{4\omega_0^2}}{2} = \omega_0 + \frac{B}{2} \tag{22.33}$$

Thus for high Q, the center frequency is about halfway between the cutoff points, in approximately the center of the passband.

Example 22.4: In the RLC series circuit we have $R = 100 \ \Omega$, $L = 0.1$ H, and $C = 0.1$ μF. Find the center frequency, the bandwidth, Q, and the approximate values of the cutoff points.

Solution: The center frequency is

$$\omega_0 = \frac{1}{\sqrt{LC}} = \frac{1}{\sqrt{(0.1)(10^{-7})}} = 10^4 \text{ rad/s}$$

and the bandwidth is

$$B = \frac{R}{L} = \frac{100}{0.1} = 1000 \text{ rad/s}$$

Therefore, we have the quality factor

$$Q = \frac{\omega_0}{B} = \frac{10^4}{1000} = 10$$

Since this is considered to be a high Q, the approximate cutoff points are, by (22.32) and (22.33),

$$\omega_{c_1} = \omega_0 - \frac{B}{2} = 10^4 - \frac{1000}{2} = 9500 \text{ rad/s}$$

and

$$\omega_{c_2} = \omega_0 + \frac{B}{2} = 10^4 + \frac{1000}{2} = 10{,}500 \text{ rad/s}$$

In hertz we have

$$f_0 = \frac{10^4}{2\pi} = 1592 \text{ Hz}$$

$$f_{c_1} = \frac{9500}{2\pi} = 1512 \text{ Hz}$$

and

$$f_{c_2} = \frac{10{,}500}{2\pi} = 1671 \text{ Hz}$$

The bandwidth is

$$B = \frac{1000}{2\pi} = 159 \text{ Hz}$$

which checks with

FILTERS

CHAPTER 22

$$B = f_{c_2} - f_{c_1}$$

RLC Parallel Circuit Example: In the case of the RLC parallel circuit of Fig. 22.8, the network function is, by (22.11),

$$\mathbf{H}(j\omega) = \frac{1}{(1/R) + j(\omega C - 1/\omega L)}$$

We may show, in the same manner as for the series circuit, that

$$\omega_0 = \frac{1}{\sqrt{LC}}$$

$$B = \frac{1}{RC}$$

$$Q = \frac{\omega_0}{B} = \omega_0 RC = \frac{R}{\omega_0 L}$$

Equations (22.30) and (22.31) also hold for the parallel circuit as well as their approximations (22.32) and (22.33) for high Q.

PRACTICE EXERCISES

22-3.1 Show by multiplying the cutoff points in (22.30) and (22.31) that

$$\omega_0^2 = \omega_{c_1} \omega_{c_2}$$

or

$$f_0^2 = f_{c_1} f_{c_2}$$

Use this result to find the center frequency if the cutoff points are 4 kHz and 16 kHz.   *Ans.* 8 kHz

22-3.2 For the RLC series circuit we have $R = 40\ \Omega$, $L = 8$ mH, and $C = 0.2\ \mu$F. Find ω_0, B, Q, ω_{c_1}, and ω_{c_2}.
Ans. 25,000 rad/s; 5000 rad/s; 5; 22,625 rad/s; 27,625 rad/s

22-3.3 For the RLC parallel circuit we have $R = 5$ kΩ, $L = 5$ mH, and $C = 0.02\ \mu$F. Find ω_0, B, Q, and the approximate values of the cutoff points.
Ans. 100,000 rad/s; 10,000 rad/s; 10; 95,000 rad/s; 105,000 rad/s

22-3.4 Show that the given circuit is a bandpass filter with $\mathbf{H} = \mathbf{V}_2/\mathbf{V}_1$, and find ω_0, B, and Q.   *Ans.* 10,000 rad/s, 1000 rad/s, 10

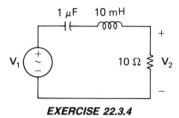

EXERCISE 22.3.4

22.4 LOW-PASS FILTERS

A *low-pass* filter is one that passes low frequencies and blocks high frequencies. A typical low-pass amplitude response $|\mathbf{H}(j\omega)|$ is that of Fig. 22.10, where ω_c is the *cutoff* frequency and the band $0 \le \omega \le \omega_c$ is the *passband*. The *bandwidth* in the low-pass case is thus $B = \omega_c$ rad/s, or $B = f_c = \omega_c/2\pi$ Hz.

As in the bandpass filter, the frequencies that pass correspond to amplitudes greater than or equal to $1/\sqrt{2} = 0.707$ times the maximum amplitude. Unlike the bandpass filter, however, in the low-pass case there is only one cutoff point and one band of frequencies that is blocked.

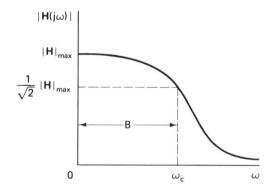

FIGURE 22.10 Low-pass filter amplitude.

An example of a low-pass filter, as we may see from its amplitude response of Fig. 22.2(a), is the circuit of Fig. 22.1(a), with network function

$$|\mathbf{H}(j\omega)| = \frac{\mathbf{I}}{\mathbf{V}_g}$$

Example 22.5: Find the network function

$$\mathbf{H}(j\omega) = \frac{\mathbf{V}_2}{\mathbf{V}_1}$$

in the circuit of Fig. 22.11, with $R = 1$ kΩ, $L = 0.1$ H, and $C = 0.05$ μF, and show that it is a low-pass filter circuit. Also find the cutoff frequency f_c.

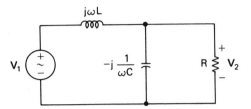

FIGURE 22.11 Low-pass filter circuit.

Solution: The parallel RC combination has impedance $\mathbf{Z}_1$ given by

$$\mathbf{Z}_1 = \frac{R\left[-j(1/\omega C)\right]}{R - j(1/\omega C)} = \frac{-jR}{R\omega C - j1}$$

Thus by voltage division we have

$$\mathbf{H}(j\omega) = \frac{\mathbf{V}_2}{\mathbf{V}_1} = \frac{\mathbf{Z}_1}{j\omega L + \mathbf{Z}_1}$$

$$= \frac{\dfrac{-jR}{R\omega C - j1}}{j\omega L - \dfrac{jR}{R\omega C - j1}}$$

which may be simplified to

$$\mathbf{H}(j\omega) = \frac{-jR}{\omega L + j(\omega^2 RLC - R)}$$

or

$$\mathbf{H}(j\omega) = \frac{R}{-(\omega^2 LC - 1)R + j\omega L}$$

The amplitude thus is

$$|\mathbf{H}(j\omega)| = \frac{R}{\sqrt{(\omega^2 LC - 1)^2 R^2 + \omega^2 L^2}}$$

or

$$|\mathbf{H}(j\omega)| = \frac{R}{\sqrt{R^2 L^2 C^2 \omega^4 + (L - 2R^2 C)L\omega^2 + R^2}} \qquad (22.34)$$

For the case we have, the amplitude is therefore

$$|\mathbf{H}(j\omega)| = \frac{10^3}{\sqrt{10^6(10^{-2})(25)(10^{-16})\omega^4 + [0.1 - 2(10^6)(5)(10^{-8})](0.1)\omega^2 + 10^6}}$$

$$= \frac{10^3}{\sqrt{25(10^{-12})\omega^4 + 10^6}}$$

or

$$|\mathbf{H}(j\omega)| = \frac{10^3}{\sqrt{10^6[1 + 25(10^{-18})\omega^4]}}$$

$$= \frac{1}{\sqrt{1 + \omega^4/4(10^{16})}}$$

Writing this last result in the form

$$|\mathbf{H}(j\omega)| = \frac{1}{\sqrt{1 + [\omega/\sqrt{2}\,(10^4)]^4}}$$

we see that the maximum amplitude is $|\mathbf{H}|_{max} = 1$, which occurs when $\omega = 0$, since the amplitude continually decreases as ω increases. Also, we see that when we have

$$\omega = \sqrt{2} \times 10^4 \text{ rad/s} \tag{22.35}$$

the amplitude is

$$|\mathbf{H}| = \frac{1}{\sqrt{1 + 1}} = \frac{1}{\sqrt{2}} = \frac{1}{\sqrt{2}}(1) \tag{1}$$

$$= \frac{1}{\sqrt{2}}|\mathbf{H}|_{max}$$

Therefore, the cutoff frequency is the value in (22.35), which in hertz is

$$f_c = \frac{\sqrt{2} \times 10^4}{2\pi} = 2251 \text{ Hz}$$

PRACTICE EXERCISES

22-4.1 Find the network function

$$\mathbf{H}(j\omega) = \frac{\mathbf{V}_2}{\mathbf{V}_1}$$

and show that the circuit is a low-pass filter by finding the amplitude and cutoff frequency.

$$Ans. \quad \frac{R}{R+j\omega L}, \frac{R}{\sqrt{R^2 + \omega^2 L^2}}, \frac{R}{L} \text{ rad/s}$$

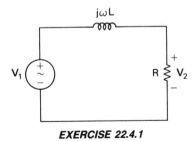

EXERCISE 22.4.1

22-4.2 Find R in Exercise 22-4.1 if the cutoff frequency is 2000 Hz and $L = 0.1$ H.

Ans. 1257 Ω

22-4.3 Find f_c in the low-pass filter of Fig. 22.11 if $R = 1$ kΩ, $L = 20$ mH, and $C = 0.01$ μF.

Ans. 11.25 kHz

22.5 OTHER TYPES OF FILTERS

Low-pass and bandpass filters are probably the most important ones, but there are many other types of filters, as well. In this section we will consider two other types, known as *high-pass* and *band-reject* filters.

High-Pass Filters: Filters that pass high frequencies and block low frequencies are called high-pass filters. An example is the circuit of Fig. 22.3(a), as we may see from its amplitude response of Fig. 22.4. Low frequencies correspond to relatively small amplitudes, and thus are blocked, and high frequencies correspond to relatively large amplitudes, and thus are passed.

In the general case the amplitude response of a high-pass filter may look like that of Fig. 22.12. The cutoff frequency is ω_c rad/s, or $f_c = \omega_c/2\pi$ Hz, which separates

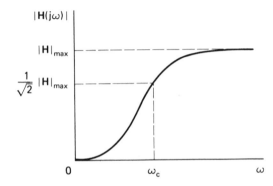

FIGURE 22.12 *High-pass amplitude response.*

the rejected band $0 < \omega < \omega_c$ from the passband $\omega \geq \omega_c$. Again, frequencies that pass correspond to amplitudes greater than or equal to $1/\sqrt{2} = 0.707$ times the maximum amplitude $|\mathbf{H}|_{max}$.

Example 22.6: Find the network function

$$\mathbf{H}(j\omega) = \frac{\mathbf{V}_2}{\mathbf{V}_1}$$

in the circuit of Fig. 22.13, with $R = 1$ kΩ, $L = 0.1$ H, and $C = 0.05$ μF, and show that it is a high-pass filter circuit. Find also the cutoff frequency f_c.

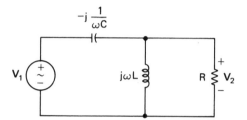

FIGURE 22.13 *High-pass filter.*

Solution: The parallel RL combination has impedance $\mathbf{Z}_1$ given by

$$\mathbf{Z}_1 = \frac{R(j\omega L)}{R + j\omega L}$$

Thus by voltage division we have

$$\mathbf{H}(j\omega) = \frac{\mathbf{V}_2}{\mathbf{V}_1} = \frac{\mathbf{Z}_1}{-j(1/\omega C) + \mathbf{Z}_1}$$

$$= \frac{\dfrac{j\omega RL}{R + j\omega L}}{-j\dfrac{1}{\omega C} + \dfrac{j\omega RL}{R + j\omega L}}$$

which may be simplified to

$$\mathbf{H}(j\omega) = \frac{j\omega RL}{(L/C) + j(\omega RL - R/\omega C)}$$

or

$$\mathbf{H}(j\omega) = \frac{jRL}{(L/\omega C) + jR(L - 1/\omega^2 C)}$$

The amplitude thus is

$$|\mathbf{H}(j\omega)| = \frac{RL}{\sqrt{(L/\omega C)^2 + R^2(L - 1/\omega^2 C)^2}}$$

or

$$\mathbf{H}(j\omega) = \frac{RL}{\sqrt{\dfrac{R^2}{C^2\omega^4} + \left(\dfrac{L}{C} - 2R^2\right)\dfrac{L}{C\omega^2} + R^2 L^2}} \tag{22.36}$$

For the circuit elements we have, this becomes

$$|\mathbf{H}(j\omega)| = \frac{10^3(0.1)}{\sqrt{\dfrac{10^6}{(25)(10^{-16})\omega^4} + \dfrac{0.1}{5(10^{-8})}\left[\dfrac{0.1}{5(10^{-8})} - 2(10^6)\right]\dfrac{1}{\omega^2} + (10^6)(10^{-2})}}$$

or

$$|\mathbf{H}(j\omega)| = \frac{100}{\sqrt{10^4\left[\dfrac{4(10^{16})}{\omega^4} + 1\right]}}$$

$$= \frac{1}{\sqrt{1 + \dfrac{4(10^{16})}{\omega^4}}} = \frac{\omega^2}{\sqrt{\omega^4 + 4(10^{16})}} \tag{22.37}$$

From this last result we see that the amplitude is zero at $\omega = 0$, and that as ω gets large, $4(10^{16})/\omega^4$ gets continuously smaller, so that $|\mathbf{H}(j\omega)|$ increases continually toward 1. Therefore, the amplitude response is like that of Fig. 22.12 with $|\mathbf{H}|_{max} = 1$. To find ω_c we note from (22.37) that when

$$\frac{4(10^{16})}{\omega^4} = 1 \tag{22.38}$$

we have

$$|\mathbf{H}| = \frac{1}{\sqrt{2}} = \frac{1}{\sqrt{2}} \cdot 1 = \frac{1}{\sqrt{2}} |\mathbf{H}|_{max}$$

Therefore, ω_c satisfies (22.38) and is given by

$$\omega_c = \sqrt[4]{4(10^{16})} = \sqrt{2} \times 10^4 \text{ rad/s}$$

In hertz the cutoff point is

$$f_c = \frac{\sqrt{2} \times 10^4}{2\pi} = 2251 \text{ Hz}$$

Band-Reject Filters: Filters that pass all frequencies except a band centered around some given frequency ω_0 are called *band-reject*, or *band-elimination*, or *notch*, filters. The frequency ω_0 is called the *center frequency*, and if the band eliminated is

$$\omega_{c_1} < \omega < \omega_{c_2}$$

then ω_{c_1} and ω_{c_2} are the *cutoff frequencies*. As in the other filter types, frequencies that pass correspond to amplitudes greater than or equal to $1/\sqrt{2} = 0.707$ times the maximum amplitude. A typical band-reject amplitude is shown in Fig. 22.14, where the rejected band has *bandwidth B* given by

$$B = \omega_{c_2} - \omega_{c_1} \text{ rad/s} \tag{22.39}$$

As in the bandpass filter, there is a quality factor Q defined for the band-reject filter. Its value is given by the expression

$$Q = \frac{\omega_0}{B}$$

which is identical to the bandpass expression.

An example of a band-reject filter is the circuit of Practice Exercise 22.1.2, with network function

$$\mathbf{H}(j\omega) = \mathbf{V}_2/\mathbf{V}_g$$

Its amplitude is

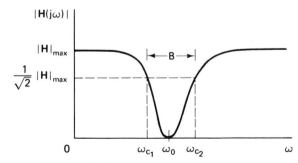

FIGURE 22.14 *Band-reject amplitude response.*

$$|\mathbf{H}(j\omega)| = \frac{|1-\omega^2|}{\sqrt{\omega^4 - 1.96\omega^2 + 1}}$$

with maximum value

$$|\mathbf{H}|_{max} = 1$$

occurring at $\omega = 0$ and at infinite frequency. Also, at $\omega = 1$, we have $|\mathbf{H}| = 0$, so that the response resembles Fig. 22.14 with $\omega_0 = 1$ rad/s.

Example 22.7: Find the network function

$$\mathbf{H}(j\omega) = \frac{\mathbf{V}_2}{\mathbf{V}_1}$$

in the circuit of Fig. 22.15, and show that it is a band-reject filter.

Solution: The impedance $\mathbf{Z}_1$ of the *LC* series portion of the network is

$$\mathbf{Z}_1 = j\omega L - j\frac{1}{\omega C}$$

so that the network function is

$$\mathbf{H}(j\omega) = \frac{\mathbf{V}_2}{\mathbf{V}_1} = \frac{\mathbf{Z}_1}{R + \mathbf{Z}_1} = \frac{j(\omega L - 1/\omega C)}{R + j(\omega L - 1/\omega C)} \tag{22.40}$$

We may write this result in the form

$$\mathbf{H}(j\omega) = \frac{1}{1 - j\left(\dfrac{R}{\omega L - 1/\omega C}\right)}$$

from which we have the amplitude response

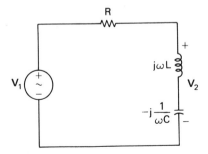

FIGURE 22.15 *Band-reject filter.*

$$|\mathbf{H}(j\omega)| = \frac{1}{\sqrt{1 + \left(\dfrac{R}{\omega L - 1/\omega C}\right)^2}}$$

or

$$|\mathbf{H}(j\omega)| = \frac{1}{\sqrt{1 + \left(\dfrac{\omega RC}{\omega^2 LC - 1}\right)^2}} \qquad (22.41)$$

Also, from (22.40) we may write

$$|\mathbf{H}(j\omega)| = \frac{|\omega L - 1/\omega C|}{\sqrt{R^2 + (\omega L - 1/\omega C)^2}} \qquad (22.42)$$

From (22.41) the maximum amplitude is $|\mathbf{H}|_{\max} = 1$, which occurs when

$$\frac{\omega RC}{\omega^2 LC - 1} = 0$$

or $\omega = 0$ and $\omega = \infty$. In (22.42) we see that $|\mathbf{H}(j\omega)| = 0$ when

$$\omega L = \frac{1}{\omega C}$$

which must occur at the center frequency ω_0. Therefore, we have

$$\omega_0 L = \frac{1}{\omega_0 C}$$

or

$$\omega_0 = \frac{1}{\sqrt{LC}} \qquad (22.43)$$

From (22.42) we see that the numerator and denominator both increase as ω increases or decreases from ω_0. Since $|\mathbf{H}(j\omega)|$ moves to 1 as ω moves to 0 or to ∞, we must have the amplitude response like that of Fig. 22.14. Thus the circuit of Fig. 22.15 is a band-reject filter.

Cutoff Points: In (22.41) we see that

$$|\mathbf{H}(j\omega)| = \frac{1}{\sqrt{2}} = \frac{1}{\sqrt{2}} |\mathbf{H}|_{\max}$$

when we have

$$\frac{\omega RC}{\omega^2 LC - 1} = \pm 1$$

or

$$\pm \omega RC = \omega^2 LC - 1 \qquad (22.44)$$

Therefore, this must be the relationship satisfied by the cutoff points. We may rewrite (22.44) in the form

$$\omega^2 LC \mp \omega RC - 1 = 0$$

which may be solved by the quadratic formula. Using the negative sign on ωRC, we have

$$\omega = \frac{RC \pm \sqrt{(RC)^2 + 4LC}}{2LC}$$

and using the plus sign, we have

$$\omega = \frac{-RC \pm \sqrt{(RC)^2 + 4LC}}{2LC}$$

The negative signs on the radicals must be ignored if $\omega > 0$, so that we have the cutoff points

$$\omega_{c_1} = \frac{-RC + \sqrt{(RC)^2 + 4LC}}{2LC} \qquad (22.45)$$

and

$$\omega_{c_2} = \frac{RC + \sqrt{(RC)^2 + 4LC}}{2LC} \qquad (22.46)$$

(We have assigned ω_{c_1} and ω_{c_2} in this way because we know that $\omega_{c_1} < \omega_{c_2}$.)
  The bandwidth B of the rejected band is given by

$$B = \omega_{c_2} - \omega_{c_1}$$
$$= \frac{RC + \sqrt{(RC)^2 + 4LC}}{2LC} - \frac{-RC + \sqrt{(RC)^2 + 4LC}}{2LC}$$
$$= \frac{2RC}{2LC}$$

or

$$B = \frac{R}{L} \tag{22.47}$$

Example 22.8: If in Fig. 22.15 we have $R = 100\ \Omega$, $L = 0.1$ H, and $C = 0.4\ \mu$F, find the center frequency, the cutoff points, the width of the rejected band, and Q.

Solution: The center frequency is

$$\omega_0 = \frac{1}{\sqrt{LC}} = \frac{1}{\sqrt{(0.1)(0.4)(10^{-6})}} = 5000 \text{ rad/s}$$

or

$$f_0 = \frac{5000}{2\pi} = 796 \text{ Hz}$$

By (22.45) and (22.46) we have the cutoff points

$$\omega_{c_1} = \frac{-(100)(4)(10^{-7}) + \sqrt{[(100)(4)(10^{-7})]^2 + 4(0.1)(4)(10^{-7})}}{2(0.1)(4)(10^{-7})}$$

$$= \frac{-0.4(10^{-4}) + 4.02(10^{-4})}{8(10^{-8})}$$

$$= 4525 \text{ rad/s}$$

and

$$\omega_{c_2} = \frac{0.4(10^{-4}) + 4.02(10^{-4})}{8(10^{-8})}$$

$$= 5525 \text{ rad/s}$$

The rejected bandwidth is

$$B = \omega_{c_2} - \omega_{c_1}$$

$$= 5525 - 4525$$

$$= 1000 \text{ rad/s}$$

This checks (22.47), where we have

$$B = \frac{R}{L} = \frac{100}{0.1} = 1000 \text{ rad/s}$$

Finally, Q is given by

$$Q = \frac{\omega_0}{B} = \frac{5000}{1000} = 5$$

In hertz the results are

$$f_{c_1} = \frac{\omega_{c_1}}{2\pi} = \frac{4525}{2\pi} = 720.2 \text{ Hz}$$

$$f_{c_2} = \frac{\omega_{c_2}}{2\pi} = \frac{5525}{2\pi} = 879.3 \text{ Hz}$$

and

$$B = 879.3 - 720.2 = 159.1 \text{ Hz}$$

PRACTICE EXERCISES

22-5.1 For the high-pass filter of Fig. 22.13 the element values are $R = 2 \text{ k}\Omega$, $L = 8 \text{ mH}$, and $C = 1 \text{ nF}$. Find the cutoff point.   *Ans.* 56.27 kHz

22-5.2 For the high-pass filter of Fig. 22.13, show that if

$$R^2 = \frac{L}{2C}$$

then the amplitude function (22.36) becomes

$$|\mathbf{H}(j\omega)| = \frac{1}{\sqrt{1 + 1/L^2 C^2 \omega^4}}$$

and the cutoff point is

$$\omega_c = \frac{1}{\sqrt{LC}}$$

Use this result to find L and R if $C = 0.01 \ \mu\text{F}$ and $\omega_c = 10,000$ rad/s.
                                                                Ans. 1 H, 7.07 kΩ

22-5.3 For the band-reject filter of Fig. 22.15, the element values are $R = 20 \ \Omega$, $L = 0.02 \text{ H}$, and $C = 0.5 \ \mu\text{F}$. Find ω_0, B, and Q.   *Ans.* 10,000 rad/s, 1000 rad/s, 10

22.6 SUMMARY

Filters are circuits that pass certain bands of frequencies and block all the other frequencies. Low-pass filters pass low-frequency signals, high-pass filters pass high-frequency signals, and bandpass filters pass a band of frequencies. Band-reject filters pass all frequencies except a certain band.

The center frequency of the band of frequencies that is passed by a bandpass filter or is rejected by a band-reject filter is the resonant frequency of the filter circuit. When the circuit is operated at this frequency it is said to be in resonance.

Low-pass and high-pass filters each have a single cutoff frequency, which separates the passband from the rejected band. In the case of bandpass and band-reject filters, there are two cutoff points that define the band that is passed in the bandpass case, or the band that is rejected in the band-reject case.

There are many examples of filter circuits. Two of the more common are the *RLC* series and parallel circuits, which are in resonance when the impedance seen at the generator is purely real. That is, its imaginary part is zero.

PROBLEMS

22.1 Find the amplitude response if the voltage of a circuit is given by

$$\mathbf{V} = \frac{1}{(j\omega)^2 + \sqrt{2}\,(j\omega) + 1}$$

22.2 Find $|\mathbf{V}|$ for

$$\mathbf{V} = \frac{(j\omega)^2}{(j\omega)^2 + \sqrt{2}\,(j\omega) + 1}$$

22.3 Find $|\mathbf{I}|$ for

$$\mathbf{I} = \frac{j\omega}{(j\omega)^2 + 2j\omega + 16}$$

22.4 Find $|\mathbf{I}|$ for

$$\mathbf{I} = \frac{(j\omega)^2 + 16}{(j\omega)^2 + 2j\omega + 16}$$

22.5 Find $\mathbf{H}(j\omega)$ and $|\mathbf{H}(j\omega)|$ where

$$\mathbf{H}(j\omega) = \frac{\mathbf{V}_2}{\mathbf{V}_1}$$

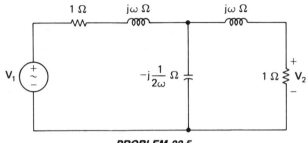

PROBLEM 22.5

22.6 The voltage **V** of Problem 22.1 is that of a low-pass filter. Find ω_c.

22.7 The circuit is in resonance at $f_0 = 5000$ Hz. If $X_L = X_C = 15$ Ω, $R = 3$ Ω, and $\mathbf{V}_1 = 30 \lfloor 0° $ V, find B, Q, and **I**. (The network function is $\mathbf{H} = \mathbf{I}/\mathbf{V}_1$.)

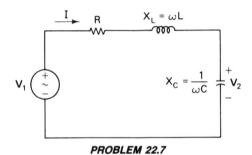

PROBLEM 22.7

22.8 In the circuit of Problem 22.7, $B = 500$ Hz, $Q = 10$, and $R = 10$ Ω. Find L and C.

22.9 In Problem 22.7, $R = 5$ Ω, $X_L = 200$ Ω, and $f_0 = 10{,}000$ Hz. Find B, f_{c_1}, and f_{c_2}.

22.10 In Problem 22.7, $L = 0.1$ H, $C = 1$ nF, and $R = 1$ kΩ. Find f_0 and B.

22.11 Find B and C if $R = 2$ kΩ, $L = 1$ mH, and $\omega_0 = 200{,}000$ rad/s. (The network function is $\mathbf{H} = \mathbf{V}_2/\mathbf{I}_1$.)

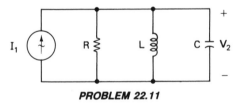

PROBLEM 22.11

22.12 In Problem 22.11, $L = 20$ mH, $C = 0.5$ μF, and $Q = 5$. Find f_0 and R.

22.13 For the low-pass filter of Fig. 22.11, show that if

$$R^2 = \frac{L}{2C}$$

then the amplitude function (22.34) becomes

$$|\mathbf{H}(j\omega)| = \frac{1}{\sqrt{1 + L^2C^2\omega^4}}$$

and the cutoff point is

$$\omega_c = \frac{1}{\sqrt{LC}}$$

Use this result to find L and R if $C = 0.02$ μF and $\omega_c = 50{,}000$ rad/s.

22.14 Find the network function

$$\mathbf{H} = \frac{\mathbf{V}_2}{\mathbf{V}_1}$$

in Problem 22.7, and note that for this case the circuit is a low-pass filter. Find ω_c if $R = 2$ kΩ, $L = 0.2$ H, and $C = 0.1$ μF.

22.15 Note that the network of Problem 22.5 is a low-pass filter and find ω_c.

22.16 The function of Problem 22.2 is that of a high-pass filter. Find ω_c.

22.17 The network function

$$\mathbf{H}(j\omega) = \frac{(j\omega)^2}{(j\omega)^2 + 20j\omega + 200}$$

is that of a high-pass filter. Show this by finding the amplitude response and ω_c.

22.18 Find the network function

$$\mathbf{H}(j\omega) = \frac{\mathbf{V}_2}{\mathbf{V}_1}$$

and show that the circuit is a high-pass filter by finding the amplitude response and ω_c.

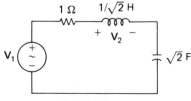

PROBLEM 22.18

22.19   Find the network function

$$H(j\omega) = \frac{V_2}{V_1}$$

for $R = 1\ \Omega$, $L = 5$ H, and $C = 0.2$ F, and show that the circuit is a band-reject filter by finding the amplitude response ω_0, B, and Q.

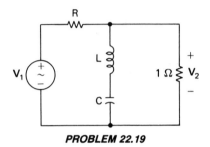

PROBLEM 22.19

22.20   In the circuit of Problem 22.19, replace the 1-Ω resistor by a 40-Ω resistor, and let $R = 40\ \Omega$, $L = 0.1$ mH, and $C = 10$ nF. Find ω_0, B, and Q.

22.21   If in Fig. 22.15 we have $L = 10$ mH, $C = 0.01\ \mu$F, and $R = 200\ \Omega$, find ω_0, B, and Q.

ANSWERS TO
ODD-NUMBERED PROBLEMS

CHAPTER 1

1.1 *abca, bdcb, abdca*
1.3 (a) 0.62137, (b) 1.6093, (c) 0.03687
1.5 (a) 21.11, (b) 122, (c) 288.71
1.7 (a) 0.001, (b) 0.127
1.9 (a) 2.5×10^3, (b) 2×10^{-5}, (c) 10^2, (d) 3.162×10^4

CHAPTER 2

2.1   (a) 20, (b) 10
2.3   (a) 150, (b) 0.06
2.5   6
2.7   12.5 V
2.9   36 W
2.11 80 V
2.13 40 mV
2.15 22, 380
2.17 (a) 60 W, (b) 1.2

2.19   25

2.21   (a) 30 W, (b) 1.2 kWh

CHAPTER 3

3.1   (a) 0.5 A, (b) 0.5 mA

3.3   (a) 0.05 ℧, (b) 120 V, (c) 720 W

3.5   0.05 A

3.7   (a) 5 W, (b) 2.5 W

3.9   (a) 20 mA, (b) 0.4 A

3.11   4 s

3.13   96 Ω, 1.25 A

3.15   5 mA

3.17   40 V

3.19   (a) 30 kΩ, 27 to 33 kΩ (b) 0.24 Ω, 0.228 to 0.252 Ω (c) 68 Ω, 61.2 to 74.8 Ω

CHAPTER 4

4.1   4 Ω

4.3   3 A, 2 V

4.5   −4, −7 A

4.7   1 and 3 kΩ

4.9   (a) 120 Ω, (b) 0.1 A, (c) 2, 3, and 7 V

4.11   0.1 A, 0.5 V

4.13   0.7 A

4.15   2 Ω

4.17   3 A, 1.5 A

4.19   5 A

4.21   72 V, 8 A, and 4 A

4.23   20 Ω

4.25   3 Ω

4.27   (a) 1 W, (b) 0.05 W

4.29   (a) 1152 W delivered, (b) 288 W absorbed, (c) 576 W absorbed, (d) 288 W absorbed

CHAPTER 5

5.1   2 A

5.3   42 Ω

5.5   2 A

5.7   18 V
5.9   1 W
5.11   1.5 A
5.13   10 A, 2.5 A, 10 V
5.15   (a) 44 Ω, (b) 0.2 A
5.17   6 Ω
5.19   4.3 V

CHAPTER 6

6.1   4 V, 12 V
6.3   4 Ω, 2 Ω
6.5   18 V, 6 V
6.7   1.4 A, 5.6 A
6.9   1, 2, 3, 4 mA
6.11   8 A, 4 A, 24 V, 8 V
6.13   2.7 A
6.15   320 Ω

CHAPTER 7

7.1   5 A, 1 A
7.3   3 V
7.5   2 A
7.7   5 A
7.9   2 A
7.11   4.5 A
7.13   3 A, 3 A
7.15   10 V
7.17   36 W

CHAPTER 8

8.1   10 V
8.3 -1 A
8.5   4 A
8.7 $V_{oc} = 14$ V, $R_{th} = 4$ Ω, $P = 12$ W
8.9 $I_{sc} = \frac{2}{3}$ A, $R_{th} = 3$ Ω, $V_2 = 1$ V

8.11 $I_{sc} = 14$ A, $R_{th} = \frac{4}{3}$ Ω, $I = 2$ A

8.13 $V_{oc} = 27$ V, $R_{th} = 3$ Ω, $I = 3$ A

8.15 $R = 6$ Ω, $V_g = 12$ V, $V_1 = 3$ V

8.17 $I_{sc} = 13$ A, $R_{th} = 2$ Ω

8.19 $V_{oc} = 27$ V, $R_{th} = 3$ Ω

8.21 $R_A = R_B = R_C = 3\ R$

CHAPTER 9

9.1   (a) 100 Ω, (b) 11.11 Ω, (c) 0.1 Ω

9.3   0.3356 Ω, 74.5 mA

9.5   333.33 Ω, 52.63 Ω, 5.025 Ω

9.7   (a) 50 mA, (b) 1.998 mA, (c) 0.1%

9.9   (a) 996 Ω, (b) 4996 Ω, (c) 19,996 Ω

9.11 9.5 kΩ, 40 kΩ, 50 kΩ

9.13 (a) 5000, (b) 2000, (c) 1000

9.15 37.5 V, 6.25%

9.17 6 V, 5900 Ω

9.19 (a) 145/9 Ω, (b) 145/3 Ω, (c) 145 Ω

CHAPTER 10

10.1   (a) 289, (b) 144, (c) 62, (d) 2601

10.3   (a) 0.008 Ω, (b) 0.0085 Ω, (c) 20 Ω

10.5   0.00204 Ω

10.7   (a) 7.94 Ω, (b) 128.35 Ω, (c) 5245 Ω

10.9   0.3997 A

10.11 (a) 8.876 Ω, (b) 143.48 Ω, (c) 5863.4 Ω

10.13 (a) 26.29 Ω, (b) 26.26 Ω, (c) 29.6 Ω

10.15 0 V, 120 V, 120 V

CHAPTER 11

11.1   150 μC, 50 V

11.3   44.28 pF

11.5   1.59 mm

11.7   (a) 5 μA, (b) −10 μA

11.9   2 mJ

11.11 20 V

11.13 3 μF

11.15 8 μF

11.17 (a) 3 μF, (b) 12 μF

CHAPTER 12

12.1 (a) 2 s, (b) $10e^{-t/2}$ V, (c) $0.5e^{-t/2}$ mA

12.3 (a) 7.36 V, (b) 2.71 V, (c) 0.13 V

12.5 (a) 0.25 s, (b) 0.4 mJ, (c) 40 μC

12.7 $20(1 - e^{-500t})$ V

12.9 $1.5e^{-20t}$ mA

12.11 $4.8e^{-10t}$ V

12.13 $30(1 - e^{-30t})$ V

12.15 $2 - e^{-5000t}$ mA

CHAPTER 13

13.1 (a) 25 T, (b) 1 T, (c) 30 T

13.3 5.305×10^6 At/Wb, 377 μWb

13.5 0.04 Wb

13.7 0.03 Wb

13.9 (a) 0.1 T, (b) 0.377 T, (c) 0.38 T

13.11 30 μWb

13.13 15.9 A

13.15 4.54×10^8 At/Wb

13.17 1.4 mWb

CHAPTER 14

14.1 25 mWb/s

14.3 $3e^{-1000t}$ V

14.5 1.26 mH

14.7 20 mH

14.9 (a) 40 mJ, (b) 0.09 μJ

14.11 (a) 40 V, 0.32 J, (b) −40 V, 0.08 J

14.13 6 mH

14.15 $20e^{-100,000t}$ mA

14.17 $3e^{-t}$ A

14.19 $20e^{-2500t}$ V, $10e^{-2500t}$ mA

14.21 $8 - 4e^{-t}$ A

CHAPTER 15

15.1   (a) $\pi/3$, (b) 4π, (c) 1.431

15.3   (a) 0.866, (b) 0, (c) 0.99

15.5   (a) 4 sin 100t mA, (b) 2 cos 200t mA

15.7   50 V, 400π rad/s, 200 Hz, 5 ms

15.9   (a) 0.5 ms, (b) 1 μs, (c) 50 ms

15.11   (a) 2000π rad/s, (b) $\pi/10$ rad/s, (c) $2\pi \times 10^6$ rad/s

15.13   15.915 V, 12.732 V

15.15 $V_m/2$

15.17   12 cos $6t$ V

CHAPTER 16

16.1   (a) $6 \underline{|90°}$, (b) $10 \underline{|53.1°}$, (c) $4 \sqrt{2} \underline{|-45°}$, (d) $15 \underline{|143.1°}$, (e) $2.236 \underline{|243.4°}$

16.3   (a) $j8$, (b) $60 - j80$, (c) $-8 - j8$, (d) $15 - j8$, (e) $-8.66 + j5$

16.5   (a) $-1 - j7$, (b) $-6 - j17$, (c) $-2 + j11$, (d) $j100$

16.7   (a) $50 \underline{|90°}$, (b) $26 \underline{|-22.6°}$, (c) $10 \underline{|278.1°}$, (d) $36 \underline{|44°}$, (e) 40

16.9   (a) $13 \underline{|120.5°}$, (b) $17 \sqrt{2} \underline{|-73.1°}$, (c) $2 \sqrt{2} \underline{|45°}$, (d) $3 \underline{|-70°}$, (e) $5 \underline{|25°}$

16.11   (a) $60 \underline{|-36°}$, (b) $10 \underline{|12°}$, (c) $70.7 \underline{|0°}$, (d) $35.35 \underline{|50°}$, (e) $20 \underline{|0°}$

16.13 $5 \underline{|-40°}$ Ω

16.15   (a) 1 kΩ, (b) $j10$ Ω, (c) $- j1$ kΩ

16.17   (a) 0, (b) $j10$ Ω, (c) $j100$ kΩ

16.19   (a) infinite (short circuit), (b) $-j0.1$ ℧, (c) $-j10$ μ℧

16.21   2 sin $(4t - 23.1°)$ A

16.23   2 sin $(4t + 83.1°)$ A

16.25   5 sin $2t$ A

CHAPTER 17

17.1 $3 \underline{|53.1°}$ Ω

17.3   (a) 1 Ω, 5 kΩ, (b) 1 kΩ, 5 Ω, (c) 62.83 Ω, 79.58 Ω

17.5   (a) leads by 42°, (b) leads by 30°, (c) lags by 10°

17.7 $\dfrac{3 + j4}{2}$ Ω, 8 sin $(t - 53.1°)$ A

17.9 $\sqrt{2}$ sin (5000t + 45°) mA, 10 $\sqrt{2}$ sin (5000t − 45°) V

17.11 1.9 $\lfloor$68.2° A, 2.69 sin (5t + 68.2°) A

17.13 5 $\sqrt{2}$ sin (5000t + 45°) mA

17.15 12 sin (6t − 90°) V

17.17 sin 2t A

17.19 sin 3t A

17.21 10 $\lfloor$−36.9° V

17.23 (a) 2 sin (4t − 36.9°) A, (b) 2.5 sin 2t A

CHAPTER 18

18.1 $\dfrac{44 - j33}{50}$ A, $\dfrac{56 + j133}{50}$ A, 2 + j2 A

18.3 8 + 6 sin (3t − 22.6°) V

18.5 10 $\lfloor$−36.9° V

18.7 $\mathbf{V}_{oc}$ = 3 + j3 V, $\mathbf{Z}_{th}$ = 3 − j3 Ω, $\mathbf{V}$ = $\dfrac{3\sqrt{2}}{5}$ $\lfloor$81.9° V

18.9 $\mathbf{V}_{oc}$ = $\dfrac{40}{17}$ (1 + j4) V, $\mathbf{Z}_{th}$ = $\dfrac{2}{17}$ (1 + j4) kΩ, $\mathbf{I}$ = 50 + j50 mA,

i = 100 sin (5000t + 45°) mA

18.11 $\mathbf{I}_{sc}$ = −j16 A, $\mathbf{Z}_{th}$ = j4/3 Ω, $\mathbf{I}$ = j2 A

18.13 $\mathbf{V}_{oc}$ = −j6 V, $\mathbf{Z}_{th}$ = 1 Ω, $\mathbf{V}$ = −j3 V

18.15 $\mathbf{Z}_A$ = 2 − j1 Ω, $\mathbf{Z}_B$ = 5 Ω, $\mathbf{Z}_C$ = −3 + j4 Ω

18.17 4 − j4 Ω

CHAPTER 19

19.1 480 W

19.3 32 W

19.5 172 W

19.7 60 W

19.9 0.6 lagging

19.11 0.8 leading

19.13 48 W, 83.1 vars, 96 VA

19.15 (a) 1, (b) 0.8 lagging, (c) 0.894 leading, (d) 0.6 lagging

19.17 5 $\lfloor$53.1° Ω

19.19 24.76 μF

19.21 10 − j6 Ω, 90 W

19.23 64 W

CHAPTER 20

20.1 20 A, 6 kW
20.3 7.2 kW
20.5 $100\sqrt{3}$ V, 10 A, $1500\sqrt{3}$ W
20.7 35.36 A, 11.25 kW
20.9 $10\underline{|36.9°}$ Ω
20.11 $21 + j18$ Ω
20.13 4.8 kW
20.15 8.1 kW
20.17 69 W, 531 W, 600 W
20.19 0, 500 W, 500 W
20.21 $7.97 - j0.48$ A, $-10.83 - j1.83$ A, $2.86 + j2.31$ A

CHAPTER 21

21.1 105 V, 40 V
21.3 135 V, −80 V
21.5 (a) 0.125, (b) 0.75, (c) 1
21.7 $-24 + j56$ V, $-24 + j48$ V
21.9 $7.5 \sin(2t - 36.9°)$ A, $1.5 \sin 2t$ A
21.11 4 J
21.13 $3\sqrt{2}\underline{|-45°}$ A, $\sqrt{2}\underline{|45°}$ A
21.15 2, $100\underline{|0°}$ V, $2\underline{|60°}$ A
21.17 100 W, 100 W
21.19 $20\underline{|36.9°}$ A, $2\underline{|36.9°}$ A
21.21 (a) 100, 32 W, (b) 50, 8 W
21.23 $15\underline{|0°}$ V, $12\underline{|10°}$ A
21.25 $25 + j25$ Ω

CHAPTER 22

22.1 $\dfrac{1}{\sqrt{1 + \omega^4}}$

22.3 $\dfrac{|\omega|}{\sqrt{\omega^4 - 28\omega^2 + 256}}$

22.5 $\mathbf{H} = \dfrac{1/2}{(j\omega)^3 + 2(j\omega)^2 + 2j\omega + 1}$, $|\mathbf{H}| = \dfrac{1/2}{\sqrt{1 + \omega^6}}$

22.7 1000 Hz, 5, $10\underline{|0°}$ A

22.9   250 Hz, 9,876 Hz, 10,126 Hz

22.11   20,000 rad/s, 0.025 μF

22.13   20 mH, 707 Ω

22.15   1 rad/s

22.17 $\dfrac{\omega^2}{\sqrt{\omega^4 + 4(10^4)}}$, 10 $\sqrt{2}$ rad/s

22.19 $\dfrac{j5(\omega^2 - 1)}{\omega + j10\,(\omega^2 - 1)}$, 1 rad/s, 0.1 rad/s, 10

22.21 10^5 rad/s, 2×10^4 rad/s, 5

INDEX

INDEX

A

Abc sequence, 508
Ac (alternating current), 24, 339–66
Acb sequence, 509
Ac circuit, 363–65
Ac generator, 340
Ac voltage, 341, 365
Admittance, 391, 396
Alternating current (ac), 24, 339–66
American Wire Gage, 218
Ammeter, 185, 192–95, 209
Ampere, 22
Ampere's circuital law, 305, 312
Ampere-turns, 298
Angular frequency, 350, 366
Anode, 31, 33
Apparent power, 477–78
Armature, 36
Atomic structure, 6–7
Autotransformer, 562
Average value, 354–59
  of a sine wave, 339, 357

B

Bandpass filter, 582–88, 601
Band-reject filter, 592, 595, 601
Bandwidth, 583, 584, 589, 595
Battery, 31, 32, 33, 34, 35
  life of, 35
  rating of, 35
B-H curve, 301
Breakdown voltage, 223, 257
Bridge circuit, 96, 180
Brush, 316

C

Capacitance, 234–35
Capacitive reactance, 411
Capacitors, 158, 223, 233–60, 262
  ceramic disk, 248
  charging of, 263
  color code, 258
  dipped, 251
  discharging of, 263, 264–65
  electrolytic, 252–54

Capacitors *(Continued):*
  energy stored in, 242–43
  Mylar, 248
  parallel, 245–47
  polystyrene, 248
  series, 244–45
  variable, 254–56
Carbon resistors, 51
Cathode, 32, 33
Cell, 32
Center frequency, 582, 595
Charge, 6, 19–25, 240
Choke coil, 317
Circuit, 64
Circuit breaker, 225, 230–31
Circular mil, 214
Coefficient of coupling, 541, 567
Color code:
  capacitors, 258
  resistors, 58, 59, 63
Complex conjugate, 387
Complex numbers, 368, 373–80, 401
  operations with, 381
  polar form, 374, 375
  rectangular form, 374
Complex power, 483, 499
Conductance, 45–46, 63, 409
Conductor, 42, 89, 212–31
  wire, 216–19
Coulomb, 21, 39
Cramer's rule, 137
Current, 5, 19–25, 39, 42, 240
  alternating, 24, 339, 366
  conventional, 23, 39
  direct, 24
  electron, 23, 39
Current division, 112, 116–21, 417, 422–23
Current source, 38–39
Cutoff frequency, 582, 589, 592, 595, 597
Cycle, 340

D

d'Arsonval movement, 185, 186–91, 209
Dc (direct current), 24
Dc meters, 185–209
Decade resistance box, 56, 57
Delta (Δ) connection, 458–59
Δ-Y conversion, 179, 182, 460–61
Determinants, 137–40
Diamagnetic material, 296
Dielectric, 223

Dielectric constant, 236–38
Dielectric strength, 223
Digital meter, 206
Direct current (dc), 24
Dot convention, 543, 567
Dual, 46, 81, 118, 325, 328, 332

E

Electric field, 234, 235–36
Electricity, 1
  history of, 2–6
Electrolyte, 31, 33
Electromagnetism, 290
Electromotive force (emf), 25, 36, 298
Electron, 7
Energy, 29–31, 39, 49, 322

F

Farad, 235
Faraday's law, 316, 339, 340, 538, 540, 567
Ferromagnetic material, 296
Filters, 571–601
  bandpass, 582–88
  band-reject, 592, 595
  high-pass, 592
  low-pass, 589–91
Flux:
  electric, 235
  magnetic, 291, 294–96
Flux density, 296
Frequency, 24, 349–51, 366
  ranges, 351
Fuse, 224, 229–30, 232

G

Galvanometer, 126, 189, 190
Generator, 36–37, 316
  principle of, 316–17, 340–44
Ground, 146

H

Half-power frequency, 583
Henry, 317, 319
Hertz, 24, 350
High-pass filter, 592, 601

Horsepower, 29
Hysteresis, 302–3
   loop, 303
   loss, 303, 312

I

Imaginary number, 369–73, 401
Impedance, 391, 392, 406–7
   of a capacitor, 394–95
   of an inductor, 393–94
   of a resistor, 393
Induced voltage, 316, 538
Inductance, 315, 319–20
Inductive reactance, 411
Inductors, 158, 233, 294, 315–36
   energy stored in, 322
   parallel, 325
   series, 324
   shielding of, 326
Inferred absolute zero, 220
Insulator, 89, 212–31, 234
Internal resistance, 171
International System of Units (SI), 11
   prefixes in, 15
IR drop, 69
Iron core transformer, 552
Iron-vane movement, 208, 209

J

j operator, 371–72
Joule, 12, 30, 39

K

Keeper, magnet, 291
Kilowatthour, 30, 39
Kirchhoff's current law, 64, 65–69, 398
Kirchhoff's voltage law, 64, 68–69, 88, 398

L

Ladder network, 103, 109, 123
Lagging, 413
Leading, 413
Leakage flux, 537
Lenz's law, 317

Linear resistor, 45
Loop, 65
Loop analysis, 132, 133–36, 430–31
Loose coupling, 542
Low-pass filter, 589–91, 601

M

Magnetic circuit, 295, 305, 306
Magnetic field, 187, 234, 290, 291–94
Magnetism, 290–314
Maximum power transfer, 494–96
Mesh analysis, 132, 154, 430–33
Metal-film resistor, 52
Meter movement, 185
Meters, 11, 185
Mho, 46, 62
Millman's theorem, 173–76, 182
Mks system, 11
Multimeters, 205, 209
Multiple-load transformer, 563
Mutual inductance, 535, 536–43, 567

N

Natural frequency, 576
Network function, 582
Nodal analysis, 132, 146–49, 154, 425–29
Node, 65
Norton's theorem, 162, 166–68, 181
   for ac circuits, 451–52, 465–66
Nucleus, 6

O

Ohm, 43, 62
Ohmmeter, 185, 201–4, 209
   series, 201
   shunt, 203
Ohm's law, 43, 62, 63, 71, 77
   for ac circuits, 396
   for magnetic circuits, 297, 299
Ohms-per-volt rating, 198

P

Parallel circuits, 64, 76–84
Paramagnetic material, 296

Passband, 583, 589, 592
Period, 349
Permeability, 295–311
    relative, 295–96
Permittivity, 237
    relative, 237
Phase angle, 351–54, 412
Phasor, 368–401
Phasor diagram, 433–39
Polarity, 9
Polyphase source, 503
Potential difference, 25
Potentiometer, 54, 55
Power, 27–29, 39, 47, 85, 243, 555
    ac steady-state, 470–99
    apparent, 477–78
    average, 360, 470–78
    reactive, 482–83, 499
Power factor, 477–81, 499
    correction of, 488–94
Power triangle, 482–87
Primary winding, 538
Proton, 7

Q

Quality factor, 585, 595

R

RC circuit, 262–89
RC time constant, 262, 268–70
Reactance, 407, 440
    capacitive, 411
    inductive, 411
Reactive power, 482–83, 499
Real number, 369, 401
Rectification, 36
Reflected impedance, 556–57
Relay, 309
Reluctance, 298–99, 311
Residual flux density, 303
Resistance, 42–61, 63, 213, 231, 407
Resistive circuit, 43, 64–91, 132–54
Resistivity, 213, 214, 231
Resistors, 42, 43, 47, 50, 262
    carbon composition, 51
    carbon film, 51
    color code, 58, 59, 63
    integrated-circuit, 53
    metal film, 52

Resistors (Continued):
    parallel, 78
    power rating, 48, 50
    series, 70–73
    variable, 54–58
    wire wound, 52
Resonance, 571, 576–81
    parallel, 579
    series, 577–78
Rheostat, 54, 55, 56
Right-hand rule, 293
RL circuit, 326–35
RL time constant, 328
Rms (root-mean-square) value, 360–62

S

Saturation, 301
Scientific notation, 13–16
Secondary winding, 538
Selectivity, 585
Self inductance, 535, 538
Semiconductor, 8, 9, 212, 231
Series circuit, 64, 70–75
Series-parallel circuit, 92–108
Siemens, 46
Sine wave, 339, 341, 345–48
Slipring, 316
Solar cell, 37
Source conversion, 169–72
Specific gravity, 33
Superposition, 158, 181, 445–50, 466
Susceptance, 409
Switch, 224–28

T

Taut-band support, 187, 188
Temperature coefficient, 213, 221–22
Tesla, 296
Thévenin's theorem, 162–65, 181, 451–52, 465–66
Three-phase circuit, 503–34
Three-phase generator, 504–7
Tight coupling, 542
Time constants:
    RC, 262, 268–70
    RL, 328
T network, 176
Transformer, 326, 503, 535–68

V

Var, 483
Varmeter, 498
Volt, 25
Voltage, 25–26, 39, 42
Voltage division, 112–16, 421–22
Voltage source, 31–35
Volt-ampere, 477
Volt-ampere reactive, 483
Voltmeter, 185, 196–200, 209
Volt-ohm-milliammeter (VOM), 185, 205, 209

W

Watt, 12
Wattmeter, 185, 496
Weber, 294
Wheatstone bridge, 112, 126–28

Y

Y-connected generator, 507–12, 532
Y connection, 176, 458–59
Y-Δ conversion, 177, 182, 459, 522
Y-Δ system, 519
Y-Y system, 512–17, 532